The biology of polar bryophytes and lichens

Studies in Polar Research

This series of publications reflects the growth of research activity in and about the polar regions, and provides a means of synthesising the results. Coverage is international and interdisciplinary: the books are relatively short and fully illustrated. Most are surveys of the present state of knowledge in a given subject rather than research reports, conference proceedings or collected papers. The scope of the series is wide and includes studies in all the biological, physical and social sciences.

Editorial Board

Other titles in the series

G. Deacon, *The Antarctic Circumpolar Ocean*
Y. I. Chernov, transl. D. Löve, *The Living Tundra**
B. Stonehouse (ed.), *Arctic Air Pollution*
G. Triggs (ed.), *The Antarctic Treaty Regime*
Sir Anthony Parsons (ed.), *Antarctica: The Next Decade*
Francisco Orrego Vicuña, *Antarctic Mineral Exploitation: The Emerging Legal Framework*
C. Lamson and D. Vanderzwaag (eds.), *Transit Management in the Northwest Passage*
Donat Pharand, *Canada's Arctic Waters in International Law*
V. Aleksandrova, transl. D. Löve, *Vegetation of the Soviet Polar Deserts*
Nigel Leader-Williams, *Reindeer on South Georgia: The Ecology of an Introduced Population*

* Also available in paperback.

The biology of polar bryophytes and lichens

R. E. LONGTON
Department of Botany
School of Plant Sciences, University of Reading

PUBLISHED IN ASSOCIATION WITH THE
BRITISH BRYOLOGICAL SOCIETY

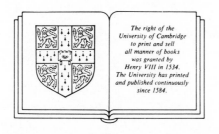

The right of the
University of Cambridge
to print and sell
all manner of books
was granted by
Henry VIII in 1534.
The University has printed
and published continuously
since 1584.

CAMBRIDGE UNIVERSITY PRESS
Cambridge
New York   New Rochelle   Melbourne   Sydney

Published by the Press Syndicate of the University of Cambridge
The Pitt Building, Trumpington Street, Cambridge CB2 1RP
32 East 57th Street, New York, NY 10022, USA
10 Stamford Road, Oakleigh, Melbourne 3166, Australia

First published 1988

Printed in Great Britain at The Bath Press, Avon

British Library cataloguing in publication data
Longton, R. E.
The biology of polar bryophytes and
lichens.
1. Polar regions. Bryophytes   2. Polar
regions. Lichens
I. Title
588'.0911

Library of Congress cataloguing in publication data
Longton, R. E.
The biology of polar bryophytes and lichens/R. E. Longton.
   p.   cm.
Bibliography.
Includes index.
ISBN 0–521–25015–3
1. Bryophytes – Polar regions.   2. Lichens – Polar regions.
I. Title.
QK533.84.P73L66   1988
588'.0998 – dc 19   88-1095

ISBN 0 521 25015 3

Contents

	Preface	page vii
1	**The polar regions**	**1**
	Topography and geological history	1
	Climate	7
	Soils	12
	Vegetation types and zones	13
	Polar bryophyte and lichen floras	23
2	**The cryptogamic vegetation**	**32**
	Vegetation classification	32
	Cryptogamic growth form types	34
	Vegetation in the cold-Antarctic	37
	Vegetation in the cold-Arctic	47
	Vegetation in the frigid-Antarctic	49
	Vegetation in mild and cool-polar regions	52
	Benthic bryophytes	64
3	**Pattern, process and environment**	**66**
	Environmental relationships of polar cryptogamic vegetation	66
	Colonisation and succession	79
	Pattern in polar vegetation	94
4	**Radiation and microclimate**	**106**
	Solar radiation	106
	Net radiation	116
	Energy budgets	117
	Annual temperature regimes	131
5	**Physiological processes and response to stress**	**141**
	Environmental control of carbon dioxide exchange	141
	Net assimilation rate under field conditions	185
	Response to plastic stress	192
	Adaptation and selection	203
6	**Vegetative growth**	**211**
	Patterns of growth in relation to net assimilation and translocation	211
	Vegetative phenology and annual growth increments	229
	Environmental and inherent control of growth	246

7	**Cryptogams in polar ecosystems**	**253**
	Energy flow	253
	Nutrient cycling	290
	Other effects of bryophytes and lichens	298
	Human influence on polar cryptogams	299
8	**Reproductive biology and evolution**	**310**
	Reproductive processes and propagules	310
	Reproduction in polar species	313
	Reproductive phenology	316
	Failure in sexual reproduction	319
	Dispersal and establishment	323
	Life history strategies	331
	Origin and evolution of polar cryptogamic floras	335
	Concluding remarks	341
	References	343
	Index of generic and specific names	383
	Subject index	386

Preface

This book is about bryophytes and lichens in the Arctic and Antarctic. It considers the evolution and adaptations of the polar floras, and the role of these plants in the vegetation and in the functioning of tundra ecosystems. The study of plant ecology in the polar regions has advanced dramatically in recent years as a result of work initiated in the Antarctic during the International Geophysical Year (1957–58), and in both Arctic and Antarctic as part of the International Biological Programme Tundra Biome investigations. Much attention has been focused on bryophytes and lichens because of their obvious abundance in local communities. The work has been broad in scope, ranging from phytogeography to physiological ecology, and from vegetation ecology to reproductive biology. The results, as synthesised here, are of relevance far beyond the polar regions, because they make a substantial contribution towards a general understanding of the environmental relationships of bryophytes and lichens. It should be noted, however, that mosses are of considerably greater ecological significance in the tundra than liverworts; they have consequently received more emphasis in research and therefore in the present text.

Authorities for most of the plant names cited in the text can be found in the following: mosses – D. M. Greene (1986), Steere (1978a); hepatics – Schuster (1966–80), Grolle (1972); lichens – Thomson (1984), Lindsay (1974); vascular plants – Greene & Walton (1975), Scoggan (1978–79). Where differences occur, the nomenclature used by the first-mentioned authority for each group has been followed.

The present author's work in polar ecology has been supported financially and/or logistically by the British Antarctic Survey, the United States Antarctic Research Program, the Royal Air Force Mountaineering Association, the National Science and Engineering Research Council of Canada, the Natural Environment Research Council (UK), and the

University of Manitoba Northern Studies Committee: it has been dependent on facilities generously provided in the Departments of Botany in the Universities of Birmingham, Manitoba and Reading. Grateful acknowledgement is made to these organisations, and to the many colleagues with whom stimulating and enjoyable discussion has helped to formulate my ideas on the topics considered here, including Drs L. C. Bliss, T. V. Callaghan, N. J. Collins, D. L. Hawksworth, D. C. Lindsay, G. A. Llano, D. M. Moore, W. O. Pruitt, R. I. Lewis Smith, J. M. Stewart, D. H. Vitt and D. W. H. Walton. I am particularly grateful to Dr S. W. Greene for advice and encouragement over many years, and to Dr D. H. Brown and Professor P. Kallio for reading a draft of the present manuscript; their comments and suggestions have materially improved the final product without being in any way responsible for whatever deficiencies remain. I also thank Dr R. I. Lewis Smith for supplying the photographs for Figs. 1.6, 2.1, 2.5, 3.3 and 6.12, J. D. Ives for Fig. 1.3, Dr F. J. A. Daniels for Fig. 2.10, and Mr S. Brooks and Mr D. J. Farmer for assistance in preparing the other illustrations.

Finally, I thank my family who have cheerfully endured long periods when I have been out of communication, either working at some remote polar station or closeted in the study. They have never complained; well, hardly ever!

R. E. Longton
Reading, December 1986

1

The polar regions

Topography and geological history

This book considers the biology of bryophytes and lichens in polar tundra and adjacent, open woodland, their contribution to the vegetation, their role in polar ecosystems, and their adaptations to the rigours of life in regions generally regarded as among the least hospitable on earth. Tundra is used here in the broad sense of treeless regions beyond climatic timberlines. It occurs in the polar regions as defined in Figs. 1.1 and 1.2, and in some alpine and oceanic areas at lower latitudes.

The polar regions are diverse in topography and climate. Arctic lands comprise substantial parts of the North American and Eurasian continents which, with Greenland and smaller islands, encircle a polar ocean (Fig. 1.1). The terrain ranges from extensive flat-bedded plains and plateaus to folded mountains, often high and imposing as in the Brooks Range, Alaska, where elevations reach over 2800 m (Figs. 1.3 and 1.4). Except in Greenland, contemporary glaciation is localised, and largely confined to Spitzbergen and other far-northern islands. In contrast, the Antarctic continent is centred over the pole and is surrounded by the vast expanse of the Southern Ocean, with a minimum width of over 850 km between the Antarctic Peninsula and Cape Horn (Fig. 1.2). The continent is fringed by coastal mountains, and rises to an inland ice-plateau at elevations of 1800 to 3800 m. Over 98% is currently buried beneath an ice sheet more than 4 km thick in places. Islands lying close by the mainland or widely scattered in the Southern Ocean thus provide the major terrestrial habitats in Antarctic regions, where vegetation is restricted to the almost universally rugged terrain of coastal regions, and to nunataks penetrating the inland ice sheet (Figs. 1.5 and 1.6). The highest, Mt Kirkpatrick, reaches an altitude of about 4500 m.

Arctic lands were formerly part of Laurasia, and although their positions have shifted as a result of continental drift, it is unlikely that the

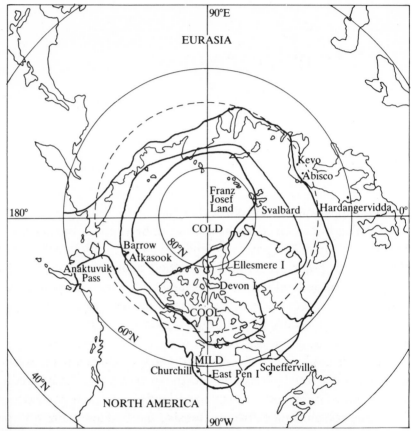

Fig. 1.1. Arctic regions showing vegetation zones and the principal localities referred to in the text. Southern boundaries of mild-Arctic and cool-Arctic after Bliss (1979); southern boundary of the cold-Arctic after Aleksandrova (1980).

major landmasses have ever been as widely separated as the southern continents. North America was connected to Europe via Greenland until the Eocene, and a Beringian land bridge united Alaska and Siberia as recently as 13 000–14 000 yr BP. Palaeontological evidence indicates that Mesozoic climates were warmer than those of today. Forests of ferns and gymnosperms, including cycads and other plants indicative of warm-temperate conditions, occurred in Alaska during the early Cretaceous, giving way in the late Cretaceous to angiosperm-dominated vegetation of cold-temperate affinity (Cox & Moore, 1980). Cooling continued through the Tertiary, but coniferous and deciduous forest appears to have persisted at far-northern latitudes until the late Pliocene (Bliss, 1986).

The northern hemisphere experienced four major glaciations in the

Fig. 1.2. Antarctic regions showing vegetation zones and the principal localities referred to in the text. Northern boundaries of mild-Antarctic, cool-Antarctic and cold-Antarctic after Greene (1964); northern boundary of frigid-Antarctic after Longton (1985a).

Pleistocene, beginning about 1 million yr BP, with Arctic conditions spreading far south of the present forest line. Moss fragments preserved in peat and lake sediments provide valuable evidence about associated vegetational changes (N. G. Miller, 1980). The most recent, the Wisconsin or Würm glaciation, reached its maximum extent about 115 000 yr BP. Its major retreat began some 14 000 yr BP and substantial areas of the Arctic are thought to have become deglaciated only during the past 5000–10 000 yr. Conversely, extensive terrain in Alaska and the Yukon, in parts of the Canadian Arctic Archipelago and in Siberia remained clear of ice, at least during the Wisconsin. A general lowering of sea levels at maximum glaciation no doubt increased the extent of these refugia, as

Fig. 1.3. Forest-tundra ecotone in Labrador-Ungava (mild-Arctic). The principal trees include *Picea glauca*, *P. mariana* and *Larix laricina*. Reproduced from Ives & Barry (1974) by courtesy of J. D. Ives and R. G. Barry.

well as reducing dispersal barriers, for example by creating the Beringian land connection. Recolonisation by plants and animals followed glacial retreat, and there is evidence that forests extended further north than at present during the hypsithermal period of maximum post-glacial temperatures 4000–6000 yr BP (Kay, 1979).

Antarctica once formed part of Gondwanaland, its present, isolated position resulting from continental drift during the Mesozoic and early Cenozoic. It comprises two geological regions separated by an ice-filled channel extending below sea level beneath the central ice-plateau. The smaller West Antarctica includes the Antarctic Peninsula and Marie Byrd Land: it has stratigraphic affinities with the South American Andes,

Fig. 1.4. Mountains in the United States Range near Lake Hazen, northern Ellesmere I, North West Territories (NWT), Canada (cool-Arctic). The muskox (*Ovibos moschatus*) in the foreground are at the edge of a stand of sedge-moss meadow surrounding the lake on the right.

whereas the larger East Antarctica is related geologically to southern Africa. Several major floras inhabited Antarctica prior to the fragmentation of Gondwanaland, while parts of the continent supported phanerogamic vegetation throughout much of the Mesozoic. Fossil assemblages

Fig. 1.5. Signy I, South Orkney Is (cold-Antarctic), with the British Antarctic Survey Station in the foreground (1965).

Fig. 1.6. Nunatak on Alexander I, west Antarctica (frigid-Antarctic).

including members of the Podocarpaceae and *Nothofagidites* occur in early Tertiary rocks in the Antarctic Peninsula region (Elliot, 1985).

Formation of continental glaciers began at least 30 million yr BP. Build-up of the Antarctic ice-cap was essentially complete by about 10–15 million yr BP, but recent evidence suggests that it may since have melted and reformed several times (Anderson, 1986). As in the Arctic, there is evidence of four major periods of ice expansion during the Pleistocene, the most recent reaching its maximum extent 17 000–21 000 yr BP (Elliot, 1985). Ice cover was at times continuous, except perhaps for isolated nunataks, throughout the frigid- and cold-Antarctic regions, and extensive in the cool-Antarctic and in Tierra del Fuego, but ice-free refugia are thought to have existed on several cool-Antarctic islands (Gremmen, 1982; Walton, 1984). Recolonisation of Antarctic regions must have been severely restricted by ocean barriers of a type not present in the Arctic (Figs. 1.1 and 1.2).

Glaciation created a range of landscape features, including broad U-shaped valleys carved by flowing ice, eskers, or gravel ridges marking the course of subglacial streams, and moraines in the form of irregularly shaped piles of till deposited by the glaciers. There are extensive lacustrine deposits of various textures laid down in the rivers and lakes that formed as the glaciers melted, and aeolian silts and clays redistributed by wind. Also widespread are raised beaches resulting from isostatic uplift as the land surface rose after being depressed by the weight of the ice. Such features are important in increasing habitat diversity, and can yield valuable information on rates and patterns of recolonisation (Chapter 3).

Climate

Polar climates are characterised by cooler summers than occur elsewhere, except on high mountains. Other factors vary widely in response to differences in energy budget, degree of oceanicity, altitude, and the resulting patterns of atmospheric circulation (Table 1.1). Annual solar radiation declines with increasing latitude leading, in general terms, to a corresponding decrease in mean annual temperature. However, seasonal variation in daylength and in daily totals of solar radiation increases with latitude. Beyond latitude 66°33′ N and S the sun remains below the horizon for 24 hours per day in midwinter, with a corresponding period of continuous daylight in summer. Maximum irradiance is relatively low, but daily totals of solar radiation reach values similar to those in temperate regions for a few weeks in midsummer. This regime may favour efficient

Table 1.1 *Climatic data for representative polar localities*

Locality and zone	Mean monthly air temperature (°C) Warmest month	Mean monthly air temperature (°C) Coldest month	Number of months with mean air temperature >0°C	Mean annual precipitation (mm rainfall equivalent)	Mean annual relative humidity (%)	Mean cloudiness (tenths)	Mean annual wind-speed (m sec^{-1})	Source
Mild-Arctic								
Churchill, Manitoba (58°45'N, 94°10'W)	12.0	−27.5	4	407	—	6.4	6.3	Hare & Hay (1974)
Coral Harbour, NWT (64°08'N, 83°10'W)	8.3	−30.4	4	249	—	6.0	5.7	Hare & Hay (1974)
Mild-Antarctic								
Evangelistas, Chile (52°24'S, 75°06'W)	8.7	4.4	12	2570	—	8.4	11.8	A. Miller (1976)
Stanley, Falkland Is (51°42'S, 57°51'W)	8.8	2.2	12	668	82	7.2	7.0	Moore (1968)
Cool-Arctic								
Barrow, Alaska (71°17'N, 156°47'W)	3.9	−27.9	3	110	79	—	5.4	Hare & Hay (1974)
Eureka, Northern Ellesmere I, NWT (80°00'N, 85°56'W)	5.7	−37.3	3	67	—	4.9	3.4	Vowinkel & Orvig (1970)
Maria Pronchistsheva, USSR (75°30'N, c 113°E)	4.0	−31.2	2	219	86	—	5.0	Matveyeva et al. (1975)

Cool-Antarctic								
King Edward Point, South Georgia (54°16'S, 36°30'W)	5.3	−1.5	9	1405	75	—	4.3	Smith & Walton (1975)
Macquarie I (54°30'S, 158°57'E)	6.7	3.1	12	926	89	—	8.6	Jenkin (1975)
Cold–Arctic								
Bukhta Tikhaya Franz Josef Land, USSR (80°19'N, 52°48'E)	1.1	−17.2	2	117	—	7.8	5.7	Vowinkel & Orvig (1970)
Mys Chelyuskin, USSR (77°43'N, 104°17'E)	0.8	−31.1	2	294	88	—	6.5	Lydolph (1977)
Cold–Antarctic								
Deception I (62°59'S, 60°43'W)	1.4	−8.0	3	398	90	8.5	6.1	Schwerdtfeger (1970)
Signy I (60°43'S, 45°36'N)	1.3	−9.0	4	400	86	8.6	6.9	Collins *et al.* (1975)
Frigid-Antarctic								
Mawson (67°36'S, 62°52'E)	−0.2	−18.5	0	—	60	7.2	10.9	Rudolph (1967a) Schwerdtfeger (1970)
McMurdo (77°51'S, 166°40'E)	−3.4	−27.8	0	119	57	—	6.5	Rudolf (1967a) Schwerdtfeger (1970)
Syowa (69°00'S, 39°35'E)	−1.0	−19.4	0	—	73	6.6	5.9	Schwerdtfeger (1970)

energy utilisation in photosynthesis, as relatively little radiation is received at irradiance far above saturation.

Net radiation is of greater significance than total solar radiation in terms of heat exchange processes and therefore the temperature of plants and their surroundings. Net radiation represents the balance between incoming and outgoing fluxes of solar radiation, and of infrared radiation upwards from vegetation and ground surfaces and downwards from cloud and other matter in the atmosphere. It is thus influenced by cloud cover and by surface reflectivity or albedo, which varies from 0.10–0.25 for vegetation, soil or rock, to as much as 0.80–0.90 for ice and snow (Chapter 4).

As a result of low total solar radiation combined with the high albedo of glaciated surfaces annual net radiation is negative over much of the Antarctic continent, whereas it is positive for most non-glaciated land areas. Typical values are $-56\,\mathrm{kJ\,cm^{-2}\,yr^{-1}}$ in the central Antarctic plateau, $67-75\,\mathrm{kJ\,cm^{-2}\,yr^{-1}}$ at the Arctic tree line and $40-60\,\mathrm{kJ\,cm^{-2}\,yr^{-1}}$ in the Arctic tundra (Barry, Courtin & Labine, 1981; Weyant, 1966). Such differences are largely responsible for driving atmospheric circulation, which results in heat exchange between areas of high and low net radiation. The Antarctic continent acts as a major heat sink.

Oceanicity also has a profound influence on seasonal temperature regimes and other aspects of climate. Water bodies with their high specific heat show a slower temperature decrease when net radiation is negative in autumn and winter, and a slower temperature rise when it increases in spring, than land areas under similar radiation regimes. Ocean temperatures influence those in coastal land areas which thus tend to experience a relatively narrow annual temperature range, particularly on coasts subject to prevailing onshore winds or surrounded by pack ice where the high albedo further retards ocean warming.

The coldest lands occur in continental regions of Antarctica. Mean air temperature for the warmest month each year remains below 0 °C throughout most of the frigid-Antarctic (page 17), where means for the coldest month may fall below −70 °C (Table 1.1). Other Antarctic lands (Fig. 1.2) comprise the west coast of the Antarctic Peninsula, parts of southern South America, and islands, some coastal and others occupying isolated oceanic positions. All lie in the path of strong, westerly winds blowing over the Southern Ocean, leading to cooler summers and milder winters than continental land areas at comparable northern latitudes (Table 1.1). High relative humidity and frequent cloud cover limit the duration of sunshine in these oceanic regions. In contrast, many Arctic land areas, particularly inland, experience short but relatively warm summers with extended periods of sunshine, combined with prolonged, severe winter frost (Table 1.1). Snow cover is continuous for several months each winter, at least in sheltered situations, throughout the polar regions, except in the mild-Antarctic and on some cool-Antarctic islands where plant growth may be possible throughout the year.

Water availability also varies widely. Annual precipitation is moderate in the mild-Arctic, commonly 125–450 mm yr^{-1} rainfall equivalent. Water stored in winter snowfall is released by the spring thaw early in the growing season, and subsequent losses are minimised by the effect of flat terrain on surface runoff, by impeded drainage due to peat and permafrost (page 12), and by low to moderate summer temperatures which restrict evaporation. Wetland and mesic habitats thus occupy much of the land surface. Summer precipitation is generally rain.

Precipitation is also moderate to high in many southern polar regions, with means greater than 1400 mm yr^{-1} on some cool-Antarctic islands and greater than 7000 mm yr^{-1} in parts of the magellanic moorland region of the mild-Antarctic (Pisano, 1983). Rain predominates in the mild-Antarctic, but the proportion of annual precipitation falling as snow increases with latitude. Some rain falls in summer as far south as the cold-Antarctic, where precipitation is light but frequent, and as cool, cloudy conditions prevail (Table 1.1) there is ample water in many habitats. Summer snowfalls, at times persisting for several days, occur throughout the southern polar regions except in the mild-Antarctic.

The continental climates in the frigid-Antarctic and much of the cool-Arctic are marked by low precipitation (Table 1.1) resulting in conditions of extreme aridity. Wind removes much of the winter snowfall from exposed habitats, and in the frigid-Antarctic the frequently subfreezing summer temperatures result in water being locked away as ice. Availability of liquid water plays a major role in controlling the extent of the vegetation in both areas.

The effects of cold and aridity are commonly accentuated by strong winds, for the influence of wind on polar cryptogams is enhanced by the scant shelter provided by taller plants. Mean annual wind speeds vary widely. They are higher in some coastal parts of the frigid-Antarctic than anywhere else in the world, reaching 18.5 m s^{-1} in places such as Cape Denison, Mawson's 'Home of the Blizzard' (Phillpot, 1985). Winds are in general no stronger than in temperate regions throughout much of the cool-Arctic, but even here storms occur in association with cyclonic disturbances. A principal component analysis of data on climate, soil and vegetation from a range of polar sites indicated that wind is a factor of overriding importance on cool-Antarctic islands (French & Smith. 1985).

Polar regions as understood here are thus characterised by cool to cold, often short, summers, with mean monthly air temperatures below 10–12 °C. The general trend is for mean temperatures to decrease and aridity to increase with latitude. Nevertheless, polar climates are diverse in winter temperature, duration of snow cover, precipitation and other factors, and particularly in the ecologically critical combination of water availability, summer temperature and duration of the annual period when mean air temperatures exceed 0 °C. Table 1.1 gives only a general impression of polar climates because many parameters vary both spatially in response to topography and from year to year at a given site, while microclimatic conditions that directly affect bryophytes and lichens differ dramatically from those indicated by standard meteorological recording (Chapter 4). Climatic variation from year to year takes on added significance in areas of climatic severity where conditions are close to the tolerance limits of many species (Hustich, 1975).

Soils

Polar soils share in common several features attributable to their youthfulness and the predominance of physical rather than chemical weathering in cold, and especially in cold, dry environments, but they vary markedly in features determined by precipitation and vegetation cover. A detailed account of polar soils is given in Tedrow (1977), and those of the Antarctic are discussed by Claridge & Cambell (1985).

The parent materials comprise, in the uplands, bedrock and scree scraped clear of soil and vegetation by Pleistocene ice movement, and in the lowlands, glacial till, aeolian silts and clays, and lacustrine deposits laid down during or since glacial retreat. The soils generally lack highly differentiated profiles and show little incorporation of organic matter below the surface horizon. Many include a layer of permafrost, i.e. soil that remains frozen throughout the summer, often with the particles cemented together by ice. The active layer above permafrost, where the soil thaws during summer, ranges from 20 to 80 cm in depth, being shallowest in the colder climates and under closed vegetation or peat. Permafrost is continuous throughout most polar regions, but occurs discontinuously in parts of the mild-Arctic, and is absent from the mild- and cool-Antarctic (Ives, 1974; Walton, 1984).

Mires overlying thick peat deposits occur extensively in mild-polar low-lands. Many other mild-Arctic soils show an A horizon with abundant organic matter over strongly gleyed lower strata of variable texture (Everett, Vassiljevskaya, Brown & Walker, 1981), and brown soils show-ing weak podsolisation occur locally beneath closed vegetation at well-drained sites (Tedrow, 1977). Where the vegetation cover becomes sparse at higher latitudes and altitudes, most soils are composed predominantly of mineral material, often coarse-textured, although superficial peat deposits may occur locally.

Organic soils formed under waterlogged conditions tend to be acidic but, depending on the nature of the parent material, the mineral soils are commonly neutral or alkaline, especially in arid regions of the cool-Arctic and frigid-Antarctic. Available nitrogen and phosphorus are pres-ent at low concentrations in many tundra soils, the former occurring prin-cipally as ammonia, although large, unavailable pools of both these elements may be present in permanently frozen organic matter (Chapter 7). However, deposition from precipitation, spray and marine animals results in non-limiting concentrations of all the major nutrients in some maritime regions (Davis, 1981).

Many types of patterned ground have developed as a result of frost heave, solifluction and related processes. They have been described in detail by Tedrow (1977), and their distribution in Antarctic regions is reviewed by Walton (1984). These features include stone circles, poly-gons, steps and stripes, and they are significant in increasing the range of microhabitats available to mosses and lichens. Substrate instability associated with frost heave, solifluction and thermocarst (page 301) is important in some areas in disrupting existing vegetation, in preventing the establishment or reestablishment of closed plant cover, and thus in providing niches for colonising species.

Vegetation types and zones
Vegetation types

The polar regions support a diversity of vegetation types, of which several classifications have been proposed (Aleksandrova, 1980; Bliss, 1981; Wielgolaski, 1972). Ten major phanerogamic vegetation units are recognised in the simplified scheme in Table 1.2, each being distinguished principally by the growth form of the dominant species. Bryophytes and lichens form an abundant component of most phanerogamic vegetation, and they are dominant in other communities in both the Arctic and the Antarctic (Chapter 2).

Table 1.2 *Major vascular plant-dominated vegetation types in polar regions*

Vegetation type	Distinctive features and distribution in polar regions	Representative genera
Wetland communities		
Wet meadow	Ombrogenous or soligenous wetland vegetation dominated by grasses, sedges and rushes (graminoids), sometimes with abundant dwarf shrubs; vascular plants usually rooted in a bryophyte understorey; often overlying peat and then classed as mire. Surface level or hummocky. Extensive in mild polar regions and on cool-Antarctic islands; widespread in small stands in the cool-Arctic.	*Arctagrostis* (N) *Carex* (N) *Eriophorum* (N) *Juncus* (NS) *Rostkovia* (S)
Cushion bog	Wetland areas dominated by low, hard cushion-forming dicotyledons, often over peat. Confined to the mild-Antarctic	*Astelia* (S) *Gaimardia* (S)
Mesic communities		
Woodland	Stands of open, often stunted arboreal vegetation. Confined to well-drained substrata in mild-polar regions, particularly in sheltered river valleys.	*Betula* (N) *Larix* (N) *Picea* (N) *Nothofagus* (S)
Scrub	Dominated by shrubs 0.5–1.0 (to 3.0) m tall, often with herb and moss layers below. Extensively developed only in the warmer parts of the mild-polar regions.	*Alnus* (N) *Salix* (N) *Cheliotrichum* (S)
Dwarf shrub heath	Dominated by erect or ascending shrubs 10–50 cm tall with a range of life forms represented in the lower strata. Widespread in mesic habitats in mild-polar regions, extending into the cool-Arctic in sheltered sites insulated by snow in winter.	*Cassiope* (N) *Ledum* (N) *Vaccinium* (N) *Empetrum* (NS) *Pernettya* (S)
Grass heath	Vegetation in which caespitose grasses, sedges and rushes are prominent on relatively dry substrata, without deep peat formation. A variety of other life forms occur between the graminoid tussocks, and creeping species of *Salix* may assume dominance.	*Carex* (N) *Kobresia* (N) *Luzula* (N) *Salix* (N)

Community	Description	Genera
Herbfield	Mesic communities in which broad-leaved herbaceous perennials are dominant, often associated with grasses and mosses. Widespread on cool-Antarctic islands but in the Arctic restricted to small stands in late-lying snowbeds or on substrata subject to nutrient enrichment by animals.	*Poa* (NS) *Festuca* (NS) *Cortadaria* (S) *Ranunculus* (N) *Saxifraga* (N) *Acaena* (S) *Cotula* (S) *Pleurophyllum* (S)
Tussock grassland	Dominated by caespitose grasses forming tussocks overtopping a basal pedestal of fibrous peat, the whole reaching heights of up to 3 m. Forms a coastal fringe around cool-Antarctic islands and, until largely destroyed by grazing, the Falkland Is.	*Poa* (S)
Xeric communities Fellfield	Communities comprising abundant cushion-forming phanerogams interspersed among plants of other growth forms: generally open, but intergrading in more mesic sites with vegetation having a closed moss/lichen layer between the vascular plants. Extensive on semi-arid substrata throughout mild- and cool-polar regions, but principally on uplands in the former.	*Dryas* (N) *Silene* (N) *Azorella* (S) *Colobanthus* (S)
Barren	Predominantly bare ground with scattered individuals of several growth forms including tufted graminoids, cushion and rosette plants. Extensive throughout the cool-Arctic.	*Draba* (N) *Kobresia* (N) *Oxyria* (N) *Potentilla* (N) *Saxifraga* (N)

N = Arctic regions
S = Antarctic regions

Vegetation zones: previous terminology

It has long been recognised that growth form spectra vary within polar regions in such a way that the vegetation tends to become progressively more open, lower in stature and simpler in structure with increasing latitude and climatic severity. The relative importance of mosses and lichens rises correspondingly, suggesting that these plants are particularly well equipped to survive in severe polar environments (Chapters 5 and 6). Variation in growth form composition provides a basis for recognising vegetation zones. Many schemes have been proposed, unfortunately with similar terms applied to areas supporting quite dissimilar vegetation, as the following examples illustrate.

Polunin (1951) regarded the extent of continuous plant cover as a major criterion in the recognition of latitudinal belts within Arctic tundra, i.e. 'the low-Arctic, in which the vegetation is continuous over most areas, the middle-Arctic in which it is still sufficient to be widely evident from a distance, covering most of the lowlands, and the high-Arctic in which closed vegetation is limited to the most favourable habitats and is rarely at all extensive.' He considered the sub-Arctic to comprise more southerly regions of open forest and forest-tundra, which form an ecotone between closed boreal forest and low-Arctic tundra. Many western authors have recognised only two tundra zones in the northern polar regions, the low-Arctic and high-Arctic, the boundary lying within Polunin's middle-Arctic (Bliss, 1981).

Soviet approaches to defining polar vegetation zones have been reviewed by Aleksandrova (1980), who placed considerable emphasis on the distribution of plakors, or plant associations occurring in favourable, mesic habitats which are protected by snow cover in winter but are exposed for much of the summer. Aleksandrova recognised three principal latitudinal zones. Her region of sub-Arctic tundra corresponds approximately with the low-Arctic as defined by Bliss (1981), rather than with Polunin's sub-Arctic. Within the high-Arctic, Aleksandrova recognised a region of Arctic tundra, and, in the far north, a region of Arctic polar desert characterised by the dominance of cryptogams. Other authors apply the term desert more widely to high-Arctic vegetation (Bliss, 1981; Porsild, 1951).

Cool temperate, sub-Antarctic and Antarctic zones are commonly recognised in the southern polar regions (Greene, 1964; Wace, 1960, 1965). The cool-temperate zone comprises parts of southern South America and oceanic islands lying near the subtropical convergence, a circumpolar boundary between subtropical surface water and cooler sub-

Antarctic water which results in a sharp change through some two to three degrees in temperatures at the ocean surface. Many cool temperate areas support stands of semiprostrate evergreen trees to 30 m tall with ferns abundant in the understorey or, in southern South America, forests formed by species of *Nothofagus*. However, arboreal vegetation is rare or absent in the Falkland Is and the magellanic moorland region of south-west Chile (Godley, 1960; Moore, 1968, 1979).

The sub-Antarctic consists exclusively of oceanic islands located near the Antarctic Convergence, where cold, north-flowing Antarctic surface water sinks beneath warmer sub-Antarctic water resulting in a further sharp change in surface temperature and associated climatic factors. The Antarctic continent, its off-lying islands and other, isolated islands in the South Atlantic comprise the Antarctic zone, within which maritime and continental regions are recognised due to differences in vegetation and climate (Holdgate, 1964a). The sub-Antarctic zone as thus defined lacks trees and dwarf shrubs, and its vegetation differs dramatically from that in the sub-Arctic as understood by either Aleksandrova (1980) or Polunin (1951). Moreover, the maritime and continental regions of the Antarctic zone are referred to as low- and high-Antarctic by some authors (e.g. Wace, 1965), although their vegetation is essentially cryptogamic and quite unlike that in much of the low- and high-Arctic. More thorough reviews of polar vegetation zones, and their subdivision on floristic and vegetational grounds, are presented by Bliss (1981), Markov *et al.* (1970) and Pickard & Seppelt (1984).

Vegetation zones: a revised terminology

Rationale Four vegetation zones are recognised in Antarctic regions and three in the Arctic as a framework for the present account (Figs. 1.1 and 1.2). They correspond closely to those of Bliss (1979, 1981) and Aleksandrova (1980) in the north, and of Greene (1964) and Smith (1984a) in the south (Table 1.3), but with a revised nomenclature. The prefixes mild-, cool-, cold- and frigid- each relates to Arctic and Antarctic regions supporting physiognomically similar vegetation. They can be used specifically, e.g. cool-Arctic, or collectively, e.g. cool-polar regions. The terms sub-, high-, low- and polar desert have been avoided in view of the variety of their past usage.

The present zonation is based essentially on vegetation, rather than a broad combination of ecological features including fauna, climate and soils. Although desirable, and probably feasible for either Arctic or Antarctic regions individually, the latter approach is fraught with difficulty when applied to both northern and southern hemispheres. This stems

Table 1.3 *Vegetation zones in polar regions*

Zone	Approximately corresponding zones of previous authors		Highest mean monthly air temperature (°C)	Characteristics of the vegetation
	Arctic	Antarctic		
Mild-polar	Low-Arctic (B, P). Part of the mid-Arctic (P). Sub-Arctic tundra subregion (A)	Those parts of the southern temperate region lacking arboreal vegetation (G, W)	6–10 (to 12)	Extensive grass heath, dwarf shrub heath, mire and other closed phanerogamic vegetation. *Sphagnum* abundant in many mires, though local in the mild-Antarctic. Fellfields on the drier uplands.
Cool-polar	Most of the high-Arctic (B, P), Part of the mid-Arctic (P). Arctic tundra subregion (A)	Sub-Antarctic zone (G)	3–7	Open fellfields and barrens predominant but mire, dry meadow and other closed angio-sperm-dominated communities locally extensive in favourable habitats. Dwarf shrub heaths of restricted occurrence or absent. *Sphagnum* seldom a major component of mires.
Cold-polar	Parts of the high-Arctic (B, P). Arctic polar desert region (A)	Maritime Antarctic (H). Low-Antarctic (W)	0–2	Closed stands of bryophytes, lichens or algae extensive where wet or mesic conditions occur, with open cryptogamic vegetation on drier ground. Herbaceous phanerogams subordinate to cryptogams or absent. Liverworts frequent.
Frigid-polar	—	Continental Antarctic (H) High-Antarctic (W)	<0	Vegetation largely restricted to scattered colonies of mosses, lichens or algae, and to endo-lithic microorganisms. Phanerogams absent. Liverworts very rare.

A = Aleksandrova (1980)　　H = Holdgate (1964a)
B = Bliss (1979)　　　　　　P = Polunin (1951)
G = Greene (1964)　　　　　W = Wace (1965)

from the effect of dispersal barriers which have restricted biota native to some Antarctic regions, and from lack of correlation between the major climatic variables (Table 1.1). Thus summer air temperatures are similar in the cool-Arctic and cool-Antarctic, but the former has an arid climate combined with severe cold in winter whereas cool-Antarctic islands experience frequent, heavy rainfall and relatively mild winters.

The vegetation zones, as determined by the range of plant growth forms, are more closely correlated with mean summer temperature than with winter temperature or precipitation (Tables 1.1 and 1.3), a fact emphasised by the present terminology. In contrast, both the extent of plant cover and the distribution of the principal polar soil types appear to be governed more strongly by precipitation than by temperature, resulting in dissimilarities between the present zonation based on vegetation and others based on soils (e.g. Tedrow, 1977).

Imperfections are inevitable, even in a system based solely on vegetational criteria, as spatial changes in plant communities often occur gradually along environmental gradients, and altitudinal zonation over short horizontal distances may resemble the broader regional pattern. The zones are by no means uniform and can be subdivided, as Weyant (1966) has done for the frigid-Antarctic. Bipolar comparison is further complicated by climatic differences between the northern and southern hemispheres that result in tundra vegetation extending to lower latitudes in the south. One consequence of this is the occurrence in southern polar regions of distinctive tundra formations, notably tussock grassland, cushion bog and herbfield (Table 1.2), which are related to tropical mountain communities and have no direct counterparts in the Arctic (Troll, 1960).

Mild-polar regions The mild-polar regions include land areas beyond the poleward limit of extensively developed woodland, an important boundary because the presence of tree cover has a profound influence on the microclimate affecting cryptogamic vegetation. Dwarf shrubs are an important component of mild-Arctic vegetation. They are regarded as woody chamaephytes, generally under 50 cm tall, but with the principal stems and branches erect or ascending, and are thus distinguished from woody perennials with the principal stems creeping along the ground surface (e.g. *Acaena* spp, *Salix arctica*), and from low cushion-forming perennials such as *Dryas* spp. The term dwarf shrub was used in the present sense by Wace (1960), although it is employed more broadly by some authors.

Mild-polar lowlands support extensive, closed stands of grass heath, dwarf shrub heath and wet meadow, but fellfields are present in the uplands. Woodland and scrub occur in sheltered valleys, and become more extensive in southern parts of the mild-Arctic. Wet meadows with an abundance of graminoids, *Sphagnum* spp and pleurocarpous mosses cover large areas of the mild-Arctic plains. Similar communities occur in the mild-Antarctic, where cushion bogs without sphagna are also extensive.

The southern boundary of the mild-Arctic is difficult to fix because forest gives way to tundra through a broad ecotone tens of kilometres wide supporting a mosaic of woodland and tundra communities. Larsen (1974) presented a useful map showing, for much of North America, the line along which woodland cover declines to approximately 50%. The forest boundary has been regarded as corresponding most closely with the 10 °C July isotherm (Köppen, 1918), the Nordenskjöld line which takes into account both summer and winter temperatures (Polunin, 1951), or potential evapotranspiration (Hare, 1950). More recently, it has been shown that the southern boundary of open tundra coincides across large parts of North America and Eurasia with the mean summer position of the Arctic front (Bryson, 1966; Krebs & Barry, 1970). This interface between cold, dry Arctic air and warmer, moist air of more southerly origin is likely to be associated with relatively sharp spatial variation in temperature and other features of the summer climate. The influence of the summer-cool water body in Hudson's Bay on the southern boundary of the mild-Arctic is clearly evident in Fig. 1.1.

The mild-Antarctic includes only the Falkland Is and the magellanic moorland region of southern Chile, and is clearly defined. The Falkland Is lack native trees, and, except in river valleys, the Patagonian *Nothofagus* forests scarcely extend into terrain underlain by dioritic rocks, which is occupied by the magellanic moorland (Godley, 1960; Moore, 1979). This raises the question whether the treeless areas considered here as mild-Antarctic should be regarded as areas of polar tundra, or of heathland reflecting oceanic or geological influences, especially as mean monthly air temperature remains above 0 °C in winter. The soils may be continuously frozen for 3–6 months each winter, however, and Pisano (1983) regarded low temperature as a major limiting factor, noting also a strong physiognomic similarity between magellanic moorland and Arctic tundra. Godley (1960) suggested that temperature was unlikely to be limiting for tree growth in the magellanic moorland region, but he nevertheless combined the present mild-Antarctic with a number of more southerly islands within his sub-Antarctic zone, from which areas with extensive arboreal vegetation were excluded. It is significant that prolonged frost and snow cover are seldom experienced on some of the latter islands, here referred to as cool-Antarctic, where mean summer air temperatures remain so low that a polar influence can scarcely be doubted (Table 1.1).

Controversy also exists as to the status of some northern tundra, for example in Iceland, northern Scandinavia and around the southwestern shores of Hudson's Bay where, as at Churchill (Table 1.1), the winters are severe but mean July temperatures reach 10–12 °C (Larsen, 1972). For present purposes, tundra regions beyond polar forest lines with mean monthly air temperatures at sea level below 12 °C are regarded as polar. It must be recognised, however, that the absence of trees results in some areas from the interaction of what might be regarded as polar and maritime influences combined with geological and historical factors such as human interference and the time available for recolonisation since the land rose above the sea by isostatic uplift.

Cool-polar regions The cool-polar regions include land beyond the poleward limit of extensive dwarf shrub heath, in which flowering plants nevertheless continue to play a significant role in the vegetation. Graminoids, cushion- or rosette-forming plants, and creeping perennial species of *Salix* are predominant among cool-Arctic phanerogams forming open fellfields, and barrens where only scattered individual plants survive. Wet meadow, grass heath and, less extensively, dwarf shrub heath occur locally where the moisture regime is favourable, but *Sphagnum* spp are rare. Grass heath, herbfield and mire are extensive in some cool-Antarctic lowlands, with tussock grassland widespread in coastal regions, but dwarf shrub heath is absent. Fellfields and barrens occur extensively on non-glaciated terrain in cool-Antarctic uplands and also predominate in some exposed lowlands.

The southern boundary of the cool-Arctic is again impossible to define with precision, as the extent of dwarf shrub heath, and of closed phanerogamic vegetation generally, decreases gradually with increasing latitude. In contrast, an ocean barrier leads to discontinuity in climate and vegetation between mild- and cool-Antarctic regions. Aleksandrova (1980) suggests that the boundary between her sub-Arctic and Arctic tundra regions, which roughly corresponds with that between the mild- and cool-Arctic (Table 1.3), coincides with the 6 °C July isotherm. Similarly, mean summer air temperatures remain below 7 °C in the cool-Antarctic (Holdgate, 1964a), although the two cool-polar regions otherwise differ substantially in climate.

Cold-polar regions   Cold-polar regions support extensive, closed communities dominated by mosses, lichens and algae, although much of the terrain is occupied by open cryptogamic vegetation. Liverworts are frequent, but *Sphagnum* is absent. Flowering plants are seldom abundant: there are two native species in the cold-Antarctic, one of them, a grass, locally forming small areas of closed turf. In the cold-Arctic, over 50 angiosperm species may grow within a given area but they occur exclusively as scattered individuals forming small tufts or cushions, with creeping willows also widespread. The vegetation is almost lacking in stratification, as the occasional flowering plants do not extend above the mosses and lichens, and they fail to form a closed root system (Aleksandrova, 1980).

Once more we find that the boundaries between the cool- and cold-polar regions are imprecise in continental Eurasia and North America, but clearly defined in the Antarctic due to ocean barriers. Aleksandrova (1980) recognises within her Arctic polar desert region a northern belt analogous with the cold-Antarctic in growth form composition and structure of the vegetation, and a southern belt which represents a transition to the cool-Arctic. Both are included in the cold-Arctic in Fig. 1.1: the northern belt includes only scattered areas in the extreme north of the Eurasian mainland, Franz Josef Land, the northernmost part of the Severnaya Zemlya Archipelago, the De Long Is, Meighen I, and a number of other small islands in the Arctic Ocean and in the Canadian Arctic archipelago. Lowland glaciation is extensive in the cold-polar regions. Mean monthly air temperatures during the warmest month of the year are generally 0–2 °C, but the winters are milder and summer conditions moister in the Antarctic than in the Arctic (Table 1.1).

The frigid-Antarctic   Sparse, open populations of mosses, lichens and algae are interspersed with more extensive areas of terrain devoid of macroscopic vegetation even in non-glaciated parts of the frigid-Antarctic. Closed cryptogamic vegetation is seldom extensive and usually comprises single-species stands, only one liverwort has been recorded, and flowering plants are unknown. Mean monthly air temperatures only locally reach 0–1 °C in summer, and aridity is a major limiting factor. Both mosses and lichens are locally abundant at scattered coastal sites, but lichens generally predominate on inland nunataks. The cold- and frigid-Antarctic zones intergrade on the west coast of the Antarctic Peninsula around Marguerite Bay.

Particularly severe conditions occur in the ice-free dry valley region of southern Victoria Land, a mountainous area several thousand square kilometres in extent that is cut off by the Transantarctic Mountains from the flow of glaciers from the continental interior. Here, maximum air temperatures never rise significantly above 0 °C and mean annual relative humidity may be as low as 45%. Precipitation is light, in the order of 150 mm yr^{-1} rainfall equivalent. It falls exclusively as snow, much of which is lost by sublimation so that little melt water is available in the absence of glaciers. The soils are highly saline, and endolithic lichens and other microorganisms occurring within the surface layers of rocks and stones constitute the principal biota (Friedmann, 1982; Wilson, 1970). The frigid-Antarctic has no northern-hemisphere counterpart.

Detailed ecological studies have been carried out at a number of sites in the polar regions defined in Table 1.3, and in open woodland in the northern boreal zone as understood by Ahti, Hämet-Ahti & Jalas (1968). Particularly important sites that are referred to repeatedly in subsequent chapters are indicated in Figs. 1.1 and 1.2 and in Table 1.4.

Polar bryophyte and lichen floras
The Arctic flora

Assessment of the size, geographical affinity and history of polar cryptogamic floras is hampered by taxonomic uncertainty and inadequate distribution data. These constraints apply especially to Antarctic regions and to confusing genera such as *Bryum* and *Lecidea*. There are moss and liverwort floras of the Soviet Arctic (Abramova, Savicz-Ljubitskaja & Smirnova, 1961; Schljakov, 1982), and publication of a moss flora of Arctic North America has begun (Mogensen, 1985). Bryophyte checklists and more detailed floristic accounts have been published for smaller areas, e.g. Arctic Alaska (Steere, 1978a; Steere & Inoue, 1978), Devon I (Vitt, 1975), northern Ellesmere I (Brassard, 1971a, 1971b, 1976) and Iceland (Jóhansson, 1983). The distribution of over 50 bryophytes in Greenland has been mapped by Mogensen & Lewinsky (1982), and a wealth of distribution data on American Arctic hepatics can be found in Schuster (1966–80). For lichens there are checklists of Truelove Lowland on Devon I (Table 1.4) and other sites, a flora of the Alaskan Arctic Slope and the first volume, covering macrolichens, of a lichen flora of the North American Arctic (Thomson, 1979, 1984; Barrett & Thomson, 1975). These and other studies have demonstrated that the Arctic cryptogamic flora is diverse, and comprises several coherent floristic elements.

Table 1.4 *Sites of major investigations into the biology of polar bryophytes and lichens*

Vegetation zone	Locality	Latitude	Longitude	Site description
Northern Boreal	Stordalen Mire, Abisco, Sweden	68°20'N	18°51'E	Sonesson (1980)
	Hardangervidda, Norway	60°17'N	7°44'E	Wielgolaski (1975a)
	Kevo, Finland	69°43'N	27°04'E	Wielgolaski (1975a)
	Schefferville, Quebec	54°48'N	66°49'W	Hicklenton & Oechel (1976)
Mild-Arctic	Churchill, Manitoba	58°45'N	94°10'W	Ritchie (1956)
	East Pen I, Ontario	56°46'N	88°46'W	Kershaw & Rouse (1973)
	Atkasook, Alaska	70°28'N	157°23'W	Komarkova & Webber (1980)
Cool-Arctic	Barrow, Alaska Truelove Lowland,	71°17'N	156°47'W	Tieszen (1978a)
	Devon I, NWT	75°33'N	84°40'W	Bliss (1977)
Cool-Antarctic	Marion I	46°55'S	37°45'E	Gremmen (1982)
	Maquarie I	54°30'S	158°57'E	Jenkin (1975)
	South Georgia	54–55°S	36–38°W	Smith & Walton (1975) Headland (1984)
Cold-Antarctic	Signy I, South Orkney Is	60°43'S	45°36'W	Collins *et al.* (1975)
Frigid-Antarctic	Cape Hallett, Southern Victoria Land	72°19'S	170°13'E	Rudolph (1966)
	Ross I, McMurdo Sound, Southern Victoria Land	77°51'S	166°40'E	Longton (1937a, 1974a)
	Syowa Station Prince Olav Coast	69°00'S	39°35'E	Kanda (1981a)

Particularly rich bryophyte floras are known from areas such as Iceland (426 moss species, 146 hepatics) and Arctic Alaska (415 mosses, 135 hepatics) which include a variety of montane and lowland habitats subject in part to mild-Arctic climatic regimes. The occurrence of glacial refugia is another factor contributing to the richness of the Alaskan flora. Steere (1978a) has suggested that the moss flora of Alaska may reach 475–500 species, noting that the present total exceeds those for many better-known states in continental USA. At least 166 moss species are known from cool-Arctic sites on northern Ellesmere I, but here diversity is greatest in inland valleys where summer conditions are unusually warm for the latitude (81–83° N). The total Arctic bryophyte flora is likely to comprise some 600–700 species, compared with about 900 flowering plants (Polunin, 1959). About 2000 lichen species have been recorded, indicating remarkable diversity even allowing for some eventual reduction to synonymy (Thomson, 1972).

Distribution patterns of Arctic bryophytes and lichens to some extent parallel those of flowering plants (Porsild, 1958), but fewer species are restricted to high latitudes. The majority are also widespread in boreal, or in boreal and temperate regions (Fig. 1.7), commonly with extensions to alpine areas, and most, perhaps 80% of the bryophytes and 70% of the lichens, have circumpolar distributions (Steere, 1978a; Thomson, 1972). There is a small cosmopolitan, and a significant bipolar element. Small amphi-Atlantic, amphi-Beringian and other disjunct elements are also represented (W. B. Schofield, 1972; Thomson, 1984), but there are few local endemics.

Of particular phytogeographical significance are the small temperate disjunct, and the larger circumpolar Arctic elements. The former comprises species having their main areas of distribution in temperate regions, with disjunct Arctic populations often centred in glacial refugia such as the north slope of Alaska and parts of northern Ellesmere I (Fig. 1.8). The circumpolar Arctic element (Fig. 1.9) comprises some 15% of Arctic bryophytes and 35% of the macrolichens (Steere, 1965a; Thomson, 1984). Many of these species are also widespread in glacial refugia, some having disjunct populations on lower latitude mountains. A number of the bryophytes have their closest relatives in warm-temperate or tropical regions.

Both Schuster (1983) and Steere (1978a) believe the bryophytes in these two elements to be remnants of an early Tertiary flora formerly existing in Arctic regions under temperate or subtropical conditions. The present species are thought to have persisted throughout the Pleistocene in refugia from which they have subsequently radiated to varying extents.

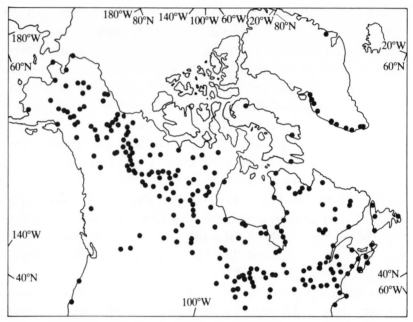

Fig. 1.7. Distribution of the lichen *Cladonia stellaris* in North America. This species has a circumpolar distribution in temperate, boreal and mild-Arctic regions. Data from Thomson (1984).

Arctic recolonisation by the more widely distributed taxa is seen as having occurred by migration from both northern refugia and regions south of the Pleistocene ice boundary. Brassard (1974) has suggested that some members of the circumpolar Arctic element evolved in the Arctic during more recent times.

Thomson (1972) considered most Arctic lichens to be of pre-Pleistocene origin. In contrast to Steere's position regarding mosses, however, he maintained that each species has had time to colonise all parts of the Arctic. Thus disjunct distributions were seen as a reflection not of historical factors, but of corresponding disjunctions in critical features of the microenvironment. Species disjunct between temperate regions, Arctic glacial refugia and scattered intervening sites in formerly glaciated areas were cited in support of this view. The Beringian flora also includes elements showing affinities with lower-latitude steppe vegetation (Murray, Murray, Yertsev & Howenstein, 1983).

The Antarctic flora
Size and affinities of the flora   Far-southern cryptogamic floras have been increasingly studied in recent years. For bryophytes, there

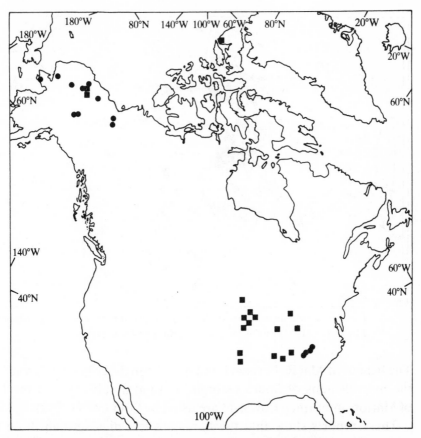

Fig. 1.8. Distribution of the lichen *Umbilicaria caroliniana* (■) and the moss
Seligeria pusilla (●) in North America. *U. caroliniana* also occurs in Japan and
eastern Siberia and *S. pusilla* in central and western Europe, the Caucasus and
Japan. Data from Thomson (1972, 1979) and Steere (1978a).

are checklists covering the cold- and frigid-Antarctic and several cool-
Antarctic islands (Gremmen, 1982; Greene, 1968; Grolle, 1972; Seppelt,
1977, 1980) and floristic accounts of some frigid-Antarctic localities
(Kanda, 1981a; Seppelt, 1984). More detailed systematic treatments
are also available for selected genera on South Georgia and in the two
southernmost zones (Greene, Greene, Brown and Pacey, 1979; Lightowlers, 1985). Several of the checklists require updating in the light of
more recent studies.

Lamb (1968) has presented detailed accounts of *Buellia, Rinodina* and
several macrolichen genera from the Antarctic Peninsula region and
Dodge (1973) has provided a lichen flora of the Antarctic continent, a
work sometimes criticised for its narrow species concept (Lindsay, 1977).

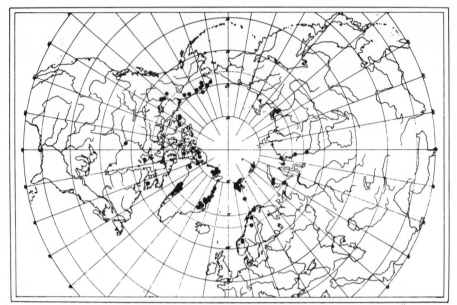

Fig. 1.9. World distribution of the moss *Cinclidium arcticum*. Reproduced from Mogensen (1973) by courtesy of the Editor of *Lindbergia*.

The lichens of MacRobertson Land have been described by Filson (1966), the macrolichens of South Georgia by Lindsay (1974), and the lichens of Marion and Prince Edward I are listed by Gremmen (1982).

These reports show that the floras are less diverse than those in the Arctic. Table 1.5 emphasises the important role of bryophytes and lichens, but shows that the numbers of species decline sharply with increasing climatic severity from the cool- to the frigid-Antarctic. There is considerable overlap in species representation between the zones and the total numbers of species so far recorded in the areas covered by Table 1.5 are probably in the order of 250–300 mosses, 150–175 hepatics and 300–400 lichens. A considerable richer flora exists in the mild-Antarctic, as indicated by floristic accounts for several parts of the zone (e.g. Boelcke, Moore & Roig, 1985; Engel, 1978; Hässal de Menéndez & Solari, 1975; Matteri, 1986; Seki, 1974).

Skottsberg (1960) recognised three major distribution patterns among southern cool-temperate vascular plants, the Fuegian type centred in southern South America, the Neozealandic type centred in Australasia and the bicentric type represented in both these regions, with bipolar and cosmopolitan elements within each one. Most Antarctic cryptogams also occur in temperate areas where they show one or other of these patterns.

Table 1.5 *Approximate numbers of native plant species in Antarctic regions*

Vegetation zone	Mosses	Liverworts	Lichens	Pteridophytes	Flowering plants	Macrofungi
Cool-Antarctic	250	150	>300	16	56	>70
Cold-Antarctic	75	25	150*	0	2	>20
Frigid-Antarctic	30	1	125*	0	0	2

Data from Smith (1984a). No comparable figures are available for the mild-Antarctic.
* Dodge (1973) recognises approximately 430 lichen species from these two zones.

Cold- and frigid-Antarctic regions The bryophyte flora of the cold- and frigid-Antarctic thus differs from that in the Arctic not only in its relative paucity, but also in the dearth of endemic taxa. Of the species apparently confined to these areas (Greene, 1968), the majority are doubtfully distinct members of difficult genera such as *Bryum* and *Ceratodon*. Indeed, *Bryum antarcticum*, is considered by Kanda (1981b) to be synonymous with the widespread species *Pottia heimii*. A few others such as *Grimmia lawiana* and *Schistidium antarctici* are better marked and may stand as endemics unless they are eventually discovered elsewhere. There are differences of opinion concerning the degree of endemism among the lichens. Dodge (1973) regarded the majority of species as endemic, often to local areas within Antarctica, but Lindsay (1977) recognised fewer endemics, notably an assemblage of fruticulose members of otherwise crustose genera centred in the Antarctic Peninsula region.

The vast majority of bryophytes, and many of the lichens, are confined, in Antarctica, to the cold-Antarctic zone, but others have circumpolar distributions embracing also the frigid-Antarctic (Fig. 1.2). Indeed, some species are more widespread and abundant there than in the Peninsula region, and future studies could confirm the presence of elements centred in particular sectors of the frigid-Antarctic (Rudolph, 1967b).

The Antarctic flora has strong affinities with the Fuegian region, but some bicentric species are represented, and there are substantial cosmopolitan and bipolar elements, particularly among lichens (Greene, 1964; Lindsay, 1977). Most Antarctic bryophytes are thought to be post-glacial immigrants (Robinson, 1972), but Dodge (1973) and Lamb (1970) have suggested that many of the lichens, particularly endemic species, survived in Antarctica throughout the Pleistocene on high-altitude nunataks. There are problems with this interpretation for, as Lamb points out, many of the Antarctic Peninsula endemics are ornithocoprophilous (Chapters 2 and 3), and it seems unlikely either that this characteristic could have evolved in the relatively short period since maximal glaciation (c 20 000 yr) or that a substantial avifauna persisted in the Antarctic throughout the Pleistocene.

The theory that the present Antarctic lichen flora may be derived in part from species surviving on nunataks is lent support by the observations of Filson (1982). Working around Mawson Station, he found that where melt water streams flow from inland nunataks to coastal outcrops the same lichen species tend to occur at both sites, but that coastal outcrops not at the termination of streams emanating from vegetated areas are devoid of lichens. Similarly, lichen fragments were found in melt streams, but only those originating in areas supporting lichen vegetation. The extent of Pleistocene survival among the Antarctic lichen flora is likely to remain controversial, at least until the degree of endemism becomes clearly established.

Cool-Antarctic islands The bryophyte and lichen floras of the cool-Antarctic appear to be composed largely of species showing Skotts-berg's Fuegian or bicentric distribution types, with the Neozealandic type well represented on Macquarie I (Fig. 1.2). A striking number of species, particularly of mosses, are widely distributed on these isolated oceanic islands and many are bipolar. Several mosses and rather more hepatics at present appear to be endemic to Marion I and other islands in the southern Indian Ocean (Gremmen, 1982), possibly reflecting persistence throughout the Pleistocene. The degree of endemism is otherwise low, and cool-Antarctic bryophyte floras, like those existing further south, are thought to comprise principally post-glacial immigrants from the Magellanic region (Smith, 1984a).

The origin of bipolar distributions, and aspects of the reproductive biology of bryophytes and lichens that influence their capacity for dispersal, establishment and evolution are discussed in Chapter 8.

2

The cryptogamic vegetation

Vegetation classification

Boundaries between communities based on the distribution and abundance of cryptogams and of flowering plants may only partially coincide, as Alpert & Oechel (1982) demonstrated in Alaska. The most useful vegetation accounts thus consider all the major plant groups. The procedures of the Braun-Blanquet school have been adopted to define vegetation types at isolated Antarctic localities (Kappen, 1985a), and more comprehensively on Svalbard (Philippi, 1973), in the Soviet cold-Arctic (Aleksandrova, in press), in Greenland (Daniëls, 1982, 1985), and on Marion and Prince Edward Is (Gremmen, 1982). In a rather different approach, some Arctic lichen communities have been delimited by principal component analysis (Kershaw & Rouse, 1973; Richardson & Finegan, 1977), or by an agglomerative technique that indicates diagnostic species (Sheard & Geale, 1983). Although appearing with increasing frequency (e.g. Bliss & Svoboda, 1984), such quantitative analyses cannot yet form the basis for a comprehensive, objective classification of polar vegetation because of the limited database and differences in methodology. Thus Brossard, Deruelle, Nimis & Petit (1984) used small sample plots (100 cm^2) to study lichen-dominated vegetation on Svalbard, and as a consequence recognised communities on a smaller scale than those indicated by the conventional, larger quadrats.

In view of these problems, the present description of Antarctic vegetation is based on a hierarchical classification, formulated subjectively, with the major units defined by growth form. The two formations (Table 2.1) include vegetation dominated respectively by vascular and non-vascular plants, while subformations are based on growth form of the community dominants. Minor units reflect floristic composition; associations are characterised by small groups of constant species, and sociations by dominant or codominant species. This is a convenient procedure in

Table 2.1 *Outline classification of vegetation in the cold- and frigid-Antarctic*

Antarctic herb tundra formation
  Grass and cushion chamaephyte subformation
  Deschampsia antarctica – Colobanthus quitensis association: 4 sociations

Antarctic cryptogam tundra formation
  Appressed lichen subformation
  Buellia – Lecanora – Lecidea association: 5 sociations
  Caloplaca – Xanthoria association: 10 sociations
  Placopsis contortuplicata association: 3 sociations
  Verrucaria association: 5 sociations

Fruticose and foliose lichen subformation
  Usnea – Umbilicaria association: 15 sociations
  Alectoria minuscula association: 1 sociation

Short moss turf and cushion subformation
Andreaea association: 6 sociations
Bryum algens associations: 4 sociations
Bryum antarcticum – B. argenteum association: 4 sociations
Bryum pseudotriquetrum association: 5 sociations
Ceratodon association: 3 sociations
Pohlia nutans association: 2 sociations
Pottia austro-georgica association: 2 sociations
Sarconeurum glaciale association: 1 sociation
Tortula – Schistidium antarctici association: 6 sociations

Tall moss turf subformation
  Campylopus association: 2 sociations
  Polytrichum alpestre – Chorisodontium aciphyllum association: 6 sociations
  Polytrichum alpinum – Pohlia nutans association: 7 sociations

Bryophyte carpet and mat subformation
  Brachythecium association: 1 sociation
  Calliergidium austro-stramineum – Calliergon sarmentosum – Drepanocladus uncinatus association: 6 sociations
  Cephalozia badia association: 1 sociation
  Cryptochila grandiflora association: 2 sociations
  Marchantia association: 1 sociation

Moss hummock subformation
  Brachythecium austro-salebrosum association: 1 sociation
  Bryum algens – Drepanocladus uncinatus association: 3 sociations

Macroscopic alga and cyanobacterium subformation
  Prasiola crispa association: 2 sociations
  Nostoc association: 1 sociation

Autotrophic microorganism subformation

The classification has been developed by Holdgate (1964a), Longton (1967, 1973a) and Gimingham & Smith (1970). Communities within the last two subformations have not been formally classified. Other sociations are listed by Longton (1979a) with additions by Seppelt & Ashton (1978), Kanda (1981a) and Smith (1985a). Modifications may be anticipated following further vegetation studies and taxonomic clarification, particularly regarding *Bryum* spp.

the Antarctic, where the most abundant taxa are easily recognised but unresolved taxonomic difficulties prevent compilation of the exhaustive species lists required for refined phytosociological analysis (Ando, 1979). Communities at all levels in the hierarchy intergrade, and they should therefore be regarded as noda within continua rather than discrete entities. Nevertheless, quadrat data from Signy I provide support for the validity of several of the vegetation units (Smith & Gimingham, 1976). The system has been widely applied in the Antarctic by Engelskjøn (1981), Furmańczyk & Ochyra (1982) and papers reviewed in Longton (1979a).

Emphasis on growth form is maintained for other regions to present a unified treatment, and because there is close correlation between growth form and habitat. It is hoped that the resulting account will thus prove more useful than one based purely on floristic composition, and that it will provide an effective background to subsequent ecophysiological discussion. It is drawn from personal field observations supplemented freely by material in the papers cited above, and in others that provide more detailed information than can be given here. These include: for cool-Antarctic islands, Ashton & Gill (1965), Davies & Greene (1976), Huntley (1971), Lindsay (1974) and Smith (1981); for the mild-Antarctic, Greene, Hässal de Menéndez & Matteri (1985), Moore (1979) and Pisano (1983); for the mild-Arctic, Hesselbo (1918), Larsen (1972), Matveyeva, Polozova, Blagodatskykh & Dorogostaiskaya (1974) and Sonesson (1970); for the cool-Arctic, Bliss (1977) and Tieszen (1978a); for Arctic cryptogamic communities, Brassard (1971a), Holmen (1955), Miller & Alpert (1984), Steere (1978b) and Thomson (1979, 1982, 1984).

Cryptogamic growth form types

The growth forms are outlined in Tables 2.2 and 2.3. Some hepatics are thallose, but in most bryophytes the gametophytes comprise stems bearing leaves and rhizoids. Bryophyte growth forms are recognised in part by branching pattern and other aspects of shoot morphology, but the greatest emphasis is placed on arrangement of shoots within a colony. The distinction between the three principal lichen growth forms is based on internal structure, plant size, shape and means of attachment. The treatment in Table 2.3 follows Hale (1983) in, for example, including leprose species under the heading crustose rather than as a separate type (Hawksworth & Hill, 1984).

In lichens, the fungal component (mycobiont) forms a narrow cortical region of densely packed hyphae surrounding a broader medulla of looser

Table 2.2 *Growth form types in polar bryophytes with examples of genera in which they occur*

Small cushion
   Systems formed by sparingly branched acrocarpous mosses with the main shoots radiating in dome-shaped colonies <5 cm in diameter. Branches adopting the same direction of growth as the main shoots. High spatial density of shoots and branches resulting in compact colonies (*Andreaea, Grimmia*).

Large cushion
   Systems formed by acrocarpous or pleurocarpous mosses with the main shoots radiating in dome-shaped colonies >5 cm in diameter. Branches adopting the same direction of growth as the main shoots. Spatial density of shoots and branches lower, and the colonies thus looser, than in small cushions (*Brachythecium, Bryum*).

Short Turf
   Systems formed by sparingly branched acrocarpous mosses with main shoots erect and parallel. Branches erect, of indeterminate growth. Colonies loose or more frequently compact, <2 cm tall (*Bryum, Pottia*).

Tall turf, branches erect
   Systems formed by sparingly branched acrocarpous mosses with main shoots parallel and erect. Branches erect, of indeterminate growth. Colonies loose or compact, >2 cm tall (*Dicranum, Polytrichum*).

Tall turf, branches divergent
   Systems formed by freely branched pleurocarpous mosses, with main shoots erect. Branches divergent, of determinate growth. Colonies compact or more frequently loose, >2 cm tall (*Drepanocladus, Sphagnum*).

Carpet
   Systems formed by pleurocarpous mosses with distal portions of shoots parallel and erect or ascending, derived from a prostrate basal region of partially denuded stems. Colonies loose or compact (*Calliergon, Drepanocladus*).

Compact mat
   Systems formed by leafy hepatics (*Cephalozia*) or mosses (*Racomitrium*) with determinate branching. Shoots prostrate or ascending, interweaving to form compact colonies.

Thallose mat
   Systems formed by dichotomously branched thallose hepatics, the thalli prostrate and often interweaving (*Marchantia*).

Weft
   Systems formed by freely branched pleurocarpous mosses, principal branching determinate, often pinnate or bipinnate, shoots typically robust, erect or ascending and interweaving to form loose colonies (*Hylocomium, Pleurozium*).

Canopy
   Systems formed by sympodial shoots, at first stoloniferous, later becoming erect. Erect portion dendroid, unbranched and bearing scale leaves below, with abundant, determinate branches above, the branches bearing photosynthetic leaves and forming loose colonies (*Climacium*).

Thread
   Shoots occurring individually, not aggregated into colonies: frequently within colonies of other species (*Cephaloziella*).

After Gimingham & Birse (1957)

Table 2.3 *Growth form types in polar lichens with examples of genera in which they occur*

Crustose
> Plants small, lacking a lower cortex, mostly intimately attached to the substratum by hyphae of the medulla, and scarcely separable from it without damage. In typical crustose forms the thalli are often circular in outline, and comprise an upper cortex, algal layer and often a narrow medulla; attached to the substratum by rhizoidal hyphae or by an endolithic, fungal hypothallus (*Lecidea*). Some species with an alga-free, usually black, marginal prothallus (*Rhizocarpon*). Several other levels of structural complexity can be recognised including:
>> Leprose: irregular mats of hyphae and loosely associated algal cells, not differentiated into layers of tissue, often forming a scurfy or powdery crust on the surface of the substrate (*Lepraria*). Endolithic forms have the vegetative hyphae and algal cells among rock crystals just under the rock surface.
>> Placodioid: similar to typical crustose forms, but with the thallus margin comprising a series of discrete lobes closely appressed to the substrate (*Caloplaca*).
>> Squamulose: discrete lobe-like structures partially or wholly free of the substrate: medulla thus exposed due to absence of lower cortex (*Psoroma*).

Foliose
> Thallus strongly flattened, prostrate, typically larger and less firmly attached to the substrate than in crustose forms. In structure, dorsiventrally symmetric with an upper, and usually a lower, cortex surrounding an upper algal layer and a lower medulla. Two principal types are recognised:
>> Laciniate: thalli lobed, loosely attached to the substrate, often by rhizines (compressed strands of hyphae) which may arise anywhere on the lower surface (*Cetraria, Peltigera*).
>> Umbilicate: thalli roughly circular, firmly attached to the substrate by a central holdfast (*Umbilicaria*).

Fruticose
> Thallus strap-shaped (*Ramalina*) or shrubby (*Cladonia*), the main axis prostrate to erect, commonly branched, attached to the substrate by basal rhizoids or free. Basically radially symmetric, though often flattened, with a dense outer cortex surrounding a thin algal layer, a medulla and either a more or less hollow centre (*Cladonia*) or a dense central cord (*Usnea*).

The descriptions of growth form type are based on Hale (1983).

texture that may facilitate gaseous exchange. Most mycobionts are asco-mycetes or deuteromycetes, but a few are basidiomycetes. Basidiolichens, particularly species of *Botrydina* and *Coriscium*, are widely distributed but little studied in boreal and Arctic regions (Heikkilä & Kallio, 1969). The former grow on soil near fruiting bodies of *Omphalina* spp and other agarics as a green, gelatinous growth formed by globules of green algae (*Coccomyxa* spp) surrounded by a colourless fungal envelope. *Coriscium* spp form squamules around similar basidiocarps.

The autotrophic component of lichens (photobiont) normally occurs in a narrow zone between the cortex and medulla, but in genera such as *Collema* and *Mastodia* it forms the principal structure of the thallus. Unicellular green algae, including species of *Trebouxia*, are most common but *Nostoc* spp and other cyanobacteria also occur. Genera such as *Pelti-gera* and *Stereocaulon* contain both green algae and cyanobacteria, the latter confined to swollen cephalodia. *Nephroma arcticum* may have three photobionts, a green alga in the main body of the thallus, and two cyano-bacteria occurring in separate cephalodia (Hawksworth & Hill, 1984).

Hale (1983) points out that the lichen growth forms are variable (Table 2.3), and intergrade to some extent due to lack of strict correlation between morphological and anatomical features. Thus *Xanthoria* spp are usually regarded as foliose lichens as most species possess a lower cortex, but the thalli are small and grow close against the substratum in the manner of crustose species while a lower cortex is lacking in the wide-spread species, *X. elegans* (Hooker, 1980a). The term appressed is applied here to all small lichens growing flat against the substratum, regardless of anatomy. In *Cladonia* spp, young plants form squamules, which have been regarded as foliose (Jahns, 1973) or crustose (Hale, 1983), and which give rise to a more prominent, fruticose phase formed by large, freely branched podetia. These are hollow; they may eventually bear apothecia, and are regarded as being part of the fruiting body. There may also be habitat-correlated or developmental variation within a species. Despite these inconsistencies, growth form is of considerable ecological signifi-cance in both mosses and lichens.

Vegetation in the cold-Antarctic
Antarctic herb tundra formation

Two native flowering plants are widely distributed in the cold-Antarctic. *Deschampsia antarctica*, a grass, forms low mats, and *Coloban-thus quitensis* (Caryophyllaceae) occurs in small, compact cushions. Both are most frequent at low altitudes as occasional components of essentially cryptogamic communities on level ground, or more typically on north- or west-facing slopes where the vegetation receives maximum heating from solar radiation. *D. antarctica* is the more widespread, and usually the more abundant species, locally forming continuous swards extending over several square metres on rock ledges and stony ground. *C. quitensis* and mosses such as *Drepanocladus uncinatus* are abundant in some of these occasional stands of angiosperm-dominated vegetation, which may overlie soil showing general similarities to temperate brown earths (Allen & Heal, 1970).

Antarctic cryptogam tundra formation

Appressed lichen subformation The cold-Antarctic supports a wide diversity of luxuriant cryptogamic vegetation, including examples of all eight subformations of the Antarctic cryptogam tundra formation (Table 2.1). The appressed lichen subformation is dominated by crustose, dwarf foliose and other small lichens (Fig. 2.1). Such species form the total plant cover in many stands, while *Ramalina terebrata* and other macrolichens, or small cushion-forming mosses such as *Muelleriella crassifolia* occur as associates in others. Many dominants are halophilous or ornithocoprophilous, and thus the present vegetation is particularly characteristic of coastal cliffs influenced by salt spray and nutrient enrichment by sea birds. Other communities occur extensively on dry, rocky substrata inland and on the shingle of raised beaches.

Vertical zonation is often evident on coastal rocks. Black or brown species of *Verrucaria* are abundant from immediately above to slightly below the high-water mark, and contrast strongly with *Caloplaca* and *Xanthoria* spp which form striking orange and yellow splashes on the rocks above. Each sociation occupies a characteristic position in relation to high-tide level, spray deposition, and proximity to bird nesting-sites. These communities are largely replaced by the *Beullia–Lecanora–Lecidea* and the *Placopsis contortuplicata* associations on exposed inland rocks, the latter being characteristic of relatively high altitudes to at least 600 m. *Buelliua* and *Verrucaria* spp are also locally abundant in freshwater pools.

Fruticose and foliose lichen subformation   Macrolichen communities are widespread and extensive from sea level to altitudes of at least 500 m, principally on inland boulders, cliffs and scree, and on finer mineral soils in dry or exposed situations. Fourteen sociations in a single *Usnea–Umbilicaria* association have been described, many with abundant *Usnea antarctica*. Other fruticose lichens, including *Himantormia lugubris* and *Usnea fasciata*, or foliose species of *Umbilicaria*, also assume dominance in different sociations. The black and yellow thalli of *Usnea fasciata* reach lengths of 15 cm, and the black apothecia, each up to 2 cm wide, make this one of the most striking of Antarctic plants (Fig. 2.2). Crustose species commonly form an understory beneath the macrolichens, and small cushion-forming mosses in such genera as *Andreaea*, *Dicranoweisia* and *Schistidium* become prominant on relatively moist rocks (Fig. 2.3).

Short moss turf and cushion subformation   Bryophyte vegetation is more varied than that dominated by lichens and shows greater ecological amplitude. Growth form is strongly correlated with water availability. Short turf- and small cushion-forming mosses predominate in the driest

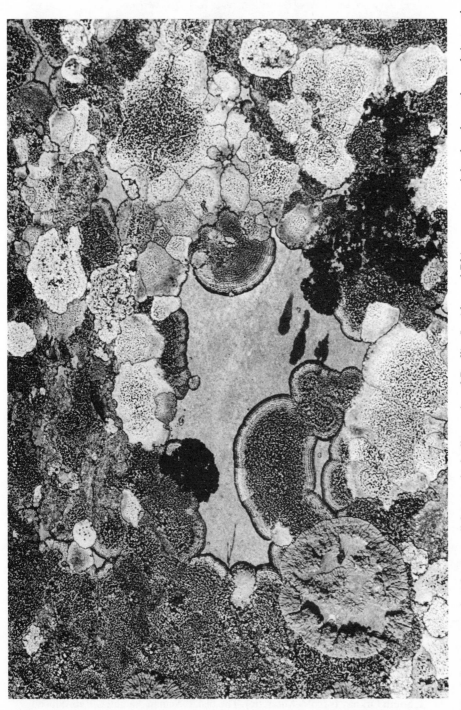

Fig. 2.1. A community of crustose lichens, principally species of *Buellia*, *Lecidea* and *Rhizocarpon*, on inland rocks on the cool-Antarctic island of South Georgia. Similar communities occur in the cold-Antarctic and elsewhere. Several thalli surrounding the central bare area show marginal zonation (Chapter 6).

Fig. 2.2. The fruticose lichen *Usnea fasciata* on an upland rock on Signy I, South Orkney Is (cold-Antarctic). The disk is 5 cm in diameter. Reproduced from Longton (1985a) by courtesy of Pergamon Press.

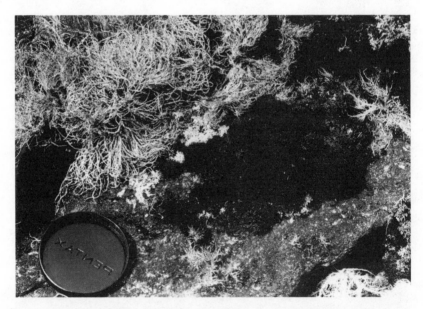

Fig. 2.3. *Usnea antarctica* with *Andreaea* sp on an exposed boulder on Signy I, South Orkney Is (cold-Antarctic). The disk is 5 cm in diameter. Reproduced from Longton (1967) by courtesy of the Royal Society.

places supporting abundant bryophytes in vegetation that may be regarded as cryptogamic fellfield. The mosses form small, discrete colonies or continuous, undulating stands several metres wide.

Communities dominated by *Andreaea* spp occur extensively on acidic substrata, including rocks, scree and mineral soil; they occur in dry to rather moist habitats, and are characteristic of exposed upland soil influenced by cryoturbatic disturbance. The dark brown or blackish cushions of *Andreaea* spp are associated with many other bryophytes and lichens. Cushions of *Dicranoweisia* and *Schistidium* spp, and mats of *Racomitrium* spp are prominent among the mosses, while permanently moist, sloping gravels support an *Andreaea*–hepatic sociation, one of the few cold-Antarctic communities in which liverworts are abundant away from the influence of geothermal heat and moisture. The hepatics include mat-forming and thread-like species of *Cephaloziella, Hygrolembidium, Herzogobryum* and *Pachyglossa*. Lichens associated with *Andreaea* spp include fruticose species of *Cladonia, Ochrolechia, Stereocaulon* and *Usnea*, foliose species of *Umbilicaria* and a variety of crustose species. Many lichens grow on stone or rock surfaces, but others, notably *Ochrolechia frigida*, are epiphytes on moss.

The *Tortula–Schistidium* association is characteristic of base-rich habitats, including marble outcrops and recent volcanic soils. Small stands also occur around piles of limpet shells deposited by gulls on otherwise acidic soil. Species of *Bryum, Ceratodon, Schistidium* and *Tortula* are dominant in various combinations. Associated species include other short turf- and mat-forming bryophytes and many lichens.

Highly porous substrata of recent volcanic origin on the South Sandwich Is and South Shetland Is form the principal habitat of the *Ceratodon* and *Pohlia nutans* associations, in which the mosses are extensively encrusted by epiphytic lichens (Fig. 2.4). Large areas of scoria-covered plains support turfs of *Pohlia nutans* so heavily colonised by a scurfy, powdery crust of *Lepraria* sp that the vegetation assumes a pale, greenish-white colour. Species of *Bryum* and *Ceratodon* are also abundant in comparable habitats, forming irregular, elongate cushions up to 5 cm wide, which are impregnated with volcanic ash and hollow in the centre, possibly as a result of frost action. They are extensively colonised by a reddish-brown species of *Psoroma*.

The *Pottia austro-georgica* association is locally distributed on young morains and ground disturbed by solifluction, frost-heave and animal-induced phenomena. The vegetation is generally open, with short turfs of the *Pottia*, or more locally of *Pterygoneurum* sp, giving 10–15% cover.

Fig. 2.4. Short turfs of *Pohlia nutans* and *Polytrichum alpinum* on a scoria-covered plain on Candlemas I, South Sandwich Is (cold-Antarctic). The disk is 5 cm in diameter.

Other small acrocarpous mosses and thallose mats of *Marchantia polymorpha* are the principal associates. Short turf- and small cushion-forming mosses are also well represented in rock crevice communities, accompanied by other growth forms including fruticose and foliose lichens and bryophyte mats.

Tall moss turf subformation   The tall moss turf subformation includes some of the most striking and distinctive Antarctic vegetation. In the dominant turf-forming species, the shoots may be loosely aggregated, as in *Polytrichum alpinum*, or more densely packed with abundant interlacing rhizoids below as in *Polytrichum alpestre* and *Chorisodontium aciphyllum*. The two latter species commonly overlie semi-ombrogenous peat up to 2 m deep. Tall turf-forming mosses grow abundantly in mesic sites over boulders, scree, and more finely divided but generally well-drained substrata below an altitude of 150 m. The deeper peats are permanently frozen below 20–30 cm from the surface.

The *Polytrichum alpestre–Chorisodontium aciphyllum* association occurs as discrete turfs or continuous banks of either or both species which may extend continuously for up to 2500 m², particularly on well-insolated north- and west-facing slopes (Figs. 2.5 and 2.6). *C. aciphyllum*

Fig. 2.5. A bank of tall turf-forming mosses, principally *Chorisodontium aciphyllum*, on Signy I, South Orkney Is (cold-Antarctic). The pole is 70 cm high. Reproduced from Longton (1985a) by courtesy of Pergamon Press.

Fig. 2.6. A tall turf of *Chorisodontium aciphyllum* with occasional *Polytrichum alpestre* (top right) adjacent to a carpet of *Calliergon sarmentosum* and *Drepanocladus uncinatus* on wet, level ground on Signy I, South Orkney Is (cold-Antarctic). The disk is 5 cm in diameter. Reproduced from Longton (1967) by courtesy of the Royal Society.

prefers wetter conditions than *P. alpestre* and thus tends to predominate on gently-sloping ground, where it forms oval, dome-shaped islands of peat. Associated bryophytes include *Pohlia nutans* and species of *Barbilophozia* and *Cephaloziella*, the moss occurring as isolated turfs and the liverworts as individual stems among the mosses or small mats on the surface. The surface of the larger banks is typically undulating, terraced or fissured due, in part, to frost-induced phenomena. Nevertheless, the moss turfs provide a sufficiently stable, well-drained substratum to support epiphytic lichens. Colonisation by crustose species of *Buellia, Ochrolechia* and *Rinodina*, and fruticose species of *Alectoria, Cladonia* and *Usnea*, occurs extensively where upper parts of the banks are blown clear of snow in winter.

The *Polytrichum alpinum–Pohlia nutans* association, although widespread, is particularly characteristic of recent volcanic substrata. *Polytrichum alpinum* is usually the most abundant species. Its colonies range from compact cushions about 20 cm in diameter and impregnated with volcanic ash, to large, loose mounds up to 1 m wide and 30 cm tall. These may merge to form continuous stands, but deep peats are not formed. The colonies develop from radiating, subterranean rhizomes, and are frequently circular in outline. *Pohlia nutans* is usually present, and sometimes codominant.

The third association of tall turf-forming mosses (Table 2.1) is confined to Deception I and the South Sandwich Is where stands of *Campylopus* spp are locally extensive on ground subject to volcanic heat and moisture (page 95).

Bryophyte carpet and mat subformation The bryophyte carpet and mat subformation occupies soil saturated with standing or slowly moving water on level or gently sloping ground at low altitudes, around lakes, or on slopes subject to melt water seepage. The characteristic species are carpet-forming pleurocarpous mosses which in places overlie soligenous peat up to 15 cm deep. *Calliergidium austro-stramineum, Calliergon sarmentosum* and *Drepanocladus uncinatus* are usually dominant or codominant (Figs. 2.6 and 2.7) and *Brachythecium austro-salebrosum* forms similar carpets more locally. There are several regularly associated bryophytes, notably *Cephaloziella varians*: the alga *Prasiola crispa* is locally frequent, but lichens are rare.

Hepatic-dominated communities are largely confined to relatively warm, moist ground around fumaroles on the South Sandwich Is. They range from shallow, compact mats of *Cephalozia badia* to luxuriant stands

Fig. 2.7. A variety of carpet-forming mosses on a moist seepage slope on Signy I, South Orkney Is (cold-Antarctic) with hummocks of *Brachythecium austrosalebrosum* in the centre. The boulder at top centre is about 75 cm wide. Reproduced from Longton (1985a) by courtesy of Pergamon Press.

of *Cryptochila grandiflora* and, more locally, *Triandrophyllum subtrifidum*, which occur as mats, carpets or even tall turfs. Thallose mats of *Marchantia berteroana* are dominant locally near the fumaroles.

Moss hummock subformation   The moss hummock subformation is confined to the margins of melt water streams, wet rock ledges subject to dripping water and other flushed sites in contact with flowing water. The growth forms have been described by Smith (1972) as a 'tall compact cushion (*Bryum algens*), a tall loose cushion (*Brachythecium austrosalebrosum* and *Tortula excelsa*), or a deep, undulating carpet (*Drepanocladus uncinatus*)'. The large cushions sometimes coalesce to form closed hummocky stands, but the communities are seldom extensive (Figs. 2.7 and 2.8).

Macroscopic alga and cyanobacterium subformation   The green alga *Prasiola crispa* is one of the most conspicuous plants in coastal regions. Its bistratose thalli are abundant over large areas subject to trampling and manuring by marine vertebrates. Few other macrophytes can

Fig. 2.8. Hummocks of *Brachythecium austro-salebrosum* by a small stream on Signy I, South Orkney Is (cold-Antarctic). The disk is 5 cm in diameter. Reproduced from Longton (1967) by courtesy of the Royal Society.

tolerate the conditions of mechanical disturbance, low pH and high concentrations of nitrogen, sodium and phosphorus which characterise such biotically influenced sites. Bryophytes are rare except on the periphery of the *Prasiola* zone. Lichens colonise some of the larger boulders and include *Mastodia tesselata*, a lichenised form of *Prasiola*. The most intensively disturbed ground within penguin colonies is devoid of visible vegetation. Elsewhere, clay soils saturated with almost stagnant, non-nitrogenous melt water locally support an abundance of *Nostoc* spp in rosettes up to 5 cm in diameter and spaced almost equidistantly some 10–20 cm apart. *Nostoc* also occurs in a lichenised form (*Leptogium*).

Autotrophic microorganism subformation   Assemblages of autotrophic microorganisms, in places occurring in the absence of macroscopic vegetation, are a feature of the cold-Antarctic and other polar regions. Most prominent are aggregations of unicellular algae which form red, green or occasionally yellow patches on snow. They are widespread in coastal areas, particularly on firn snow during thaws in mid- to late-summer when the coloration may extend continuously over several hundred square metres.

Extent of the major vegetation types

Signy I is a small, relatively well vegetated and intensively studied member of the South Orkney Is. Approximately 30% of the total area of about 20 km² is glaciated: of the rest, 12% supports the bryophyte carpet and mat subformation and 6% the tall moss turf subformation, particularly banks of *Polytrichum alpestre* and *Chorisodontium aciphyllum*. The fruticose and foliose lichen subformation, and *Andreaea*-dominated stands of the short moss turf and cushion subformation extend over 50% of the ice-free land surface, but they are often open with cover less than 50%. Some 26% of the ice-free surface is either bare or colonised by appressed lichens, and a further 5% is so strongly influenced by birds and seals as to support only *Prasiola crispa*. Estimated cover of phanerogamic vegetation is less than 0.001% (Tilbrook, 1970).

Extent and diversity of cryptogamic vegetation decrease from the South Orkney Is southwards along the west coast of the Antarctic Peninsula until at Marguerite Bay (latitude 67–69° S) plant cover is locally distributed among arid substrata devoid of macrophytes. Candlemas I, Deception I, Bouvetøya and other recent volcanic islands are also less extensively vegetated than the South Orkney Is, from which they differ further in a rarity of *Andreaea* spp, *Polytrichum alpestre*, and *Chorisodontium aciphyllum*, and greater abundance of *Polytrichum alpinum*. The dry, east coast of the Antarctic Peninsula is sparsely vegetated, principally by lichens, and is frigid-Antarctic in character.

Vegetation in the cold-Arctic

This account of cold-Arctic vegetation is drawn largely from Aleksandrova (1980, 1988), who stresses the dominant role of cryptogams in communities comparable in structure with those in the cold-Antarctic. Flowering plants comprise exclusively species with distributions centred in cool- and cold-Arctic regions, and they occur in isolated tufts or small cushions. Aleksandrova recognises three longitudinal provinces in the cold-Arctic, marked by differences in oceancity. The Canadian Province is arid, with annual precipitation 50–150 mm, compared with 100–230 mm in the Siberian Province, and up to 300 mm in the Barents Province of western Eurasia. Winter temperatures are higher in the more oceanic regions, but mean monthly temperatures in summer everywhere remain below 2 °C.

Vegetation is best developed in the Barents Province, for example in the Franz Josef Land archipelago. These islands are 85% glaciated but, as in the cold-Antarctic, melt water, light but frequent summer precipitation and the prevalence of cloud cover, fog and high relative humidity ensure adequate water availability in some habitats. Plant distribution is strongly influenced by the depth and duration of winter snow cover.

Distinctive communities occur on finely divided soils in which frost action causes small-scale polygonal structure, the polygons ranging from 20 to 50 cm wide. Aleksandrova (1980) considers optimal conditions to occur where snow reaches 10–25 cm in depth and melts by mid-June. Loamy soil on the surface of the polygons is then almost completely covered by a colourful mosaic of crustose lichens including species of *Collema* (black), *Ochrolechia* (yellow) and *Pertusaria* (white), with mats of *Cephaloziella* and *Tritomaria* spp and other liverworts together forming a crust about 5 mm thick. The channels support an assemblage of tall turf-forming mosses such as *Ditrichum flexiacule* and *Polytrichum alpinum*, with *Cetraria* spp and other fruticose lichens and colonies of *Nostoc* spp. Scattered flowering plants, including *Cerastium, Draba, Papaver* and *Saxifraga* spp, grow among the mosses, and on otherwise bare ground between the mosses and crustose lichens. Flowering plant cover is only 1–6%, compared with around 75% for the lichens.

Where winter snow accumulates to depths greater than 25 cm, it persists until beyond mid-June and the vegetation is sparse, with the grey fruticose lichen *Stereocaulon rivularum* predominating in mesic habitats. Distinctive communities in wetter places include an assemblage of the mat-forming mosses *Orthothecium chryseum* and *Campylium stellatum* with turfs of *Bryum* spp in small mires by streams, and a black film of hepatics including *Cephaloziella* and *Lophozia* spp on the surface of polygons. The liverworts are marked by tightly overlapping leaves and large cells with thick walls: they are associated with cyanobacteria and the black mucilaginous lichen *Collema ceraniscum*.

There is no macrophytic vegetation where snow cover persists beyond mid-July, but further communities occur in exposed areas where snow is removed by wind resulting in a severe winter environment but a long growing season. Flowering plants are virtually absent, but bryophytes and lichens form sparse open communities. Boulders support, first, an assemblage of fruticose lichens including *Neuropogon sulphureus* and *Alectoria* spp with foliose species of *Umbilicaria* and occasional cushions of *Andreaea rupestris* and, second, a community dominated by species of *Lecidea, Lecanora, Rhizocarpon* and other crustose lichens. Between the boulders, fruticose lichens are associated with turf- and mat-forming bryophytes, notably *Racomitrium lanuginosum*. Crustose lichens predominate on the finer sands and gravels.

Most plant communities in Franz Josef Land could thus be accommodated within the appressed lichen, fruticose and foliose lichen, tall moss turf, and more locally the bryophyte carpet and mat subformations as recognised in the cold-Antarctic. The same appears to be true of other parts of the cold-Arctic, although the vegetation is more sparsely developed under arid conditions in parts of northern Siberia and Canada, and similarities in growth form belie considerable longitudinal variation in floristic composition.

Vegetation in the frigid-Antarctic
Appressed lichen, and fruticose and foliose lichen subformations
  The impression presented by ice-free terrain in the frigid-Antarctic is of bare gravel, rock and scree: flowering plants are absent and cryptogamic vegetation is only locally prominent. The occasional stands of macrophytic vegetation are referable to the appressed lichen, fruticose and foliose lichen, short moss turf and cushion, and macroscopic alga and cyanobacterium subformations. They are most widespread in coastal regions, but mosses extend on inland nunataks to beyond 84°S, lichens and cyanobacteria to beyond 86°S, and other bacteria and yeasts to 87°21′S (Cameron, Lacey, Morelli & Marsh, 1971; Wise & Gressitt, 1965).

Lichen communities are more widespread and variable than those dominated by mosses, particularly inland. Thus Kappen (1985a) recorded 33 lichens and 5 mosses from Birthday Ridge, a coastal area of northern Victoria Land, compared with 20 lichens and 2 mosses from similar granitic substrata 70 km from the coast. The bipolar, fruticose lichen *Neuropogon sulphureus* is characteristic of epilithic communities at Birthday Ridge, associated with species of *Umbilicaria* (foliose), *Pseudephebe* (fruticose) and *Lecidea* (crustose). A further lichen assemblage occurs on moribund moss cushions, with crustose species of *Candelariella, Rhizoplaca* and *Rinodina* prominent. Elsewhere, *Buellia* spp and other crustose lichens, with foliose and fruticose species of *Alectoria, Umbilicaria* and *Usnea*, are locally frequent (Llano, 1965), and *Neuropogon sulphureus* is dominant in some communities near the east and southwest coasts of the Antarctic Peninsula (Gimingham & Smith, 1970).

Vegetation in some of the driest sites is restricted to crustose lichens on the sides or lower surface of pebbles, and in holes and fissures in stones and rocks. Translucence of the pebbles becomes an important ecological factor in such situations (Cameron, 1972). In contrast, the greatest variety of lichens at Kappen's inland, montane site occurs on exposed, north-facing rocks where the plants are thought to derive moisture from cloud.

Short moss turf and cushion subformation

Lichens occur both on rock surfaces and on more finely divided sands and gravels, but the small, acrocarpous mosses are found principally on the latter. They are most abundant on the margins of gulleys carrying melt water during the warmest summer weather, and in seepage areas below persistent snow banks. Small, isolated turfs occur on drier ground and, at Birthday Ridge, scattered green shoots of *Bryum algens* occur to depths of 3 cm in the soil (Kappen, 1985a). No tall turf-, carpet-, or hummock-forming mosses have been recorded and there is only one known hepatic, a species of *Cephaloziella* (Seppelt, 1983a). Perhaps surprisingly, *Andreaea* spp appear to be absent. Floristic composition varies between the widely separated localities so far studied.

Plant abundance in the moss communities is variable. On Ross I, *Bryum antarcticum* locally gives 85% cover on moist sandy soil whereas cover values are less than 5% in moss communities on dry, stony ground. In some cases mosses such as *Bryum argenteum* and *Sarconeurum glaciale* grow as scattered, short turfs having their shoot apices flush with the substratum, projecting stones affording additional protection from wind. Elsewhere, particularly in the wetter sites, other species of *Bryum* form compact, often elongate and contorted cushions up to 10 cm wide and 5 cm high. These dimensions are clearly attained by persistence rather than rapid growth, as the green, photosynthetic layer commonly extends only to 1–2 mm, and studies of branching pattern and rhizoid banding suggest ages of 25–100 years for the larger cushions (Matsuda, 1968; Seppelt & Ashton, 1978). Many of the mosses are colonised by lichens such as *Buellia grimmiae* and, in wetter habitats, by cyanobacteria. Both could inhibit moss growth, but the latter may provide a source of nitrogen (Chapter 7).

Macroscopic alga and cyanobacterium subformation

Algal vegetation is locally more variable and extensive at coastal sites in the frigid-Antarctic than in the cold-Antarctic. *Prasiola crispa* is again abundant around penguin colonies, and regularly spaced colonies of *Nostoc* spp are particularly extensive in several localities. Algae are also prominent in some seasonally wet habitats in and beside shallow melt streams and pools. This abundance of algae could be related to lack of competition from carpet-forming mosses that occupy similar habitats in the Antarctic Peninsula region.

Autotrophic microoganism subformation

The proportion of ice-free land surface supporting visible vegetation declines with increasing altitude and distance inland, until the ground surface appears lifeless. The terrain also appears barren under conditions of unusually low precipitation and high soil salinity in the ice-free dry valley region of southern Victoria Land (Chapter 1). However, fungi, bacteria and algae have colonised many sites lacking visible vegetation, and in places the microorganisms support arthropods and other soil animals. The poleward limit of soil biota is thought to be determined by soil temperatures above 0 °C failing to penetrate to sufficient depths to liberate ice-bound water (Janetschek, 1970). Beyond this limit, in much of the central ice-plateau region, there is no known life.

Friedman (1982) considers that endolithic microorganisms occurring beneath the surface of rock comprise the most widespread biotic communities in the frigid-Antarctic. Chasmoendoliths grow in minute fissures, whereas cryptoendoliths penetrate between the crystals of relatively porous rocks such as granite and marble. Primary producers occurring as endoliths include lichens, cyanobacteria and green and yellow-green algae, and they are commonly associated with colourless bacteria. Melt water absorbed in the surface layers of rock gives moisture and maintains relative humidity in air between the rock particles above ambient for several days following summer snowfalls.

Chasmoendoliths are widespread in coastal regions of the frigid-Antarctic, and they are also known from the dry valleys and the cold-Antarctic. They grow in fissures running perpendicular to the rock surface and under thin flakes parallel with the surface, in places extending beneath 20% of the rock surface. Zonation has been reported, with cyanobacteria in the innermost vegetated region where light intensity and oxygen concentrations are likely to be low, and green algae and lichens nearer the surface (Broady, 1981).

Cryptoendolithic lichens are the dominant form of life in the dry valley region (Friedmann, 1982). The rocks commonly show an abiotic crust 1–2 mm wide, inside which occur successive zones appearing black (about 1 mm wide), white (2–4 mm) and green (2 mm). The three inner zones each support filamentous fungi and unicellular green algae, the latter being most abundant in the innermost, green zone and scarce in the white zone. The organisms include species of *Trebouxia*, and of previously undescribed genera in the Chlorophyta and Hyphomycetes (D. L. Hawksworth, personal communication; Tschermak-Woess & Friedmann, 1984).

Algae in the innermost zone are unlichenised. Elsewhere there is ample evidence of a symbiotic, lichen association between the two types of organism, although the mycobiont lacks the organised thallus found in epilithic lichens. Many fungal hyphae are closely associated spatially with algal cells, being appressed to their surface or penetrating by means of haustoria: concentric bodies occur in the mycobiont cytoplast, as in other lichens, and lichen substances such as norstitic acid are produced. Where cryptoendolithic lichens become exposed on the rock surface by weathering (Chapter 3) it appears that they may, under favourable conditions, produce small crustose thalli bearing apothecia or pycnidia. Genera such as *Acarospora*, *Buellia* and *Lecidea*, all strongly represented in the Antarctic epilithic flora, have been recognised on this basis (Hale, 1987), but the link between the epilithic lichens and the endoliths remains to be confirmed (D. L. Hawksworth, personal communication). Comparable endolithic assemblages of autotrophic and dependent microorganisms could prove to be widespread in polar regions.

Vegetation in mild- and cool-polar regions

Mild- and cool-polar regions support a greater range of plant species and vegetation types than do the climatically more severe areas so far considered, and no comprehensive classification of their vegetation comparable with that in Table 2.1 has yet been attempted. This account provides an impression of the role played by cryptogams in the principal angiosperm-dominated vegetation types (Table 1.2), followed by a consideration of bryophyte and lichen communities in habitats not freely colonised by flowering plants.

Wetlands

Wet meadow   Marsh and bog communities dominated by graminoids occur extensively on level and gently sloping ground in lowland valleys and at sites of impeded drainage, particularly in the mild-Arctic and on cool-Antarctic islands. They are partially replaced by cushion bogs in the mild-Antarctic, and although widespread in the cool-Arctic they are limited in extent by the prevailing aridity. Wet meadows range from relatively homogeneous stands of graminoids to more varied assemblages showing pronounced patterns of microrelief caused by cryoturbatic disturbance, and hummock formation by mosses and plants such as *Eriophorum vaginatum*.

Arctagrostis latifolia, Carex aquatilis and *Dupontia fisheri* are among characteristic plants of the more homogeneous meadows throughout much of the Arctic. These and other rhizomatous sedges and grasses form leafy shoots at regular intervals, among which are interspersed tussocks of *Eriophorum* spp with *Pedicularis* spp and other small herbs. *Betula nana* and low-growing species of *Salix* are common in some stands. Rushes in the genera *Marsippospermum* and *Rostkovia* are major dominants in Antarctic mires with species of *Juncus* and *Agrostis* in flushed sites. Many of the communities overlie peat.

In most mires the flowering plants are rooted in a continuous understorey of bryophytes: lichens are seldom abundant, but foliose species of *Lobaria* and *Peltigera* are common locally. The mosses include pleurocarpous species of *Calliergon* and *Drepanocladus*, which form carpets or tall turfs with divergent branches, and acrocarpous species of *Cinclidium, Meesia* and *Tortula* occurring as tall turfs with erect branches (Table 2.2). Hummocky turfs of *Sphagnum* spp are abundant in mild-Arctic mires and occur more locally in other areas. Hepatics, occurring as turfs (e.g. *Blepharidophyllum* spp), mats (*Riccardia* spp) and thread-like forms (*Cephaloziella* spp) are also locally frequent.

Many wet meadow communities show characteristic patterns of surface relief. Some of the component species occur also in homogeneous mires, but the diversity of species and growth forms is greater, due to the wider range of microhabitats. Thus short turfs of the moss *Seligeria polaris* occur in association with cyanobacteria, other bryophytes and occasional sedges in open vegetation on the mineral surface of hummocks in frost boil sedge meadows on Devon I, the more luxuriant assemblage of sedges and tall turf-forming mosses being restricted to peat in the intervening channels. Over 40 bryophytes have been recorded from sedge meadows on Devon I, but *Riccardia pinguis* is the only common hepatic (Vitt & Pakarinen, 1977).

Crustose lichens predominate on frost boils in some Arctic sedge meadows, and lichens become prominent in relatively dry meadows. Thus in a *Carex aquatilis–Poa arctica* community in Alaska the cryptogam layer comprises a mosaic of tall turf-forming mosses, notably *Dicranum elongatum* and *Polytrichum alpinum*, with *Cetraria, Cladonia, Dactylina* and *Thamnolia* spp and other fruticose lichens.

Cushion bog Cushion bog is a southern hemisphere vegetation type occurring throughout the mild-Antarctic. The dominants are broadleaved dicots (Table 1.2). They form low, hard cushions which hold water in a dense covering of old leaves and branches, and cushion bogs can develop on slopes, as well as on level ground. The vegetation is typically waterlogged and overlies peat. Bryophytes are usually present, but are seldom conspicuous due to the abundance of flowering plants, species of *Breutelia* and *Dicranoloma* being among the taxa represented.

Mesic communities

Woodland   The mild-Arctic and boreal regions merge in a broad ecotone comprising a mosaic of tundra and open woodland. The latter extends furthest into the tundra on well-drained river banks and in sheltered valleys, both in the Arctic and in the magellanic moorland region of southwestern Chile. Birch (*Betula* spp) predominate in the most northerly woodlands in Fennoscandia, Greenland and Iceland, whereas spruce (*Picea* spp) and larch (*Larix* spp) play this role across much of North America and Siberia, with willows (*Salix* spp) as common associates. Mild-Antarctic woodlands include *Pilgerodendron uvifera* and other conifers, and broad-leaved evergreens such as a southern beech, *Nothofagus betuloides*.

The trees in Arctic woodland are mostly less than 10 m in height and are widely spaced so that a closed canopy seldom develops. A dwarf shrub stratum overlying a more or less continuous layer of mosses and lichens is characteristic. Where the tree canopy is relatively dense, as on moist sites not recently subject to fire, the dominants are robust, weft-forming, feather mosses, notably *Hylocomium splendens* and *Pleurozium schreberi*. They are associated with less abundant, tall turf-forming species (e.g. *Dicranum* spp), mats of leafy hepatics (*Barbilophozia* and *Ptilidium* spp), and *Peltigera* spp and other foliose lichens.

Almost closed stands of *Cetraria, Cladonia, Stereocaulon* spp and other fruticose lichens are a feature of more open lichen woodland (Fig. 2.9) on drier, sandy or gravelly soils, and during secondary succession (Chapter 3). On drying, the lichen cover frequently cracks into polygonal areas between which mosses become established. Kershaw (1985) suggests that dense lichen cover is dependent on winter snow accumulation combined with a low level of competition from vascular plants, and that the open nature of the woodland is maintained in part through inhibition by the lichens of tree seedling establishment (Chapter 7). A variety of lichens occur as epiphytes in the northern woodlands, with freely branched, fruticose species of *Alectoria, Evernia* and *Usnea* often prominent in festoons hanging from tree branches. Bryophytes are a striking feature of woodland in the high-precipitation region of southwestern Chile:

hepatics predominate, as thick mats on tree trunks and branches, and in almost continuous carpet on the ground.

Scrub Scrub formed by birch, willow and alder (*Alnus* spp) seldom extends beyond the mild-polar regions. An abundant water supply in summer and snow accumulation in winter are essential for its development in the Arctic (Porsild, 1951), and thus it is most common in damp, sheltered depressions, along water courses and on lake shores. The shrubs reach 3 m in height but are most commonly 0.5–1 m. Few associated plants occur in dense scrub, but open stands support a variety of dwarf shrubs, forbs and cryptogams including *Cladonia* and *Stereocaulon* spp and the mosses *Polytrichum commune* and *Tomenthypnum nitens*.

Berberis, Chiliotrichum, Hebe and *Pernettya* are among the characteristic genera of mild-Antarctic scrub, found on gently sloping foreshores and by lakes and streams. Mat-forming hepatics, notably species of *Frullania* and the Lejeuneaceae, with cushion-forming species of *Macromitrium, Ulota* and other mosses, occur as epiphytes in southwestern Chile, and a variety of hepatics and robust mosses including *Acrophyllum, Breutelia* and *Brachythecium* spp grow on the ground. Bryophytes are less important in scrub under drier conditions on the Falkland Is.

Dwarf shrub heath Dwarf shrub heath is extensive in the mild-Arctic on well-drained slopes where snow accumulates in winter. The most conspicuous plants are erect-growing or ascending evergreen shrubs 10–50 cm tall. Ericoids, particularly species of *Ledum* and *Vaccinium*, predominate, but *Empetrum* and *Salix* spp are also abundant. Cool-Arctic dwarf shrub heaths are seldom extensive, are strongly restricted to areas of deep winter snow cover, and are commonly dominated by *Cassiope tetragona* or *Vaccinium uliginosum*.

Cryptogams are abundant, particularly in the more open stands. Fruticose lichens predominate in dry areas where species of *Alectoria, Cetraria, Cladonia, Dactylina* and *Thamnolia* may form a layer 5 cm deep. Tall turf- and mat-forming mosses including *Aulacomnium, Dicranum, Ditrichum, Hylocomium, Hypnum* and *Polytrichum* spp largely replace the lichens under moister conditions. In the mild-Arctic, graminoids and *Sphagnum* spp are prominent under wet conditions in stands intermediate between dwarf shrub heath and mire, and towards the south the heathlands support bryophytes and lichens resembling those of the boreal forest and temperate heaths. Thus in Iceland, wefts of *Hylocomium* and *Pleurozium* spp occur alongside mats of leafy liverworts such as *Frullania, Lophozia* and *Ptilidium* spp. Dwarf shrub heath occupies both peat and

Fig. 2.9. Lichen woodland with spruce (*Picea* sp) and *Cladonia stellaris* in the Canadian mild-Arctic. Reproduced from Kershaw (1985) by courtesy of Cambridge University Press.

well-drained substrata in the mild-Antarctic (Table 1.2), with tall turf-
and mat-forming species of *Campylopus, Dicranoloma, Polytrichum*, and
Racomitrium prominent in the bryophyte stratum on the Falkland Is.

Grass heath Grass heath dominated by relatively short, caespi-
tose grasses and sedges (Table 1.2) covers considerable areas of mild-
and cool-polar terrain, particularly on well-drained slopes and in the
raised centres of frost-heave polygons in wetter areas. Taller grassland
dominated by *Cortadaria pilosa* occurs extensively on the Falkland Is.
Herbs, cushion plants and dicots with creeping, woody stems occur
between tufts of the graminoids. Cryptogams are abundant in the more
open stands, where their cover may exceed that of the flowering plants.
Tall turf-forming mosses in the Dicranaceae and Polytrichaceae, together
with mat-forming species of *Racomitrium*, are particularly characteristic.
The canopy-forming *Climacium dendroides* occurs in association with
wefts of *Hylocomium* and *Rhytidiadelphus* spp on grassy slopes in Iceland.
Species of *Cetraria, Cladonia, Stereocaulon* and other fruticose lichens
are also abundant in grass heath communities, the lichens commonly
becoming established by invading moss turfs.

Another widespread series of Arctic communities is dominated by
creeping species of *Salix*, usually with associated grasses, cushion plants
and forbs including species of *Alopecurus, Dryas, Pedicularis* and *Saxifraga*. They occupy patterned ground, the vascular plants growing principally in troughs with mat- and tall turf-forming bryophytes such as
Hypnum revolutum and *Ditrichum flexicaule*, while elevated parts of the
polygons and frost boils remain largely bare, or become colonised by
species of *Collema*, with *Polyblastia* spp and other crustose lichens.

Herbfield Herbfields are dominated by broad-leaved, herbaceous perennials and, in the polar regions, they reach their maximum
development on cool-Antarctic islands. They occupy a range of habitats
from scree to gently sloping ground, being most extensive on well-drained
substrata. The *Cotula* and *Crassula* herbfields on Marion I are halophytic
or prefer sites influenced by marine animals and are thus restricted to
coastal areas. The dominants assume several growth forms. *Acaena* spp
produce a dense cover of erect, herbaceous leafy shoots from perennial,
woody stolons, while *Crassula moschata* forms erect shoots to 10 cm high
with opposite, succulent leaves, again arising from prostrate basal stems.
Grasses, ferns and small herbaceous dicots occur as associates. Lichens
are seldom prominent and bryophytes are most abundant in *Acaena-*

dominated communities and locally in association with *Crassula moschata*. On South Georgia, the stolons of *A. magellanica* permeate an almost continuous turf of *Tortula robusta*, which, on Marion and Prince Edward Is is replaced by mats of *Drepanocladus uncinatus* and *Brachythecium rutabulum*. *Aceana* herbfield occurs in contrasting habitats, on scree and in wet flushes, in both areas.

There are no Arctic counterparts of these southern herbfields. Broad-leaved herbaceous perennials are widespread in many communities, but become dominant only locally in areas of deep or late-lying snow, or of nutrient enrichment by animals. The plants are smaller than those in Antarctic herbfields, the rich assemblages including species of *Oxyria, Pedicularis, Polgonum, Ranunculus* and *Saxifraga*, with grasses and creeping willows. Tall turf- and large cushion-forming mosses such as *Bryum pseudotriquetrum* and *Philonotis fontana* are abundant, particularly in late snow fields.

Tussock grassland   Tussock grassland is another characteristic feature of mild- and cool-Antarctic islands, where it forms an almost continuous coastal fringe on hillsides, cliffs and raised beaches to altitudes of at least 225 m. The dominants are species of *Poa*, which form pedestalled tussocks 0.5–1.0 m high in the cool-Antarctic, and up to 3.0 m high on the Falkland Is. The tussocks comprise fibrous stools surmounted by dense crowns of leaves, and they commonly overlie peat. Few associated species occur where the leaves of adjacent tussocks interweave to form a closed canopy. In open stands on steep slopes, or in areas subject to erosion by seals and penguins, the ground between the *Poa* tussocks is colonised by small herbs and grasses and by turf-forming mosses such as *Bryum argenteum, Campylopus introflexus* and species of *Tortula*. On South Georgia, banks of *Chorisodontium aciphyllum* and *Polytrichum alpestre* have developed in open stands on steep hillsides and in areas subject to grazing by introduced reindeer. Distinctive bryophyte assemblages, including *Mielichhoferia eckloni, Orthodontium lineare* and *Philonotis scabrifolia*, also occur on the walls and ceilings of burrows excavated in the peat by petrels and prions.

Xeric communities
Fellfields   Varying combinations of strong wind and low precipitation prevent the establishment of closed vascular plant cover over extensive mild- and cool-polar terrain, leading to the development of fellfields

and barrens. Fellfields are characterised by cushion-forming vascular plants (Table 1.2), usually interspersed with areas of bare ground. Xero-phyllous sedges, forbs and creeping woody perennials occur between the cushions, and cryptogams are also abundant. Total plant cover ranges from about 20% on dry ground to almost 100% under more mesic conditions.

Fellfields generally support both mosses and lichens, their relative abundance varying in relation to exposure and other factors. The mosses comprise turf- and cushion-forming acrocarps including *Andreaea, Ditrichum* and *Holodontium* spp, with mat-forming pleurocarps such as *Hyloco-mium, Hypnum, Racomitrium* and *Tomenthypnum* spp and leafy liverworts. A wide variety of lichen growth forms is represented. Matveyeva *et al.* (1974) report that species of *Lecidea, Ochrolechia* (crustose), *Hypo-gymnia, Parmelia, Solarina* (foliose) and *Sphaerophorus* (fruticose), with mats of the liverwort *Gymnomitrion corraloides*, occur abundantly between *Dryas* cushions in a lichen–*Dryas* fellfield on the Taimyr Peninsula.

The ground often has a hummocky surface resulting from frost action, and some communities, particularly in the USSR, are regarded as spotty tundra due to the presence of round or oval patches 50–100 cm wide with particularly sparse vascular plant cover. The spots are surrounded by raised borders about 50 cm wide, and they are separated by a network of troughs. Variation in microrelief leads to high species richness, as in a *Dryas*–sedgemoss spotty tundra in the Taimyr that supports 57 angiosperms, 46 mosses and 50 lichens (Matveyeva *et al.*, 1974). In this example, crustose lichens predominate in the dry, open vegetation of the spots, and mosses and angiosperms in the more or less closed plant cover of the borders and troughs.

Barrens Large tracts of arid gravel, scree and clay in the cool-Arctic carry only scattered individuals of cushion- and rosette-forming dicots, tufted graminoids, creeping willows and occasional mosses and lichens (Table 1.2). Total plant cover in these barrens is generally less than 10% and often under 5%. The principal mosses form short turfs or mats, e.g. *Ditrichum, Hypnum* and *Racomitrium* spp, while crustose species such as *Thamnolia subobscura* predominate among the lichens. Some lichens, e.g. *Polyblastia bryophila*, are epiphytic on mosses, but cryptogams are rare or absent in barrens on highly saline or unstable substrate.

Communities dominated by mosses and lichens

Saxicolous communities Essentially cryptogamic vegetation types occur in mild- and cool-polar habitats unfavourable for vascular plants, or in small stands where the habitat differs markedly from the norm for a major community as on muskox dung within a wet meadow. Many of the crytogamic communities resemble those already described from the cold-Antarctic, physiognomically and to some extent floristic-ally, as in the case of epilithic lichen communities. Others resemble the cryptogamic component of some of the angiosperm-dominated vege-tation.

Similarities to the cold-Antarctic are seen in lichen communities on South Georgia, where one finds crustose species of *Verrucaria* low on coastal rocks, while inland rocks support, in addition to crustose lichens (Fig. 2.1), fruticose species of *Usnea* or the foliose lichens *Leptogium* and *Umbilicaria* spp, associated with abundant crustose species and small cushion-forming mosses. Similar saxicolous lichen associations occur widely, as in southeast Greenland. Here *Xanthoria elegans* is dominant on exposed rocks subject to bird influence, *Umbilicaria* spp on other dry, exposed rocks, the fruticose species *Pseudephebe pubescens* in steeply sloping areas, and foliose species of *Parmelia* in moister sites such as the upper surfaces of sheltered boulders. *Rhizocarpon geographicum* and other crustose lichens extend throughout this series of communities (Daniëls, 1975). The habitat relationships of other cool-Arctic lichen vegetation are discussed by Brossard *et al.* (1984) and Link & Nash (1984).

Rocks in some cool-Arctic areas are too dry to support abundant epi-liths. Vitt & Pakarinen (1977) note that epilithic mosses are rare on cal-careous substrata on Devon I, but that small cushion-forming species of *Amphidium*, *Andreaea*, *Dicranoweisia* and *Grimmia* are common on acidic rocks. In contrast, small mat- and turf-forming mosses predominate in the open vegetation of sheltered rock crevices. Examples include a *Cyrtomnium hymenophylloides* community on northern Ellesmere I, where associated mosses include species of *Fissidens*, *Isopterygium*, *Pohlia* and *Myurella*, and an assemblage of *Leskeela*, *Seligeria* spp and other mosses from Alaska. Cliff vegetation is best developed at oceanic, mild-Arctic sites where communities resembling those of temperate, montane regions are found.

Moss and lichen heath Level to moderately sloping ground where conditions do not favour angiosperms may support closed moss or lichen heaths. On South Georgia, lichen-rich *Festuca* meadow gives way on substrata subject to frost disturbance to a community dominated by *Cetraria, Cladonia* and *Stereocaulon* spp, with only scattered flowering plants among the fruticose lichens and their occasional bryophyte associates. Lichen heaths are locally extensive in the Arctic, for example on raised beaches near the west coast of Hudson's Bay. Like lichen woodland, many fruticose lichen heaths are strongly chionophytic, being covered by thick snow until late July, as Daniëls (1982) noted in Greenland (Fig. 2.10). See and Bliss (1980) recognised three assemblages of fruticose and foliose lichens ordered along pH and moisture gradients at alpine sites in the Yukon, and compared them with lichen communities reported from other Arctic and alpine sites.

Fig. 2.10. Lichen heath dominated by *Cladonia* and *Stereocaulon* spp, Angmagssalik District of southeast Greenland. The polygonal cracking in the lichen cover is drought-induced. The area covered by the photograph is about $0.75\,\text{m}^2$.

Moss heaths are also widespread, many being dominated by mat-forming species of *Racomitrium*, notably *R. lanuginosum*. The latter is frequent in dwarf shrub heath and other vascular plant communities, and becomes dominant in dry, rocky sites. Extensive *Racomitrium* heath is also characteristic of islands where a combination of strong winds and porous volcanic substrata restricts vascular plant cover, but bryophytes whose colonies absorb and retain precipitation and dew can thrive. Extensive stony ground on Iceland and Jan Mayen support closed *R. lanuginosum* heath, often with only scattered phanerogams and a small admixture of other bryophytes (Russell & Wellington, 1940), and similar, closed communities dominated by *Racomitrium* spp and *Drepanocladus uncinatus* occur on Svalbard.

Cryptogamic fellfields   Open cryptogamic fellfields are widespread on dry or windswept gravels and scree. Small cushions and other short, compact bryophyte growth forms are prominent: crustose and fruticose lichens may also be abundant, with angiosperms restricted to scattered individuals. Brassard (1971a) lists a wide range of small acrocarpous mosses as constituents of a *Bryum–Encalypta* community on dry plains of sand, silt and clay on northern Ellesmere I. In Iceland, large stones on dry rocky flats in montane regions support scattered mats, small cushions and short turfs of mosses such as *Andreaea rupestris*, *Dicranoweisia crispula* and *Racomitrium fasciculare*, and a similar assortment of growth forms is represented in an open, pioneer community on porous substrata in young lava fields. Moss fellfields form the major plant cover above altitudes of 300 m on cool-Antarctic islands where cushions formed by species of *Andreaea, Bartramia* and *Ditrichum* occur abundantly with lichens and other small, compact mosses.

Bryophyte communities in snow patches   A further series of bryophyte communities occurs where snow persists until well into summer. As well as having an unusually short growing season with freely available water, such sites are characterised by shelter from wind, and by protective winter snow cover. Distinctive assemblages of both vascular plants and bryophytes are found, but Mårtensson (1956) noted that many bryophytes that are confined to late snow areas in the low alpine belt in Lapland occur more widely in high alpine regions. Late snowbeds in the cool-Arctic support large, luxuriant cushions and tall turfs of *Bryum cryophilum, Philonotis fontana* and other mosses, in which are rooted scattered herbaceous flowering plants. The hepatic *Anthelia juratzkana* is abundant in late snowbeds throughout the Arctic, and it is one of the few hepatics to assume dominance in Arctic plant communities.

Hygrophytic bryophyte communities   Vegetation resembling the snowpatch communities in the abundance of large, luxuriant cushion-forming mosses such as *Bryum cryophilum* and *Philonotis fontana* is widely distributed on marshy ground by streams. At some sites rocks in streams support few mosses (Holmen, 1955), but communities with abundant *Hygrohypnum* and *Schistidium* spp occur locally, particularly by waterfalls. *Scapania* and *Fontinalis* are among the genera typical of such habitats in the mild-Arctic. *Calliergon, Drepanocladus* and *Scorpidium* spp are examples of mosses that occur widely in shallow, calcareous ponds, and bryophytes are also dominant in some bogs and flushes. Thus the hepatic *Blepharidophyllum densifolium* is the principal component of some ombrotrophic mires on Marion I, forming, with mosses, a continuous layer 2–5 cm thick over the peat beneath a sparse vascular plant cover.

Most bryophytes are intolerant of salinity, but an assemblage of 36 hepatic species, mostly widely distributed, weedy taxa such as *Jamesoniella colorata*, grow among rocks in the intertidal zone, or in habitats directly exposed to marine spray, in the magellanic moorland region of southern Chile (Engel & Schuster, 1973). Their occurrence is attributed to the ameliorating influence of the almost continuous, heavy rain (page 11) and runoff from adjacent coasts.

Communities of nutrient-enriched sites   The Arctic is notable for the number of bryophyte communities associated with substrata subject to nutrient enrichment by animals. Most striking are assemblages comprising members of the Splachnaceae found on faeces, bone and other animal remains, which occur throughout the Arctic.

In Scandinavia, it has been shown that sporophytes of splachnaceous mosses characteristic of animal remains and dung differ from those of related species growing as epiphytes or on soil in the presence of an expanded, often brightly coloured apophysis, and in the detailed morphology of peristome and spores (Koponen, 1978). Attraction of diptera and other insects to coprophilous species has been demonstrated, both by observation under natural conditions and by trapping experiments (Cameron & Troilo, 1982; Koponen & Koponen, 1978). Visible stimulation by the apophysis, and chemical attraction by volatile octane derivatives released from the sporophyte, are thought to be involved (Pyysalo, Koponen & Koponen, 1983).

Thus adaptations favouring insect dispersal of spores appear to have evolved within the Splachnaceae, with the advantage to the mosses that, having visited a sporophyte, an insect is likely to land in a habitat suitable for the establishment of the moss species concerned. Noting that most members of the Splachnaceae are monoecious, Cameron & Troilo (1982) suggested that insect dispersal of spores may also reduce the incidence of inbreeding by tending to deposit spores from several parent colonies at a given point.

Soil around the mouths of lemming burrows and rocks habitually used as bird perches also support distinctive bryophyte communities composed of small acrocarpous mosses. Several of those associated with lemming burrows are apparently Arctic endemics, e.g. *Desmatodon leucostoma* and *Funaria polaris* (Steere, 1978b; Vitt, 1975). In contrast, the cosmopolitan *Bryum argenteum* is among the mosses colonising sites habitually frequented by birds, while Mårtensson (1956) noted that some mosses with southern distributions extend to alpine sites in northern Sweden only around bird perches. A characteristic series of nitrophilous lichens, including species of *Parmelia, Physcia* and *Xanthoria*, occurs in similar habitats.

Benthic bryophytes

Bryophytes growing permanently submerged in lakes are a characteristic feature of both Arctic and Antarctic regions (Kaspar *et al.*, 1982; Mårtensson, 1956; Welch & Kalff, 1974; and other references in Seppelt, 1983b). They are particularly abundant under oligotrophic conditions where little incoming radiation is absorbed by phytoplankton (Priddle, 1980). Mosses are rare in shallow water, probably due to winter ice-scour, but they have been recorded at depths greater than 30 m and cover 40% of some lake bottoms, the associated biota including epiphytic algae and rotifers. Individual moss stems reach 40 cm in length in these benthic communities, whose luxuriance in the frigid-Antarctic contrasts strongly with the sparsity of the terrestrial vegetation (Light & Heywood, 1975). The moss genera include *Campylium, Dicranella, Distichium, Fontinalis* and *Marsupella*. Some species are exclusively aquatic, but most are more characteristic of dry ground and their morphology is strongly modified by submersion: internode length is typically increased and the leaves may be larger, as in *Calliergon sarmentosum*, or smaller as in *Bryum* spp, than in terrestrial forms, sometimes leading to taxonomic confusion (Seppelt & Selkirk, 1984).

Benthic bryophytes exist under continuously low temperature and irradiance, particularly in the frigid-Antarctic where the lakes are ice-covered throughout the year. Positive net assimilation has been demonstrated during all except the three darkest months in cold-Antarctic sites, although the lake surfaces are frozen for 8–12 months annually; light compensation points are low, and an estimate of 40 g carbon m^{-2} yr^{-1}

was obtained for net productivity in closed stands of *Calliergon* and *Drepa-nocladus* spp (Priddle, 1980). This relatively low figure suggests that the luxuriance of the benthic communities results from stability rather than rapid growth.

It is clear from the preceding account that the diversity of environmental regimes in polar regions is reflected in a wide variety of cryptogamic vegetation. In the next chapter we shall begin to consider the factors controlling the distribution of the various vegetation types, and to investigate dynamic aspects of the communities as evident from their successional relationships.

3

Pattern, process and environment

Environmental relationships of polar cryptogamic vegetation

Evidence concerning environmental control over the distribution of polar bryophytes and lichens is largely circumstantial as it is derived principally from observed correlations between vegetation and environment, which do not necessarily imply direct control by the factors concerned. Moreover, distribution is influenced by complex combinations of climatic and other variables so that correlations may be difficult to elucidate.

Ordination provides a means of alleviating the latter difficulty, as demonstrated by Webber's (1978) study of Alaskan tundra. Webber showed that species richness among both bryophytes and lichens is highest under mesic to dry conditions with moderate soil phosphate levels, but that species richness for lichens shows a stronger correlation with good soil aeration than does that for mosses (Fig. 3.1). Different patterns were shown by species indices based on frequency and cover (Fig. 3.2), which reveal that bryophytes are most abundant under relatively wet conditions with poor soil aeration and low phosphate, whereas lichen abundance is greatest on drier, well aerated soils with moderate phosphate.

The analysis also provided valuable insight into the habitat preferences of individual species, but such techniques have not been widely applied to polar vegetation. In some instances, however, relationships between plant distribution and environment stand out so clearly that a controlling influence for the factors concerned may reasonably be inferred from simple observation. Other evidence comes from physiological studies, as discussed in Chapters 5 and 6.

Temperature

Cool, generally short, summers are the most significant unifying feature of the varied polar climates, and the major vegetational zonation

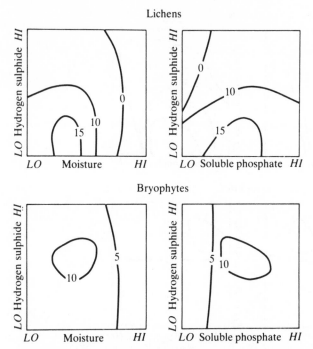

Fig. 3.1. Distribution of bryophytes and lichens in relation to edaphic factors within an ordination of Alaskan tundra vegetation, showing isolines of the number of species per $10\,m^2$ plot. Low hydrogen sulphide concentration is indicative of good soil aeration. Data from Webber (1978).

is more closely correlated with mean summer temperature than with other climatic factors (Table 1.3). How far temperature acts directly through its control of plant metabolism and growth, or indirectly by influencing rates of nutrient cycling and other ecosystem processes remains to be determined (Chapter 7).

Cryptogams are strikingly predominant in cold- and frigid-Antarctic regions, where mean monthly air temperatures remain at or below 2 °C throughout the year. Water is plentiful in summer, snow cover well developed in winter, and the major mineral nutrients freely available in parts of these regions. The cold-Antarctic has milder winters than many areas supporting angiosperm-dominated tundra, coniferous forest or grassland, and it extends north to latitude 54°26′S on Bouvetøya (Fig. 1.2), where the photoperiodic regime resembles that in temperate regions. Low summer temperatures and a short growing season thus appear to be the most potent factors determining the essentially cryptogamic facies of cold-polar vegetation.

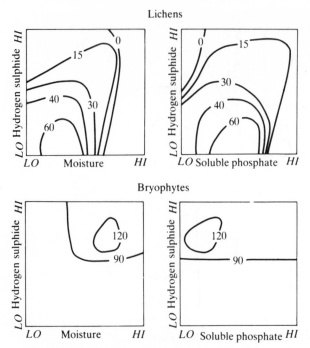

Fig. 3.2. Distribution of bryophytes and lichens in relation to edaphic factors within an ordination of Alaskan tundra vegetation showing isolines of species index values. A species index is the sum of relative frequency and relative cover, and the indices for individual species have been summed for bryophytes and for lichens. Data from Webber (1978).

Geographical isolation since Pleistocene glacial maxima has perhaps restricted immigration, particularly of vascular plants, to the cold-Antarctic, and the associated discontinuity in environmental gradients may have prevented the progressive evolution of adaptation to climatic severity (Aleksandrova, 1980; Longton & Holdgate, 1967). However, the two phanerogams native to the cold-Antarctic are rarely abundant except in sunny, sheltered sites warmed by insolation, and most flowering plants experimentally introduced from Arctic, or from mild- and cool-Antarctic localities, have failed to become established (Brown, 1912; Edwards, 1979; Edwards & Greene, 1973; Holdgate, 1964b). *Poa annua*, which survived for several years following its accidental introduction to Deception I (Longton, 1966a), is a notable exception. Some additional flowering plants might well survive locally under the climatic conditions prevailing in the cold-Antarctic, but it seems unlikely that angiosperms would challenge bryophytes and lichens as dominants in the vegetation, even in the absence of dispersal barriers, any more than they have in the cold-

Arctic with its continuous land connection to warmer areas.

That temperature is important in restricting the cryptogamic flora in polar regions is demonstrated by an examination of sites of geothermal heat. Hot springs on Iceland and Greenland are surrounded by concentrically zoned vegetation supporting bryophytes such as *Archidium alternifolium* and *Sphagnum palustre* that are not recorded elsewhere on the islands (Halliday, Kliim-Nielsen & Smart, 1974; Hesselbo, 1918; Lange, 1973). Similarly, 19 bryophytes have been recorded from the South Sandwich Is (Fig. 1.2) only near fumaroles and several, including *Campylopus* spp and the hepatic *Cryptochila grandiflora*, are abundant at these local sites of volcanic heat and moisture but are unknown elsewhere in the cold-Antarctic (Longton & Holdgate, 1979). Since water is freely available in other habitats in Greenland, Iceland and parts of the cold-Antarctic, the restriction of significant assemblages of species to the localised and possibly transient fumaroles is likely to reflect a temperature limitation in the general environment.

The influence of temperature on local distribution of other polar vegetation is difficult to assess, as this factor seldom varies in isolation. It is notable that cold-Antarctic moss carpet and tall moss turf communities are most extensive at relatively warm, low-altitude sites of northerly exposure. Comparable relationships between plant distribution and aspect can be observed throughout the polar regions. For example in Alaska, Steere (1978b) noted a contrast between the bryophyte floras of shaded, north-facing and sunny, south-facing slopes. The former comprise principally taxa with distributions centred in polar regions while the latter include species of temperate affinity, but the drier nature of the south-facing slopes complicates interpretation.

Growth form and water relations

While temperature may be of prime importance in determining the dominance of cryptogams in cold- and frigid-polar regions, water availability is of greater importance in controlling the distribution of cryptogamic growth forms, both locally and regionally. The three major lichen growth forms (Table 2.3) are well represented, while among bryophytes relatively compact turfs, mats and small cushions predominate, with carpets widespread in the cold-Antarctic. Looser growth forms, especially the weft and the canopy, are of local occurrence in tundra but wefts are abundant in boreal forests.

At a regional level, water availability is generally regarded as the most significant cause of the vegetational sparseness of the frigid-Antarctic and much of the cool-Arctic (Billings, 1974; Llano, 1965; Rudolph, 1971). Schuster (1959) has stressed the 'extremely restricted development of the bryophte vegetation' in arid parts of the Arctic. In the frigid-Antarctic, mosses are only very locally abundant by lakes, streams, and snow banks which melt gradually throughout the summer, occurring less extensively in hollows where snow accumulates during light summer flurries. Lichens in frigid-Antarctic lowlands are also most prominent in depressions where snow accumulates and in other habitats irrigated by melt water (Fig. 3.3), but the greatest abundance of crustose species such as *Buellia frigida* in the dry valley region and other parts of Victoria Land is at relatively high altitudes (about 600 m) where cloud cover and summer snow flurries are most frequent (Llano, 1965; Kappen, 1985a,b; E. Schofield, 1972). On parts of Ellesmere I, also, the most luxuriant vegetation occurs at altitudes of 300–500 m where melt water is more freely available than in the arid lowlands.

At a more local level, relationships between topography and growth form among cool-Antarctic cryptogams (Chapter 2; Fig. 3.4) also appear to be determined primarily by water availability. The highest, most wind-swept cliffs and scree are largely devoid of macrophytic vegetation. Lichens, particularly appressed forms, tend to occupy drier, more exposed habitats than mosses, but the best development of epilithic, fruticose lichens is on relatively cool, moist cliffs rather than those most strongly warmed by sunshine. The bryophyte growth forms can be regarded as a series from compact short turfs and small cushions in dry habitats through tall turfs of acrocarps and carpets of pleurocarps in mesic and wet habitats respectively to large, loose cushions by running water.

Similar relationships apply in other areas. Crustose lichens and compact cushion and short turf-forming mosses predominate in dry exposed habi-tats, on both rocks and soil, in the frigid-Antarctic and parts of the cool-Arctic, and on porous, volcanic substrata in wetter areas such as parts of Iceland and some cold-Antarctic islands. Small acrocarpous mosses and mat-forming pleurocarps are well represented in dry to mesic Arctic habitats, as are macrolichens and tall turf-forming acrocarpous mosses with erect branches (Table 2.2) comparable with those abundant on mesic slopes in the cold-Antarctic.

Sphagnum spp and pleurocarpous mosses forming tall turfs with diver-gent branches predominate in the bryophyte stratum beneath graminoids in Arctic and cool-Antarctic mires, which thus contrast with cold-Antarctic wetlands with their abundance of carpet-forming species. The family

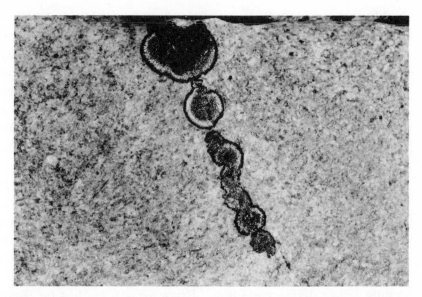

Fig. 3.3. The crustose lichen *Buellia frigida* at a frigid-Antarctic site in Wilkes Land, growing in a minute rock fissure which acts as a snow-melt drainage channel. The largest (?oldest) thallus is being colonised by the foliose species *Omphalodiscus decussatus*.

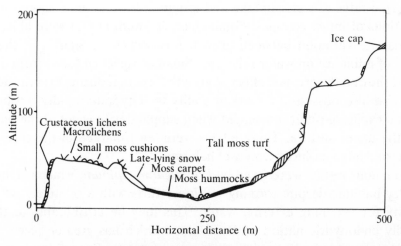

Fig. 3.4. Diagram indicating the distribution of major cryptogamic growth form types in an idealised valley in the cold-Antarctic. Data from Longton (1982).

Amblystegiaceae is well represented in both types of habitat, and a careful comparison of these carpet- and turf-forming pleurocarps might show that the carpet form results in part from the pressure of winter snow on essentially erect-growing but weak-stemmed mosses lacking protection and support from vascular plants. Finally, large, loose moss cushions are characteristic of stream banks and other sites influenced by flowing water in both northern and southern polar regions (Chapter 2), while the contribution of hepatics to the vegatation cover is everywhere disproportionally low in relation to the numbers of species involved.

The impact of growth form on bryophyte distribution stems in large measure from water relations. Lacking an effective mechanism for absorbing soil moisture and transporting it internally, most bryophytes are ectohydric, being forced to rely largely on water absorbed directly into actively photosynthetic leaf and stem cells. Such cells seldom possess a significant cuticle that would restrict water entry, but in consequence they lose water rapidly under drying conditions. Bryophytes may thus be regarded as poikilohydrous, as their water content varies widely in response to changes in ambient environment. In compensation many bryophytes, especially mosses, show greater tolerance of cytoplasmic dehydration than is typical in vascular plants (Longton, 1980; Proctor, 1982).

Aggregation of bryophte shoots into colonies, particularly in the more compact growth forms, enhances capillary uptake of soil water and retention of precipitation, as well as reducing air movement over the shoots and thus evaporation from their surfaces (Chapter 4). These effects on external water will lead indirectly to higher internal water contents. By comparing rates of water uptake and loss in individual moss shoots and in portions of intact colonies, Gimingham & Smith (1971) were able to demonstrate variation between growth form types on Signy I in their degree of influence on water relations. Small compact cushions from dry ground showed the greatest effect of growth form in reducing evaporation rates, and also possessed a marked ability to rehydrate rapidly through external capillary water movement when supplied with water at the base or at the upper surface. These effects were least marked in the case of carpets and large cushions from wet habitats.

Anomalous results were given by *Polytrichum alpinum*, which extends into dry habitats despite forming tall loose turfs with a relatively weak capacity for retaining external water. This may be attributable to the partially endohydric nature of *P. alpinum* which has greater access to soil water than ectohydric species due to the presence of water-conducting hydroids. It also has a higher resistance to loss of internal water, resulting from a relatively thick cuticle and from changes in leaf shape and orientation on drying (page 172).

These effects are illustrated in Fig. 3.5, which shows the decline in water content (internal plus external) in isolated shoots, and in portions

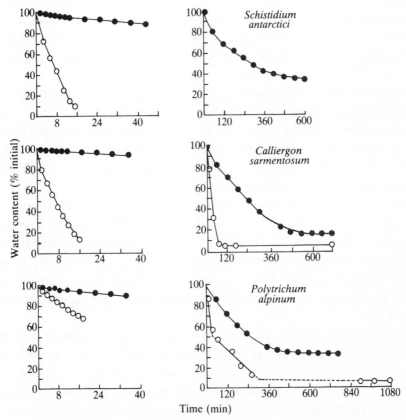

Fig. 3.5. Decline in water content of single shoots (○) and portions of intact colonies (●) in three cold-Antarctic mosses at 45% RH and 18–20 °C in still air. On the left, short-term experiments with weighings at short time intervals; on the right, long-term experiments with less frequent weighings. Broken line indicates a period over which the angle of slope is uncertain. Data from Gimingham & Smith (1971).

of intact colonies, initially water-saturated and exposed to 45% relative humidity at 18–20 °C. Water loss from individual shoots reached 90% in 16 minutes in the ectohydric mosses *Schistidium antarctici* and *Calliergon sarmentosum* compared with only 30% in *Polytrichum alpinum*. However, water retention by portions of intact colonies after 600 minutes was 35% in small cushions of *S. antarctici* but only 15% in carpets of *C. sarmentosum*.

Similar factors may account for the restricted distribution of weft- and canopy-forming mosses in the tundra. These loose growth forms are typical of sheltered, humid habitats under trees and shrubs, and they would hardly be expected to prosper under exposed, often arid, polar conditions, especially in the case of wefts which have few anchoring rhizoids and are easily disrupted by wind. W. B. Schofield (1972) noted that species such as *Hylocomium splendens* that form wefts in boreal forests normally occur in the Arctic as compact mats.

Lichens are also poikilohydrous, and it was traditionally believed that water uptake and loss are essentially physicochemical processes over which the plants exert little control. However, Blum (1973) suspects that active processes may sometimes be involved in uptake, and recent studies of Arctic lichens have shown that thallus morphology may significantly influence rates of both uptake and loss (Larson, 1979, 1981; Larson & Kershaw, 1976). Among features increasing evaporative resistance are low surface area to volume ratio, rugosity which decreases the exposure of part of the thallus surface, and, as in mosses, aggregation of individuals into mats or cushions. The rate of imbibition also varies widely, being greatest in plants such as *Alectoria* and *Usnea* spp with high surface area to volume ratios, a factor which would compensate for rapid water loss in such forms.

In experiments with *Umbilicaria* spp, Larson (1981) showed that rhizines, the lower surface of the thallus, and cortical lamellae on the lower surface of *U. muhlenbergii* are all important sites of water uptake, but that surface papules and isidia are not. In contrast, rhizines of *Peltigera canina* apparently play little role in water uptake and storage. The difference is habitat-related: *Umbilicaria* spp on steeply sloping rocks are irrigated principally by water flowing down the rock surface in contact with the underside of the thallus, but *P. canina* occurs on flat rocks or soil with its main water supply from above.

The foliose lichens in this study were collected in southern Ontario, and although few comparable data on water uptake are available for polar species, Larson's work emphasises that, as in mosses, morphological features may have a considerable influence on lichen water and habitat relations. It must be stressed, however, that growth form and other morphological attributes that influence water uptake and loss are unlikely to prevent periodic desiccation of mosses and lichens in the more arid polar habitats. Their main significance is believed to lie in extending periods when the plants are sufficiently hydrated to maintain positive net assimilation and growth (Chapter 5).

Wind and snow

The effects of wind and snow on polar vegetation are powerful, and they are interrelated as wind strongly influences the distribution of snow cover which, in turn, reduces the impact of wind. Strong winds are important through their cooling and desiccating effects. They also redistribute surface material, resulting in the accumulation of fine soil particles in sheltered depressions, and thus creating a favoured habitat for plants such as frigid-Antarctic mosses. Wind acts as an agent of erosion, particularly during winter when particles of sand and hard snow can cause severe abrasion of frozen vegetation. Webber's analysis (page 66) suggested that wind has only a minor influence on plant distribution at Barrow, due to the moist soil, high relative humidity and lack of major topographic relief. Its effects are more pronounced in arid or hilly regions. Thus mosses and lichens occur principally on sheltered, southwest-facing slopes at some frigid-Antarctic localities (Lindsay & Brook, 1971; Nakanishi, 1977), despite the warmer conditions associated with northerly aspect in regions of consistently low temperature.

Winter snow cover protects subnivean vegetation from wind-induced erosion and provides insulation against minimum temperatures and repeated freezing and thawing during periods of fluctuating air temperature. Late-lying snow gives summer moisture, and light summer snowfall accumulating in sheltered hollows also provides a significant source of water in arid regions. However, deep winter snow persisting into summer can inhibit growth by reducing the growing season. Accumulation of wind-blown sand in sheltered hollows also restricts bryophyte vegetation in some late-snow areas: in the Arctic Holmen (1955) considered *Polytrichum alpinum* and *Timmia austriaca* two of the species best able to overcome this stress. Conversely, decomposition of wind-blown organic matter deposited by melting snow could be beneficial to some snowbed vegetation by increasing availability of nitrogen and other elements (Brassard & Longton, 1970; Miller, 1982).

Relationships between vegetation and summer snow cover are thus highly variable, depending in part on the length of the summers at a given locality, and on the preferences of individual species. It was noted in Chapter 2 that luxuriant bryophyte communities are characteristic of some late snowbeds in the cool-Arctic, but that on Franz Josef Land, where the summers are particularly short and cool, the best developed bryophyte vegetation occurs where the snow melts relatively early. Koerner (1980) showed that lichen cover at some Arctic sites is inhibited where snowfields persist into July. Conversely frigid-Antarctic lichens may be most abundant on ground covered by snow throughout much of the growing season, and Gannutz (1971) regarded the subnivean environment in summer as favourable for lichen metabolism and growth (Chapter 5).

Concerning winter snow cover, some fruticose lichen-heaths in the Arctic are chionophytic (Chapter 2), while interspecific preferences are evident in lichen woodland in Manitoba (Fig. 2.9) where full development of *Cladonia stellaris* and *Stereocaulon paschale* cover is dependent on deep winter snow, while other fruticose lichens, such as *Bryoria nitidula*, are able to colonise virtually snow-free sites (Kershaw, 1977; Chapter 5). Correlation between winter snow cover and plant distribution was also demonstrated by Smith's (1972) observations on low knolls in coastal areas of Signy I. The windswept, largely snow-free summits of the knolls support a community characterised by the fruticose lichen *Usnea antarctica*. This gives way to other lichens and cushions of the moss *Andreaea depressinervis* in more sheltered locations below the summit, while slopes receiving deeper snow accumulation, and moist level ground at the foot of the knolls, support turfs of *Polytrichum alpinum* and carpets of *Drepanocladus uncinatus*. The gradual transition between the communities is emphasised in Fig. 3.6.

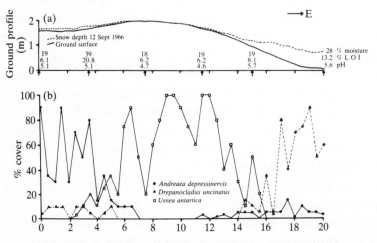

Fig. 3.6. Relationship of ground profile to soil data and winter snow cover (a), and to percentage cover of the principal species (b) along a 20 m transect across a low knoll on Signy I. LOI is loss on ignition. Redrawn from Smith (1972).

Edaphic factors

Many factors related to physical and chemical properties of the substratum influence the distribution and composition of polar vegetation. Lichen abundance is greatest on dark rock in some localities, as discussed on page 204. A preference of lichens for dolerite as opposed to sedimentary rocks in the Theron Mountains in the interior of Antarctica has been attributed to the greater stability of the dolerite, with its low susceptibility to frost shattering (Lindsay & Brook, 1971). This could be an important influence elsewhere in environments that favour rapid physical weathering by wind and frost action, as on Ellesmere I (Brassard, 1971a).

Instability of scree and finer mineral soils due to frost action, solifluction and related phenomena also has a major influence, as in frost boil sedge meadows on Devon I where graminoids and tall turf-forming mosses in wet depressions give way to algae and smaller mosses on the drier, unstable mounds resulting from cryoturbatic disturbance. Gannutz (1971) demonstrated that lichen cover was most extensive on large talus blocks and gentle inclines where the substrate was subject to minimum downslope movement at a frigid-Antarctic site. Comparable examples from the cold-Antarctic are discussed on pages 97–8.

Many cryptogams can be regarded as either calcicoles or calcifuges with others broadly tolerant, but it remains unclear how far control is exerted directly by calcium, or indirectly by pH or associated factors. Thomson (1979) indicated the preferences shown by many of the 500 lichen species on the Alaskan North Slope, and Steere (1978b) discussed the distribution of calcicoles and calcifuges in the Alaskan bryophyte flora. Species of *Sphagnum* and the Polytrichaceae are abundant on acid, boggy soils near Barrow, but are rare or absent further east near Prudhoe Bay where the substrata comprise alluvial deposits of calcareous silt and sand. Calcium levels and pH influence distribution principally at the species level, rather than the growth form level, as indicated by the distribution of short turf- and small cushion-forming mosses on Signy I (Chapter 2). A few Arctic mosses, notably *Coscinodon cribrosus* and *Mielichhoferia* spp, are indicators of copper and other metals (Brassard, 1969; Steere, 1978b).

Availability of major nutrient elements is unlikely to be seriously limiting for plant growth in the cold-Antarctic or on cool-Antarctic islands due to marine influence and the presence of large bird and seal colonies (Allen & Heal, 1970; Jenkin, 1975), although experimental fertilisation increased vascular plant production on South Georgia (Smith & Walton, 1975). In contrast, inland soils in the cool-Arctic and frigid-Antarctic are strongly deficient in certain elements, notably available nitrogen and phosphorus (Tedrow, 1977), and experimental studies suggest that net assimilation and growth may be nutrient-limited in some mosses and lichens (Chapter 6).

Other elements occur in high or even toxic concentrations in some of these dry soils due to restricted leaching or a prevalence of upward water movement. Salinity is regarded as crucial in restricting macrophytic vegetation on gravel soils in the frigid-Antarctic, mosses and lichens being less tolerant than algae (E. Schofield, 1972). Ugolini (1970) suggested that there is 'an inverse relationship between pedologic development and life succession' in these areas because of the high mineral content of the most highly developed soils, and on Ross I extensive stands of *Bryum antarcticum* by shallow gulleys are commonly moribund and coated with a white crystalline deposit attributed to evaporation of mineral-rich water flowing down the channels (Longton, 1973a). The deposit has a high sodium content (35%) and may originate as wind-blown salt of marine origin (Greenfield & Wilson, 1981). Ugolini (1977) speculated that mosses concentrate minerals from highly saline soils and that their death and subsequent removal by wind may restore salt contents to sublethal levels.

Biotic factors

The effect of severe manuring and trampling by marine animals restricts vegetation around their breeding colonies, but some cryptogamic communities show a positive association with sites of less intense animal activity (Chapter 2), and many such relationships are believed to involve nitrogen enrichment. Some of the plants concerned are distinctly nitrophilous, showing a preference for ammonia and other reduced nitrogen compounds, and they are restricted to enriched sites even in areas where available nitrogen levels are not unusually low. Examples include the alga *Prasiola crispa*, epilithic lichens such as *Xanthoria elegans* and, at least in some areas, the cosmopolitan moss *Bryum argenteum* (Nakanishi, 1977; Steere, 1978b). Local nutrient enrichment, for example around animal carcasses in the Arctic, also leads to the occurrence or increased luxuriance of species with more moderate demands in areas of generally low availability of nitrogen and phosphorus.

There is a striking correlation between the occurrence of lichens and nesting colonies of petrels and other birds on some frigid-Antarctic nunataks. Near Syowa Station, a species-poor *Buellia frigida–Rhizocarpon flavum* community occurs on slopes without bird colonies, whereas rock below nests support a richer community comprising *B. frigida* and a range of other lichens including *Caloplaca, Umbilicaria* and *Xanthoria* spp: mosses are also most frequent near bird nests here, and elsewhere near Syowa Station (Matsuda, 1968; Nakanishi, 1977). However, there is no comparable relationship between vegetation and bird colonies at some other frigid-Antarctic localities (Filson, 1966), while in Marie Byrd Land, Siple (1938) found that the abundance but not the diversity of lichens was greatest on outcrops influenced by birds. The degree of nitrophily clearly varies among Antarctic lichens, and the specific composition of the flora in a particular mountain system may influence the extent to which the local vegetation is associated with nesting sites.

In summary, short cool summers appear to be primarily responsible for the insignificant role of phanerogams and consequent dominance of cryptogams in cold- and frigid-polar vegetation, while water availability is crucial in determining the distribution and abundance of the major cryptogamic growth form types. Variation in water supply thus plays a vital role both in restricting the abundance of cryptogams in arid regions and in controlling their distribution in relation to the habitats available at a given site. Both temperature and water availability are influenced by other variables such as wind and snow cover, and this whole complex of factors is strongly modified by topography and microtopography which thus exert major if indirect control over local distribution patterns. Other factors, such as the chemical nature of the substrata and degree of biotic influence, may have an important bearing on the luxuriance of the vegetation, and on plant distribution, particularly at the species level.

Colonisation and succession
Theoretical background
Existing vegetation has resulted from colonisation of bare surfaces by pioneer organisms, followed by the gradual and progressive displacement of pioneer species by others in a succession of communities which leads, in time, to the development of relatively stable, climax vegetation. Species in each successional or seral community are visualised as modifying edaphic and microclimatic conditions in such a way as to favour the establishment of species in the next community with which, as a result, they become unable to compete. The climax community is viewed as being essentially in equilibrium with its environment. Succession at a given site reflects the interaction of autogenic control, i.e. control by plants through competition and their influence on the environment, and allogenic control by abiotic factors (Tansley, 1935). Biogenic control by animals is a major influence locally, e.g. by bird perches and penguin

breeding colonies. Muller (1952) considered some tundra vegetation to have developed by auto-succession, which he defined as 'a succession consisting of a single stage, in which pioneer and climax species are the same' due to the prevalence of allogenic control.

The climax concept exists in several forms. Its most strict adherents visualise the vegetation in each region as progressing, in the absence of disturbance, inexorably towards a single or monoclimax community, determined by climate which is regarded as exerting an overriding influence over the complex relationships between climate, soils and plant life form (Clements, 1916). Others, recognising that variation in bed rock, topography and other factors can permanently influence the interaction between soil and vegetation in a climatically uniform region, accept that such a region will support a range of climax vegetation types.

This polyclimax concept (Tansley, 1935) was modified by Whittaker (1953), who emphasised that vegetational change in space, as well as in time, can be continuous along environmental gradients and who thus preferred the concept of climax pattern. It is often expedient to refer to communities, or noda, while recognising the frequently continuous nature of variation in plant cover (Webber, 1978), as was done in Chapter 2. Further, it is now appreciated that climax vegetation is by no means static, showing cyclic and other changes at a particular point under uniform conditions, and directional change in response to any climatic change that is sustained over a period of years. Succession in polar regions is commonly so slow that it must inevitably be influenced by climatic change as well as autogenic processes. The time scale is such that the successional sequences are to a large extent inferred rather than observed, but strong if indirect evidence is provided by zonation of communities around retreating glaciers and on raised beaches of different age.

Much polar vegetation exists on terrain that has become available for colonisation only since the retreat of Wisconsin ice sheets or subsequent minor readvances, a period ranging from some 15 000 years to a matter of months where glacial retreat is currently in progress. Other sites were subject to marine incursions during the Tertiary or Pleistocene. Therefore, it is not everywhere clear whether existing vegetation has reached a climax, and primary succession on rock, glacial moraines and other areas that have not previously supported plant cover is more prevalent in polar regions than in areas where vegetation has been present continuously for many millions of years. However, secondary succession is also widespread, where plant cover has been destroyed by frost action, fire or other causes.

The role of bryophytes and lichens

Bryophytes and lichens are effective colonisers during early stages of succession on dry land. They were traditionally believed to have an important influence on rock weathering, but this point is currently somewhat controversial. Both groups are likely to be beneficial in terms of increasing the availability of nitrogen in immature soils (Chapter 7). The role of lichens in colonising rock surfaces has been discussed by Syers & Iskandar (1973), Ugolini & Edmonds (1983), and by Topham (1977) who noted that the adaptations facilitating colonisation include tolerance of desiccation and extreme temperatures, longevity and low growth rates. Physical weathering is accelerated by rhizine penetration in foliose species, and by expansion and contraction of appressed, partially endolithic, crustose thalli in response to changes in water content. Lichens may induce chemical weathering by liberating oxalic acid, various lichen acids and carbonic acid formed when carbon dioxide released by respiration combines with water, but organic acids excreted by the mycobiont and acting as chelating agents appear to be more significant.

Scanning electron microscopy and other techniques have confirmed that chemical weathering takes place beneath crustose lichens on a range of rock types, creating irregular surfaces susceptible to physical weathering (Ascaso, 1985; Jones & Wilson, 1985). Penetration of lichen thalli into rock, with incorporation of rock fragments into the thalli, has been demonstrated on Signy I quartz mica-schist (Walton, 1985). In southern Victoria Land the surface of rocks containing cryptoendolithic lichens and other organisms has been observed to peel off periodically, apparently due to the cementing material between rock crystals being dissolved by substances released by the organisms (Friedmann, 1982).

However, Williams & Rudolph (1974) reported that free-living fungi showed greater ability than either fungi or algae isolated from lichens to chelate ferric iron in culture, and Brodo (1973) considered rock weathering, humus formation and entrapment of wind-blown particles by lichens to be extremely slow. Indeed, a well-developed cover of lichens, or of bryophytes, may under certain circumstances retard weathering by protecting rock surfaces from erosion by wind-blown particles, by providing insulation against freeze-thaw cycles, or by absorbing precipitation and thus further reducing the incidence of frost shattering (Holdgate, Allen & Chambers, 1967; Lindsay, 1978). Mosses appear to be more effective than lichens in paving the way for colonisation of rock surfaces by vascular plants. Crustose lichens are commonly the first colonisers, but their presence is not always necessary to permit establishment by mosses, often in cracks or depressions. Moss rhizoids and associated fungi penetrate at least 5 mm into some rocks, permitting water entry and accelerating physical weathering (Hughes, 1982). Mosses have higher growth rates than lichens, and their colonies have a greater capacity to trap wind-blown material and contribute organic matter to developing soil (A. J. E. Smith, 1982). Oosting & Anderson (1939) considered these processes of greater significance in temperate regions than the influence of mosses on rock weathering.

Polunin (1936) defined stages of a xerosere in northern Lapland as dominated by: (1) cyanobacteria, (2) appressed lichens, (3) foliose lichens, (4) mosses and fruticose lichens, (5) herbs, (6) xeric dwarf shrubs, (7) taller shrubs and (8) birch climax. He regarded stages 1–3 as proseral, in that the mosses of stage 4 were in places able to colonise bare rock directly. The larger mosses, and species of *Cladonia* and *Stereocaulon* that grow on them, were considered primarily responsible for creating conditions favouring higher plants by trapping dust and accumulating as humus. Comparable processes have almost certainly occurred widely in mild- and cool-polar regions during the development of angiosperm-dominated tundra.

The traditional view that mosses and lichens precede flowering plants during colonisation is less generally valid for deposits of sand, gravel and till than for rock. Plant succession in frigid-Antarctic soils begins with the appearance of algae, followed by bacteria and other micro-organisms, and finally by mosses and lichens (Llano, 1965). Recently deglaciated rocks in the cold-Antarctic commonly show a lichen trimline at distances up to 2.5 m from the retreating ice. Intervening rocks lack colonising lichens, whereas beyond this boundary there is a progressive increase in species diversity, and in colony size in the first established species. However, *Drepanocladus uncinatus* and other mosses become established on local deposits of moist sand, even within the lichen trimline. Similarly on moraines, the first macrophytic colonisers appear to be lichens on rock surfaces, and mosses or the grass *Deschampsia antarctica* in pockets of soil (Corner & Smith, 1973; Lindsay, 1971; R. I. Lewis Smith, 1982a). On South Georgia, some moraines subject to cryoturbatic disturbance are colonised by grasses, with mosses becoming abundant later in association with older grass plants. Crustose lichens may appear first on more stable soils, giving way to bryophyte colonies in which flowering plants later become established (Heilbronn & Walton, 1984; Smith, 1984a).

A range of plant types can thus act as pioneer macrophytic colonisers in the Antarctic depending on the nature of the substratum and other factors. Worsley & Ward (1974) likewise concluded that mosses and grasses become established more or less simultaneously on newly formed moraine ridges near a retreating glacier in northern Norway. The different patterns of colonisation on rock and finer substrata were evident on Surtsey, an island formed in 1963 by volcanic activity off southern Iceland. Colonisation of lava flows was effected primarily by mosses, whereas sandy beaches were initially invaded by both mosses and flowering plants with only minor association between the two (Fridriksson, 1975). Kuc (1970) considered that pioneer bryophytes on finely divided substrata are commonly dwarf acrocarpous mosses forming short-lived colonies of high fertility (Chapter 8).

Although cryptogams do not invariably precede the pioneer flowering plants, these plants undoubtedly influence pedogenesis and therefore the course of succession on immature, mineral soils by contributing to the development of organic matter, and through their influence on mineral contents (Chapter 7). Dawson, Hrutfiord & Ugolini (1984) have demonstrated that usnic acid and other compounds liberated by *Cladonia mitis* in Alaska are sparingly water-soluble and mobile within the soil profile, and they suggested that such compounds may contribute significantly to podsolisation and other aspects of profile development.

Mosses also play a part in hydroseres. *Sphagnum* spp form floating turfs which extend from the shore towards the centre of pools. As the turfs increase in thickness other plants characteristic of mires, and eventually of mesic communities, can become established. In a rather different manner, mosses such as *Drepanocladus* and *Scorpidium* spp, growing submerged or partly so, influence early stages of other hydroseres by accelerating the build-up of organic matter and sediments on the bottoms of ponds and lakes (Polunin, 1935).

Directional succession

An analysis of lichen succession on two types of ultra-basic rocks in montane tundra in the northern Urals (Magomedova, 1979) illustrates several of the preceding points. Plant cover was recorded in more than 950 quadrats, each assigned to a stage of succession according to degree of rock weathering, silt accumulation and other features. Table 3.1 summarises data for pyroxenites on which lichens colonise microindentations between rock particles and a high cover is eventually attained. Dunites (Table 3.2) differ both physically and chemically: colonisation occurs principally in crevices, the surface is unstable, and percentage lichen cover

Table 3.1 *Succession in lichen communities on pyroxenites in the northern Urals*

Stage of succession	Number of species				Species per unit area	Number of species of different growth form types			Total percentage cover
	Total	Remain	Disappear	Intrude		Crustose	Foliose	Fructicose	
1. *Lecanora polytropa – Rhizocarpon geographicum*	3	–	–	3	2	3	–	–	5
2. *Lecidea pantherina – Rhizocarpon geographicum*	9	3	–	6	4	9	–	–	30
3. *Lecidea flavocaerulescens – Lecidea pantherina – Rhizocarpon geographicum*	24	8	1	16	6	23	1	–	41
4. *Umbilicaria proboscidea – Rhizocarpon geographicum – Lecidea pantherina*	38	20	4	18	7	23	15	–	62
5. *Parmelia centrifuga – Haematomma ventosum – Rhizocarpon geographicum – Lecidea pantherina*	54	31	7	23	10	21	33	–	68
6. *Xanthoparmelia centrifuga – Lecidea pantherina – Haematomma ventosum*	43	30	24	13	11	15	23	5	72
7. *Alectoria ochroleuca*	47	23	20	24	9	11	11	25	74
8. *Cetraria laevigata – Cladonia arbuscula*	44	17	30	27	8	2	5	37	82
9. *Cladonia rangiferina – Cladonia arbuscula*	15	15	29	–	6	–	–	15	81

Data from Magomedova (1980)

Table 3.2 *Succession in lichen communities on dunites in the northern Urals*

Stage of succession	Number of species				Species per unit area	Number of species of different growth form types			Total percentage cover
	Total	Remain	Disappear	Intrude		Crustose	Foliose	Fructicose	
1. *Lecidea pantherina – Aspicilia caesiocinerea*	8	–	–	8	3	8	–	–	5
2. *Gasparrinia elegans – Placynthium nigrum – Lecidea pantherina – Aspicilia caesiocinerea*	12	8	–	4	4	12	–	–	16
3. *Lecidea pantherina – Placynthium nigrum – Aspicilia caesiocinerea*	20	12	–	8	5	18	2	–	34
4. *Placynthium nigrum – Aspicilia caesiocinerea*	27	16	4	11	8	16	10	1	48
5. *Gasparrinia elegans – Placynthium nigrum – Aspicilia caesiocinerea*	16	14	13	2	7	14	1	1	37
6. *Caloplaca* spp – *Placynthium nigrum – Aspicilia caesiocinerea*	19	13	3	6	6	12	1	6	10
7. *Alectoria ochroleuca – Cladonia arbuscula – Cetraria laevigata*	14	5	14	9	6	–	–	14	13
8. *Cladonia arbuscula – Cetraria laevigata – Alectoria ochroleuca*	22	13	1	9	7	2	3	17	54
9. *Cladonia arbuscula – Alectoria ochroleuca – Cetraria laevigata*	20	17	5	3	6	–	2	18	42

Data from Magomedova (1980)

seldom exceeds 50%. Many species were restricted to one rock type, with others common on both.

Succession was marked by progressive intrusion of species during early stages, with significant loss of pioneers not occurring until the fourth stage recognised (Tables 3.1 and 3.2). Thereafter, succession involved both addition and loss of species. The latter predominated in later stages so that the total number of species, and the number per unit area, declined, particularly in the closed communities on pyroxenite. Certain species remained prominent during several stages, e.g. *Placynthium nigrum* on dunite. Crustose lichens such as *Rhizocarpon geographicum* predominated early in succession but were replaced first by foliose and later by fruticose species.

The nature of the rock surface as it changed during weathering was regarded as important in controlling the early succession. Thus total lichen cover on dunite fell from 48% in stage 4 to 10% in stage 6, a change attributed to crumbling of the substratum, whereas total cover increased progressively on the more stable pyroxenite. Later, control was increasingly exerted by competition and water relations, particularly on pyroxenite with its dense lichen cover. As in Fig. 3.3, foliose lichens colonised the thalli of crustose species, which in time became displaced. Intrusion of fruticose species also appeared to involve competition, but this became less significant by stage 9 on pyroxenite where extensive cover was provided by relatively few species. Increasing thickness of the lichen cover created relatively moist conditions. The effect can be seen in the replacement of the xerophytic *Alectoria ochroleuca* and *Cetraria laevigata* by mesophytic species of *Cladonia* on pyroxenite (Table 3.1), a substitution not occurring on the more sparsely vegetated dunite (Table 3.2). The lichens on both rock types eventually became unable to compete with grasses and low shrubs as angiosperm-dominated tundra developed.

In a comparable lichen succession on raised beaches on the South Shetland Is, the storm beach was devoid of macrophytes while the second ridge supported a sparse growth of halophilous, appressed lichens including *Acarospora*, *Caloplaca* and *Verrucaria* spp. Proceeding inland, *Usnea antarctica* and crustose species were abundant on the third ridge, with scattered mats of *Drepanocladus uncinatus*. The crests of the older ridges were dominated by *Usnea fasciata* associated with other fruticose lichens, turf- and cushion-forming mosses and *Deschampsia antarctica*. As well as autogenic factors, reduced spray deposition inland is probably involved in regulating this sequence (Lindsay, 1971). The fruticose lichen *Usnea antarctica* is able to colonise some Antarctic rock surfaces coincidentally with *Rhizocarpon geographicum* and other pioneer crustose species (Lindsay, 1978), possibly because this species, unlike many fruticose lichens, is firmly attached to its substratum.

One of the best documented cases of succession in Antarctic bryophyte vegetation involves the distinctive banks of tall turf-forming mosses (Collins, 1976a; Fenton, 1982a). The banks can be initiated either through colonisation by *Poltrichum alpestre* of *Andreaea* spp and other cushion-forming mosses on rocky slopes, or by *Chorisodontium aciphyllum* invading carpet-forming mosses on moist, level ground followed by the establishment of *P. alpestre* on the higher, drier areas of the resulting *Chorisodontium* turf. Individual stands reach areas of 2500 m², probably through coalescence of originally discrete turfs as indicated by sex distribution patterns (Longton & Greene, 1967). The surface of the larger banks becomes undulating, terraced or fissured as a result of wind, solifluction and frost-induced phenomena. Growth of the banks eventually results in the upper surfaces being blown clear of winter snow, thus setting in train colonisation by both crustose and fruticose lichens, followed by death, erosion and in many instances reestablishment of the mosses. Thus formation of the banks appears to involve directional succession, and their maintenance cyclic succession.

Despite periodic surface erosion, the banks have accumulated peat deposits up to 3 m deep, and their longevity is emphasised by radiocarbon dates of 5000 yr BP for peat at the base. Peat formation is unusual under aerobic, non-waterlogged conditions such as prevail in the upper layers of the moss banks, its occurrence here being attributable to permafrost and low rates of decomposition in the active layer (Chapter 7). Fenton (1980) has calculated that approximately half the original primary production becomes incorporated into the permanently frozen peat. Similarly, upward growth of the banks, averaging around 1 mm yr^{-1}, is about half the annual shoot extension, the difference being accounted for by erosion and by compression following partial decay. Many banks have a vertical or overhanging face along the lower edge (Fig. 3.7). This is due not to erosion, but to other factors including downslope movement of the upper parts of the bank over the rigid, permanently frozen core, and inhibition of moss growth along the lower edge by persistent snow banks or permanent ice.

Auto-succession

These examples comply with the traditional concept of directional succession involving species intrusion and replacement as vegetation at a site develops towards climax, but many polar communities appear to develop by auto-succession (page 80). Muller (1952) regarded auto-succession as occurring in environments so severe that only restricted

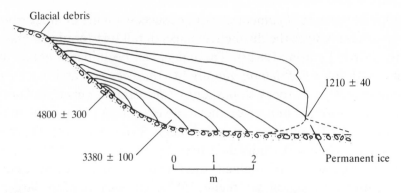

Fig. 3.7. Cross-section of a moss bank on Signy I dominated by *Chorisodontium aciphyllum* showing radiocarbon dates (yr BP) and hypothetical moss bank surfaces at 500-yr intervals. Reproduced from Fenton (1982a) by courtesy of the Regents of the University of Colorado.

suites of species can become established. Allogenic control over vegetational composition thus predominates, since autogenic control through competition and habitat modification is reduced to a minimum, and species replacement consequently fails to occur. Two types were recognised: in selective auto-succession some components of the climax community become established before others, which may be dependent on the pioneers, whereas in the more extreme case of non-selective auto-succession no dependence of one species on another is involved.

These concepts were initially applied to succession following disturbance in northern Scandinavia (Muller, 1952). Secondary succession involving pioneer species being replaced by those of the climax community occurred in birch woodland, but selective and non-selective auto-succession were found to characterise the middle and upper alpine zones respectively. Webber (1978) suggested that the concept of auto-succession may apply to drained lake basins at Barrow. Here, rapid colonisation by mosses and other plants occurs, but once cover is established change appears to be very slow, and to form part of a cyclic process driven by geomorphological change (page 99), rather than by interaction between plants and environment.

A similar view was expressed by Smith (1972), who summarised his concept of the development of cold-Antarctic cryptogamic communities as:

> when a particular habitat becomes available for plant establishment it
> is colonized only by species typical of a sociation favouring the conditions

provided by that habitat. In most cases one or a few species are more successful than the rest and form the basis of the community, usually as the dominant or codominant components. Various sociations which lie adjacent to one another show little evidence of serious competition or the displacement of one by another, but merely form a heterogeneous zone where the stands merge.

This position was supported by an examination of plant remains beneath all the widespread bryophyte communities in non-fumarolic areas of Candlemas I, which suggested that each vegetation type had developed independently of the others (Longton & Holdgate, 1979). R. I. Lewis Smith (1982a) has since described a successional sequence near a retreating, cold-Antarctic glacier that appears to conform to Muller's view of selective auto-succession, as the pioneers, principally *Bryum*, *Ceratodon* and *Pohlia* spp, were also dominant members of the most highly developed community, in which additional species had become established.

Cyclic succession

Cyclic succession comparable with that involving bryophytes and lichens in Antarctic moss banks occurs widely in polar vegetation. A very simple case is represented by appressed epilithic lichens such as *Parmelia saxtilis* and *Placopsis contortuplicata*, where invidual thalli expand to form roughly circular colonies up to 30 cm in diameter. The centre of the colonies commonly die and become eroded, leaving rings of lichen surrounding areas of bare rock which may then be recolonised by propagules of the same or different species (Lindsay, 1978). In the first case, the sequence at a given point may thus be described as species A → bare surface → species A.

A rather more complex process occurs in the *Polytrichum alpinum* sociation on volcanic ash slopes in the South Sandwich Is (Longton & Holdgate, 1979). *P. alpinum* forms hummocky turfs up to 15 cm diameter associated with smaller turfs of *Pohlia nutans* and mats of *Brachythecium* and *Drepanocladus* spp. The larger turfs are subject to lichen encrustation, and ultimately to erosion, leaving areas of bare mineral soil. A small cushion-forming species of *Ceratodon* is the principal coloniser of the bare areas, and shoots of *P. alpinum* are commonly present in the larger cushions. Analysis of soil and peat profiles beneath the living vegetation confirmed that reestablishment of *P. alpinum* frequently involves colonisation of *Ceratodon* cushions, giving the sequence species A → bare ground → species B → species A. Other members of the community may later become established over the *Polytrichum* (Fig. 3.8). Further cases of cyclic succession are considered in relation to pattern on pages 99 and 101.

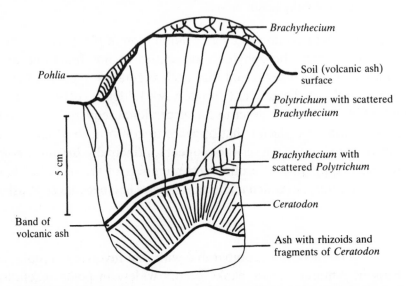

Fig. 3.8. Profile through vegetation, and the undecayed plant remains beneath, from a stand of the *Polytrichum alpinum* sociation on Candlemas I, South Sandwich Is. Redrawn from Longton & Holdgate (1979).

There are thus examples among cold-Antarctic vegetation of directional succession involving replacement of communities, as in the lichen vegetation considered by Lindsay (page 86), and banks of tall turf-forming mosses (page 87). Present evidence suggests, however, that processes akin to auto-succession as envisaged by Muller (1952) play an important role in the development of cryptogamic vegetation under the severe climatic conditions prevailing there and in other polar regions. It further appears that environmental variation resulting from microtopographic and other features has resulted in polyclimax communities, or of climax pattern depending on the intensity of local environmental gradients, and that cyclic succession is widespread in the communities thus established. An alternative hypothesis is that species replacement is occurring widely in what appear to be stable communities, but at rates too low to be readily detectable because of environmental limitation on nutrient cycling and consequent edaphic change.

Secondary succession

Secondary succession following destruction of vegetation is also of wide occurence. Lightning-induced fires cause local damage to tundra vegetation (Racine, 1981), and are a widespread and recurrent feature of boreal and mild-Arctic woodland. The trees are normally killed, but many other vascular plants survive all but the most severe fires and subsequently resprout, so that changes in species composition during the post-fire succession may be quantitative rather than qualitative (Black & Bliss, 1978; Johnson, 1981). Some mosses, notably *Polytrichum alpestre*, also survive moderate fires and regenerate from the charred remains of former turfs, but most bryophytes and lichens are readily killed and orderly successions of cryptogamic communities have been reported. Ahti (1977) considered the following a typical pattern for lichens:

(1) Bare soil stage; 1–3 years after fire.
(2) Crustose lichen stage; 3–10 years; characterised by *Lecidea* spp.
(3) Cup lichen stage; 10–30 (–50) years; characterised by species of *Cladonia* subgen *Cladonia*, e.g. *C. crispata*.
(4) First reindeer lichen stage; 30(–50)–80(–120) years; characterised by *Cladonia* subgen *Cladina*, e.g. *C. rangiferina*.
(5) Second reindeer lichen stage; beginning after 80(–120) years; characterised by *Cladonia* (*Cladina*) *stellaris*.

Numerous variations on this theme have been reported (Black & Bliss, 1978; Fig. 3.9). Thus on drumlins in the Abitau Lake region, NWT, *Polytrichum piliferum* is an abundant associate of *Lecidea* spp during the first 20 years of recovery: *Cladonia* (*Cladina*) spp are the principal species during a second phase from 20 to 60 years post-fire, then giving way to *Stereocaulon paschale* (Maikawa & Kershaw, 1976). The geographical distribution of *Cladonia* and *Stereocaulon* spp as dominants in mature lichen woodland in Canada is discussed by Kershaw (1977a).

Once established, *Cladonia stellaris*, *Stereocaulon paschale* and other fruticose lichens remain dominant throughout the lichen phase. Where closure of the tree canopy occurs the lichens are largely replaced by bryophytes, notably the weft-forming mosses *Hylocomium splendens* and *Pleurozium schreberi*, associated with foliose lichens such as *Peltigera* spp. Maikawa & Kershaw (1976) view canopy closure as part of a successional process leading to climax spruce forest with a bryophyte-dominated ground layer, which is thought to develop about 200 years post-fire. According to this interpretation, the existence of lichen woodland as an extensive vegetation type is dependent on the regular occurrence of fire

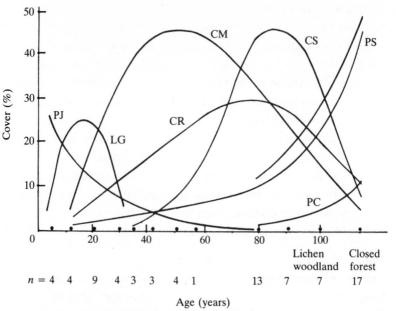

Fig. 3.9. The pattern of revegetation of the dominant cryptogams after fire in southeastern Labrador. Species abbreviations (left to right): PJ, *Polytrichum juniperinum*; LG, *Lecidea granulosa*; CR, *Cladonia rangiferina*; CM, *Cladonia mitis*; CS, *Cladonia stellaris*; PC, *Ptilium crista-castrensis*; AN, *Alectoria nigricans*; PS, *Pleurozium schreberi*: n = number of stands examined. Data from Foster (1985).

(Klein, 1982), which recurs with a mean frequency of once in about 100 years in the Abitau Lake region.

An alternative view was presented by Johnson (1981) whose analysis indicated only limited relationship between time of canopy closure and stand age. Johnson suggested that mosses become established after 30–50 years on sites where closed tree cover is permitted by mesic conditions and reasonable nutrient availability, but he and Ahti (1977) consider that fruticose lichens remain abundant in open, climax woodland on dry, oligotrophic soils in continental regions. Coversely, in Labrador and other areas with oceanic climate and longer intervals between fires, the fruticose lichens normally give way to *H. splendens* and *P. schreberi*, which may in turn be overgrown by *Sphagnum* spp (Fig. 3.10). The latter ultimately blanket the forest floor, causing reductions in pH, decomposition rates, nutrient availability and forest productivity (Foster, 1985).

Environmental conditions change dramatically during post-fire suc-

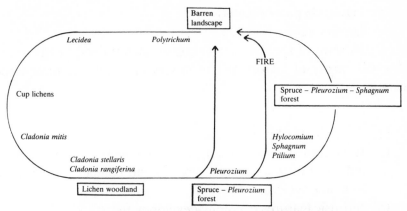

Fig. 3.10. Schematic representation fire-induced succession in *Picea mariana–Pleurozium schreberi* forest in southeastern Labrador. Data from Foster (1985).

cession. Mineral nutrients are released as the vegetation burns, and much of the nitrogen may be lost from the system. Other elements are retained, leading to higher concentrations than usual in some soil horizons, although greater nutrient flushes occur after fire in temperate communities. Leachate from charred lichen is acidic and lower in nutrients than that from spruce needles and other vascular plant material. Dubreuil & Moore (1982) suggested that this may tend to perpetuate the distribution of nutrients beneath former trees, and thus to some extent the organisation of the vegetation. Similarly, tree seedlings were more frequent among bryophytes such as *Ceratodon purpureus* and *Marchantia polymorpha* colonising moist hollows in the rooting zone of fallen trees than among the dry, charred lichens covering most of a burnt area at Churchill some ten years post-fire (Longton & Stewart, unpublished data).

Microclimatic conditions at the ground surface are severe during early stages, with wide diurnal temperature fluctuations and hot daytime conditions in summer, and there is experimental evidence that *Stereocaulon paschale* cannot withstand the resulting thermal stress (but see page 202). Colonisation by *Polytrichum* and later by *Cladonia* spp increases surface albedo thus reducing net radiation, and as organic matter accumulates it retains moisture so that energy during periods of strong irradiance is increasingly lost from the ground surface as latent heat. Maximum surface temperatures therefore decline, favouring the establishment of macrolichens, whose elimination following closure of the canopy apparently results from light limitation on photosynthesis (Kershaw, 1985; Chapter 5).

Pattern in polar vegetation

Determinants of pattern

Patterns of species distribution are determined by the interaction of a wide array of factors and processes of which the most important include:

(1) Availability of propagules.
(2) Environmental variation, itself complex, acting through its influence on the survival, growth and competitive ability of the available species.
(3) Intrinsic features of colony development.
(4) Succession, both directional and cyclic.
(5) Interaction between species.

In simple, crustose lichen communities such as that described on page 89, environmental conditions, availability of propagules, and possibly relative competitive ability determine which species become established. The general environment is also responsible for the absence of other life forms that might otherwise compete with the lichens, and these factors could be modified during directional succession. Distribution within the community will be influenced by local environmental variation, and also by cyclic successional processes, which in this instance are largely controlled by the intrinsic pattern of centrifugal colony expansion accompanied by death and erosion in the centre.

The *Polytrichum alpinum* community discussed on page 89 comprises a mosaic of *Polytrichum* hummocks in different stages of development, lichen colonisation and erosion, with small stands of other mosses and areas of bare ground in places being colonised by *Ceratodon*. Lichen colonisation and erosion of the dominant moss begin on the windward side of the hummocks, suggesting that this phase of the cycle is initiated by exposure, and wind-blown ash deposits also appear to lead to local species replacement (Fig. 3.8). Other examples of mosaic pattern in Antarctic cryptogamic vegetation are thought to be controlled by several environmental factors varying along intersecting gradients (Smith, 1972). The relative importance of the five sets of factors identified above thus varies in different situations, and under certain circumstances their interaction gives rise to distinctive distribution patterns.

Zonation

Because of its simplicity, zonation is one of the most striking forms of vegetational pattern, the many polar examples showing great variation in scale. Zonation can frequently be related to environmental gradients, as in the case of the major vegetation zones defined in Table 1.3. Smaller scale zonation along environmental gradients occurs with increasing height on coastal cliffs (page 38), near melt streams and late-lying snowbeds, and around sites of geothermal heat and moisture.

In concentrically zoned vegetation around a fumarole in the South Sandwich Is (Fig. 3.11), the vents through which steam was emitted sup-

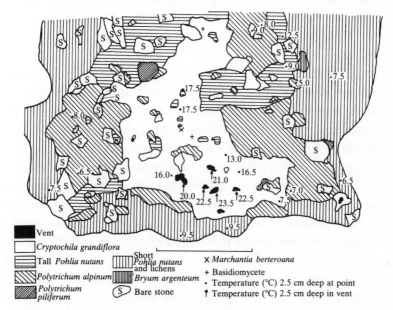

Fig. 3.11. Map of vegetation and soil temperature surrounding a series of vents at a fumarole on Candlemas I, South Sandwich Is. Redrawn from Longton & Holdgate (1979.

ported algae with low mats of *Cephaloziella* sp and mosses, including a slender form of *Pohlia nutans* with prostrate stems and small, distantly spaced leaves. The hepatic *Crypotochila grandiflora* formed carpets with tall turfs of *Campylopus* spp, *P. nutans* and other bryophytes in the zone surrounding the vents. Tall turfs of *P. nutans* extended into the third zone where they were associated with *Polytrichum alpinum*. This was surrounded by short, compact turfs of *P. nutans* and other mosses giving way gradually to almost bare ash and scoria under cooler, drier conditions at distances greater than 2–3 m from the vents. *P. nutans* illustrates the wide ecological amplitude of some species in species-poor Antarctic vegetation, as it was prominent in all four zones, and occurred in three growth forms.

Environmentally induced zonation also occurs on almost a microscale, as among endolithic microorganisms (Chapter 2), and epilithic crustose lichens near Mawson Station where rock steps about 1 cm high shelter on the leeward side a narrow zone of *Xanthoria elegans* which gives way to *Buellia frigida* and eventually to bare rock, the belt of vegetation being only 2–3 cm wide (Seppelt & Ashton, 1978). Environmentally related vegetation zones commonly intergrade, and the occurrence of sharp boundaries is often a function of discontinuities in the associated environmental gradient, as noted in Chapter 1 in connection with the major southern hemisphere zonation.

Intergrading vegetation zones also result from succession around ponds and lakes, in the wake of retreating glaciers, and on raised beach systems where environmental effects on plant distribution associated with changes in relief are superimposed on the temporal sequence of communities occupying beaches of different age. Complex patterns can result as on beach ridges along the Hudson's Bay coast in northwestern Ontario.

Each of these ridges is only a few metres high but this creates major differences in soil moisture, and the most obvious feature is environmentally induced alternation between lichen heath on the dry sand and gravel of the ridges, and wet sedge moss meadow on peat in the intervening depressions. Principal component analysis revealed the existence of further topographically related variation, for example between the lichen heath communities on the exposed crest and more sheltered slopes of a given ridge, and also of successional sequences. Thus succession on ridge slopes ran from an open *Cetraria islandica–Arctostaphylos rubra* association on the coastal ridge, estimated on geomorphological evidence to be about 300 years old, through lichen heath characterised by dwarf shrubs and fruticose lichens such as *Alectoria ochroleuca* and *Cetraria nivalis*, to closed stands of *Cladonia stellaris* on ridges 1.5 km from the coast and around 1000 years old. Succession was regarded as proceeding towards lichen woodland like that on ridges further inland. It had resulted in, and was no doubt in large measure driven by, a progressive increase in soil moisture and organic matter, indicating a considerable degree of autogenic control (Kershaw, 1974; Kershaw & Rouse, 1973; Pierce & Kershaw, 1976).

Pattern associated with moss hummocks

Many Arctic vegetation types show hummocky relief due to differential growth of mosses or other plants, and this may create considerable small-scale environmental variation. Compared with intervening depressions, hummocks commonly show lower pH, nutrient availability and substrate water content, greater depth to permafrost, shallower winter snow cover, and greater exposure throughout the year. These and other factors influence distribution within the community, as in a hummocky sedge moss meadow on Devon I where moss hummocks support *Salix arctica* with a variety of forbs and graminoids, while *Carex aquatilis* predominates on the wet hollows (Muc, 1977). Conversely, the distribution of bryophytes in some mires is related to the age of *Eriophorum* spp tussocks and their associated microtopography (Alpert & Oechel, 1984).

Some of the most striking examples of hummock–hollow microtopography are found in *Sphagnum* mires such as the Stordalen mire at Abisco (Table 1.4), which has been investigated by principal component analysis (Kvillner & Sonesson, 1980). Mosses such as *Sphagnum fuscum*, as well as *Cetraria nivalis* and other fruticose lichens, were associated with low substrate water-content on hummocks, while other mosses including *Drepanocladus schulzei* and *Sphagnum balticum* were characteristic of the wetter hollows. The moss *Dicranum elongatum* was most frequent on the drier hummocks but its productivity was greatest under moister conditions. Among flowering plants, some occurred with a similar frequency throughout the hummock–hollow complex, e.g. *Andromeda polifolia*, while *Vaccinium uliginosum* and *Carex rotundifolia* were among species associated respectively with moss hummocks and moist depressions.

Patterned ground

Polar terrain shows many microtopographic features associated with frost heave, solifluction and related phenomena. The associated plant distribution patterns reflect variation in soil moisture, substrate instability and other environmental variables as in examples described by Smith (1972) on Signy I. The transect in Fig. 3.12(a) crossed three frost-heave polygons in which instability had prevented plant establishment. Bands of *Usnea* spp occurred on the peripheral stone fringes, with *Andreaea* spp prominent among the mosses and fruticose lichens on finer, relatively stable inorganic soil between the polygons. Fig. 3.12(b) shows less regular pattern in which lichens were abundant on frost-induced stone hummocks, giving way to bryophytes on finer, moister material in intervening hollows.

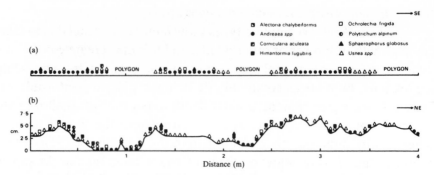

Fig. 3.12. Line transects through two *Andreaea*-lichen stands on a plateau at an altitude of 120 m on Signy I. The distribution of species is related to microtopography, (a) on level ground disrupted by three soil and stone polygons, and (b) on undulating ground subject to irregular frost heaving. Data from Smith (1972).

In a profile across soil and stone stripes on a steeper slope (Fig. 3.13) the central region of wet clay was moving downslope by solifluction faster than the surrounding stripes of stone and gravel. The former was barren centrally, but mosses and a hepatic (*Anthelia* sp) became abundant under moist conditions along the less mobile margins. Other bryophytes and lichens had colonised drier, relatively stable stones between the soil stripes.

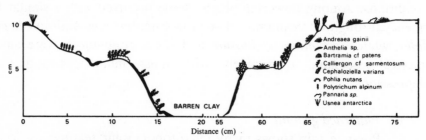

Fig. 3.13. Distribution of species across marginal ramps of a narrow soil and gravel stripe at an altitude of 50 m on Signy I. Data from Smith (1972).

Features such as stone circles, frost boils and ice wedge polygons give significant microrelief in many Arctic regions. Polygon formation on the Arctic Alaskan coastal plain has been attributed to a combination of permafrost and the waterlogged condition of the overlying active layer, which causes contraction cracks to open suddenly in the frozen ground during the severe cold of winter. The cracks form lines, some straight and parallel, others intersecting at various angles to give a polygonal network. Snow, and in summer melt water, enter the cracks and turn to ice below the level of the active layer. This process leads to the formation of ice wedges (Fig. 3.14), comprising vertical layers of ice, each

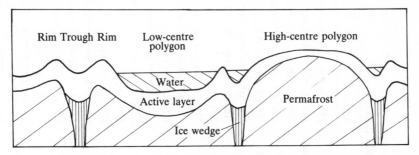

Fig. 3.14. Diagram of low- and high-centre polygons, which in the Barrow, Alaska, area are commonly 5–12 m in diameter. After Tieszen (1978a).

layer thought to represent one year's accumulation. As a wedge expands, the outer margins become pressed against the surrounding soil which is forced upwards to form paired, raised rims around the polygons. In low-centre polygons the central region is depressed and often water-filled, while high-centre polygons have the centre raised above rims (Billings & Peterson, 1980; Peterson & Billings, 1980; Fig. 3.14).

Webber's (1978) ordination defined several intergrading vegetation noda associated with different habitats in polygonal tundra near Barrow (Table 3.3). Pleurocarpous mosses, particularly *Calliergon, Campylium* and *Drepanocladus* spp are abundant in wet habitats in low-centre polygons and interpolygonal troughs while, as at Stordalen, *Dicranum elongatum* and fruticose lichens occur principally in drier habitats. Some species are largely restricted to a narrow range of habitats, e.g. *Racomitrium lanuginosum* in mesic *Salix rotundifolia* heath and *Cornicularia divergens* in the two mesic noda, while others such as *Polytrichum alpinum* show a broader range. More detailed information on the distribution of cryptogams in Alaskan tundra is given by Rastorfer (1978) and Williams, Rudolph, Schofield & Prasher (1978), while Zoltai & Tarnocai (1975) note that *Sphagnum* spp are abundant in interpolygonal troughs in tundra and boreal regions of western Canada.

As with raised beach ridges, distribution patterns in polygonal tundra are thought to be underlain by long-term successional change, in this case cyclic in nature and controlled by environmental rather than autogenic factors. An example is the thaw lake cycle (Billings & Peterson,

Table 3.3 *Vegetation of polygonal tundra near Barrow, Alaska*

Vegetation nodum	Habitat	Characteristic (c) or abundant (a) species	
		Lichens	Mosses
Dry *Luzula confusa* heath	High-centre polygons	*Alectoria nigricans* (a,c) *Cladonia cryptochlorophaea* (c)	*Dicranum elongatum* (a) *Polytrichum alpinum* (a) *Polytrichum juniperinum* (a,c) *Psilopilum cavifolium* (c)
Mesic *Salix rotundifolia* heath	Low-centre polygons that drain readily	*Cornicularia divergens* (c)	*Brachythecium* spp (a) *Dicranum elongatum* (a) *Racomitrium* spp (c)
Mesic *Carex aquatilis* – *Poa arctica* meadow	Hummocky polygon rims and more or less level, slightly polygonised terrain	*Masonhalea richardsonii* (c) *Dactylina arctica* (a)	*Bryum* spp (a) *Dicranum elongatum* (a) *Polytrichum alpinum* (a)
Moist *Carex aquatilis* – *Oncophorus wahlenbergii* meadow	Flat polygon centres and drained, shallow polygon troughs	—	*Bryum* spp (a) *Calliergon sarmentosum* (a) *Oncophorus wahlenbergii* (a) *Polytrichum alpinum* (a)
Wet *Carex aquatilis* – *Eriophorum russeolum* meadow	Low-centre polygons, poorly drained	—	*Bryum* spp (a) *Calliergon sarmentosum* (a) *Drepanocladus brevifolius* (a) *Polytrichum alpinum* (a)
Wet *Dupontia fisheri* – *Eriophorum angustifolium* meadow	Wet polygon troughs	*Peltigera canina* (c)	*Bryum* spp (a) *Calliergon sarmentosum* (a) *Polytrichum alpinum* (a)

Data from Webber (1978)

1980) where it is envisaged that the pools in low-centre polygons merge through wind-induced erosion of the intervening rims to form series of elliptic lakes orientated in line with the prevailing wind. The low shore-lines are eventually eroded by streams and the lakes drain. Some ice wedges persist through the lake phase, and in the drained lake basins the barren sediments appear as large high-centre polygons. Revegetation is slow, with initial colonisation by mosses followed by other species char-acteristic of the different habitats in polygonal tundra. With renewed growth of ice wedges, the high-centre polygons revert to the low-centre type, thus completing a cycle thought to last 2000–3000 yr. Other types of patterned ground such as palsas, peat plateau and ridged fens develop in non-permafrost terrain (Zoltai & Tarnocoi, 1975).

Circles, stripes and balls

Circular, linear or spherical patterns may also be found among polar cryptogamic vegetation. Their development is related to frost heave (Figs. 3.12 and 3.13), or to intrinsic aspects of colony development, often associated with species interaction. Pattern related to intrinsic features of the development of *Polytrichum alpinum* colonies has been described from cold-Antarctic sites where this species forms low, circular stands which gradually expand as underground rhizomes grow outwards from the periphery and give rise to aerial shoots (Smith, 1972). Translocation of photosynthate from older shoots along rhizomes to young growing points has been demonstrated in Arctic *P. alpinum* (Chapter 6). On Signy I individual colonies reach at least 1 m in diameter, but peripheral expan-sion of the larger colonies is accompanied by progressive death of plants from the centre followed by colonisation of the decaying *Polytrichum* remains by other species. *P. alpinum* may eventually reinvade the central area so that a cyclic succession is established comparable with that in some lichen communities (page 89).

Circular patterns associated with centrifugal expansion of parasitic fungi are widespread in polar mosses. Some are caused by basidiomycetes which spread from a central point over several years. Each autumn moss near the advancing edge of the fungal infection dies and it shows no subsequent recovery. Concentric rings of dead moss up to 2 m in diameter may thus be formed, as in *Racomitrium canescens* on Jan Mayen (Wilson, 1951). Similar concentric rings up to 5 m in diameter occur in *Chorisodontium aciphyllum* on Signy I but are associated with ascomycetes which undergo several periods of rapid advance during a growing season, each resulting in a ring of dead or discoloured moss (Fenton, 1983). Other 'fairy rings', such as those in *Drepanocladus uncinatus* on Signy I (Fig. 3.15), seldom exceed 20 cm wide (Longton, 1973b): intracellular penetration of the moss by ascomycetes occurs near the leading edge of the fungal colony, and the moss shoot apices in this region become white in colour and die. Recovery follows through production of lateral shoots below the dead apices. Death and recovery proceed through the growing season, so that single, progressively expanding rings of white shoots are produced. The moss rings thus appear to be caused by fungal parasitism, but it is possible that the fungal infection follows initial damage by other factors.

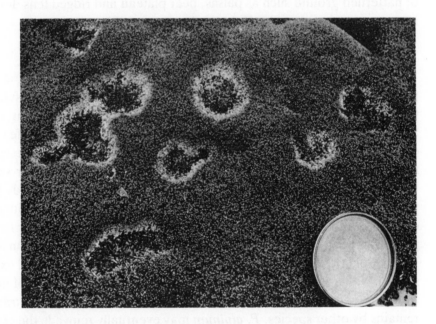

Fig. 3.15. Rings of white, moribund moss associated with fungal infection in a carpet of *Drepanocladus uncinatus* on Signy I. The disk is 5 cm in diameter. Reproduced from Longton (1985a) by courtesy of Pergamon Press.

Interaction between a moss and a dense, cushion-forming flowering plant, and their response to wind action, results in the formation of stripes comprising parallel bands of each species in fellfields on Macquarie I (Fig. 3.16). They are oriented at right angles to the prevailing wind, and move slowly over the gravel substrate as growth on the leeward side compensates for erosion on the windward. Development is initiated by *Racomitrium crispulum* which becomes established in the shelter of small stones. It forms mats which grow upwards into the strong wind

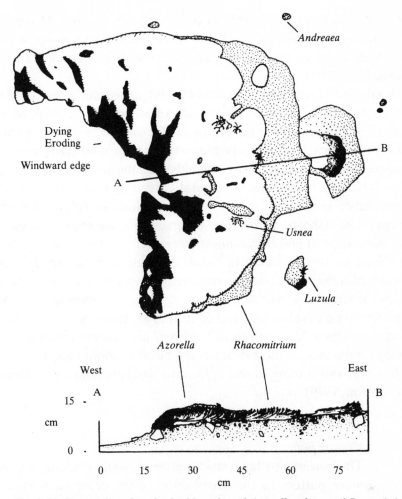

Fig. 3.16. Plan and profile of a double stripe of *Azorella selago* and *Racomitrium crispulum* on Macquarie I showing the pattern of die back on the windward side (shaded). After Ashton & Gill (1965).

and begin to die on the windward side, with cyanobacteria and crustose lichens colonising the eroding surfaces. Expansion continues on the lee-ward side of the colonies, which gradually assume an elongate, crescentic form. Seedlings of *Azorella selago* may later become established within *Racomitrium* mats. The moss at first surrounds the *Azorella*, but growth of the latter and continuing growth and erosion of the moss results in the pattern shown in Fig. 3.16. The stripes develop over 20–80 yr as estimated from leaf scar measurements. Vascular plants such as *Luzula campestris* may persist within the stripes, but small cushions of *Andreaea* and *Ditrichum* spp are unable to compete when overrun. However, erosion leaves bare ground available for recolonisation by these, and by the dominant species so that cyclic succession is again in operation (Ashton & Gill, 1965).

Some polar cryptogams undergo a mobile phase during colony development. At Mawson Station in the frigid-Antarctic, colonies of *Bryum algens* and *Grimmia lawiana* originate as small cushions, with green shoots all over the surface, which may be blown over the ground. They later become attached to the substrate by rhizoids and the lower shoot apices die. The larger cushions have a mineral core, suggesting that cryoturbatic disturbance is involved in the sedentary phase of colony development (Seppelt & Ashton, 1978). The reverse sequence characterises balls of *Bartramia patens* and *Ditrichum strictum* on cool-Antarctic islands, young cushions up to 2.5–5.0 cm in diameter being attached to the substrate from which they later become removed by frost action. Shoots then form on what was the lower side of the cushion, and mobile moss balls up to 10 cm wide develop (Ashton & Gill, 1965; Huntley, 1971). The Arctic fruticose lichen *Masonhalea richardsonii* also adopts a mobile lifestyle, being unattached to the substrate and blowing about in a curled-up state when dry, and coming to rest in hollows and spreading open when moist (Thomson, 1979).

Cryptic pattern

The patterns so far considered are visually evident, but in many communities pattern in the distribution of the component species can be detected only by statistical analysis of cover values within quadrats, or comparable data. Usher (1983) demonstrated well-developed pattern with scales of heterogeneity at about 20 cm and at 1.2–2.0 m in the distribution of *Chorisodontium aciphyllum* and *Polytrichum alpestre* in Antarctic moss banks. He speculated that the smaller-scale pattern was associated with species interaction and the larger with the distribution of rocks, and thus of permafrost, beneath the moss cover. Usher pointed out that these results from one of the world's simplest plant communities support Greig-Smith's (1979) contention that patchiness in vegetation is 'almost universal'.

The composition and distribution of polar cryptogamic vegetation is thus controlled by interaction of a wide range of factors and processes, both biotic and abiotic. Climate undoubtedly exerts an overriding influence, acting indirectly through its effects on soil-forming processes

(Chapter 7) and thus on rates of succession, as well as directly on plant growth and survival. Before the direct effects can be discussed, we must first consider the microclimatic regimes experienced by the mosses and lichens.

4

Radiation and microclimate

Solar radiation

Variation with latitude

Solar radiation exerts a pervasive influence on plant–environment relationships, supplying the energy available for photosynthesis and controlling temperature and water regimes in the microclimate of low-growing cryptogamic vegetation. Microclimatic conditions differ dramatically from those indicated by standard meteorological recording, and must be analysed in many types of ecophysiological investigation. Walton (1984) has recently provided a critical review of microclimatic studies in the Antarctic.

All energy received at the earth's surface, apart from a small amount of geothermal heat, originates as solar radiation. The solar constant, i.e. the irradiance of a plane perpendicular to the sun's rays at the outer edge of the atmosphere and at the earth's mean distance from the sun, is approximately 8.4 J cm^{-2} min^{-1} (= 2.0 cal cm^{-2} min^{-1} or 1402 W m^{-2}). Radiation receipt at the earth's surface varies widely in response to changes with latitude in the angle of solar elevation, in daylength regimes, and in attenuation of radiation during passage through the atmosphere.

The angle of solar elevation (Fig. 4.1), and therefore maximum irradiance, fall progressively from the tropics towards the poles. Daylength varies with latitude because the earth's axis of rotation is inclined in relation to its plane of revolution around the sun (Fig. 4.1), resulting in pronounced seasonality and relatively low diurnal fluctuation in solar irradiance in polar regions. At the northern summer solstice, regions north of the Arctic circle lie in the path of direct solar radiation 24 hours per day whereas south of the Antarctic circle there is no direct insolation, the converse being true at the northern winter solstice. Fig. 4.2 shows

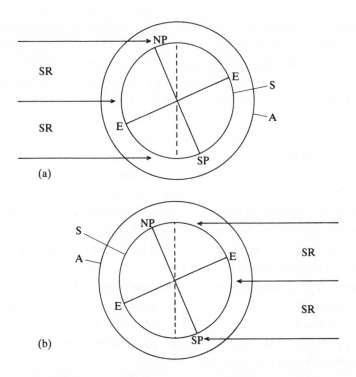

Fig. 4.1. Tilting of the earth in relation to solar radiation flux (SR) at (a) the northern summer solstice and (b) the northern winter solstice. NP = north pole, SP = south pole, E = equator, S = earth's surface, A = outer edge of atmosphere. The dotted line represents the plane of revolution of the earth around the sun. The thickness of the atmosphere is exaggerated.

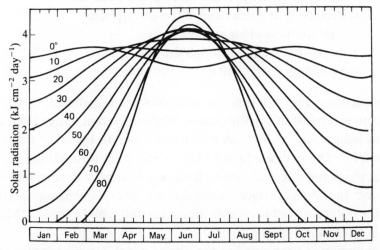

Fig. 4.2. Daily totals of undepleted solar radiation received on a horizontal surface at different latitudes in the northern hemisphere as a function of season. After Gates (1972).

the resulting daily totals of solar radiation incident upon a horizontal surface just outside the atmosphere at different latitudes: it emphasises the seasonal variation in radiation receipt in the Arctic, but also demonstrates that daily totals of undepleted solar radiation are slightly higher in the polar regions than elsewhere for a few weeks in midsummer when low maxima are offset by continuous irradiance.

As radiation passes through the atmosphere some is absorbed or reflected by cloud, dust particles and atmospheric gases, and the remainder transmitted to the earth's surface as direct insolation. The absorbed fraction is reradiated as infrared, with some passing out to space and some downwards to the surface. Of the reflected or scattered component, some again is lost to space with some reflected downwards without change in wavelength as diffuse solar radiation. The latter reaches the surface from many directions simultaneously, and is of considerable biological significance as it penetrates plant communities more effectively than does direct insolation. It is particulary important in polar regions where, at low solar elevation, it may comprise more than 40% of the solar flux (Rosenberg, 1974). The sum of direct plus diffuse solar radiation is referred to as shortwave radiation.

Radiation depletion increases exponentially with distance travelled through the atmosphere, the relationship approximating to Beer's law:

$$S = S_o e^{-cx} \tag{4.1}$$

where S = irradiance after passage through the atmosphere
S_o = irradiance at outer edge of the atmosphere
e = base for natural logarithms
x = distance travelled through the atmosphere
c = extinction coefficient of the atmosphere

It is evident from Fig. 4.1 that the path of solar radiation through the atmosphere is longest at high latitudes, while depletion varies both spatially and temporally with cloud cover and other factors that influence the value of c. The extended path of solar radiation through the atmosphere increases the effect of atmospheric water vapour in reducing irradiance and therefore surface temperatures, thus enhancing in polar regions the difference between maritime and continental climates (Milankovitch in Sørensen, 1941).

Solar radiation is distributed in the ultraviolet (UV), visible and near-infrared parts of the spectrum, whereas radiation emitted from clouds, vegetation and the earth's surface, commonly referred to as longwave

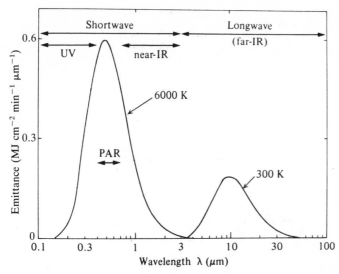

Fig. 4.3. Spectral distribution of radiation emitted from black bodies at temperatures approximating those of the sun (6000 K) and the earth (300 K). After Jones (1983).

radiation, is almost exclusively in the far-infrared (Fig. 4.3). Approximately 50% of shortwave radiation, corresponding roughly with that in the visible part of the spectrum, is regarded as photosynthetically active radiation (PAR) because it can be fixed by plants in photosynthesis, although only a small proportion is utilised. Absorbed infrared radiation is transformed to heat, while UV, particularly the shorter wavelength or biologically active UV, is absorbed by proteins and nucleic acids, and is damaging to living organisms at high irradiance.

Attenuation of solar radiation on passage through the atmosphere differs with wavelength, and the spectral composition of the transmitted radiation varies with latitude. Biologically active UV irradiance is lower in the Arctic than would be expected at comparable solar elevation in temperate lowlands, an effect attributed to depletion by relatively high concentrations of atmospheric ozone (Caldwell, 1972). Yellow lichen pigments such as usnic acid are thought to shield the algal cells from UV (Rundel, 1978), and Thomson (1982, 1984) has speculated that this may be one reason why yellow lichens are abundant at relatively low latitudes in mild-Arctic woodland and tundra, giving way to predominantly brown species in the far north where UV screening is less important.

Five-day means for shortwave radiation at representative polar sites throughout a year are shown in Fig. 4.4. They emphasise that seasonal

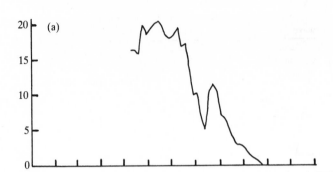

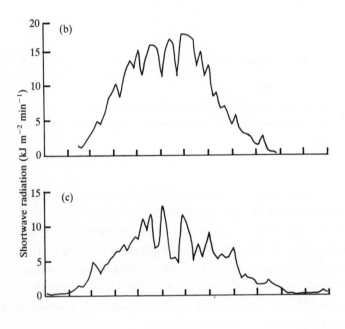

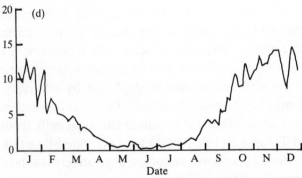

Fig. 4.4. Five-day mean solar radiation for 1972 at (a) Devon I (cool-Arctic, latitude 76° N); (b) Barrow (cool-Arctic, latitude 71° N); (c) Abisco (northern-boreal, latitude 68° N); (d) Signy I (cold-Antarctic, latitude 61° S). Data from Barry, Courtin & Labine (1981) and Walton (1977).

variation in mean irradiance increases with latitude, and that annual totals decline towards the poles. However, the highest mean irradiance in summer was recorded at the most northerly site considered, on Devon Island in the Canadian cool-Arctic. Diurnal fluctuation in shortwave radiation at contrasting polar sites on sunny days shortly after the summer solstice is compared in Figs. 4.5 and 4.6. These data reflect the low diurnal amplitude at high latitude, as the frigid-Antarctic site at latitude 78° S received continuous illumination but with a lower maximum irradiance than the mild-Arctic site at latitude 54° N where there was several hours' darkness at night. Many crustose lichens are divided into more or less flat areolae, but in Antarctic species the areoles tend to be convex or hemispherical, a feature which Dodge (1973) interprets as maximising irradiance of algal cells during continuous illumination at low solar elevation.

Influence of aspect, snow cover and vegetation

Figs. 4.5 and 4.6 refer to exposed, horizontal surfaces. Irradiation of polar cryptogams is modified by slope and aspect, and in many communities by attenuation during passage through taller vegetation or, for much of the year, through snow. The effects of slope and aspect on direct insolation are discussed by Sørenson (1941) and can be calculated from tables in Garnier & Ohmura (1968), but diffuse radiation is dependent on factors such as cloud cover and is thus less readily predictable. Daily radiation receipt in summer is greatest on moderate (40°) slopes of southerly aspect in the Arctic and northerly in the Antarctic. The influence of aspect on the distribution and composition of cryptogamic communities was noted in Chapters 2 and 3.

The subnivean environment may be favourable for metabolism in mosses and lichens, given adequate transmission of PAR (Chapter 5). Transmission is controlled by surface albedo of the snow cover and by depletion as radiation passes through the snow. Attenuation was until recently assumed to vary exponentially with snow depth according to Beer's law, as demonstrated by Weller & Holmgren (1974). They reported that the proportion of radiation transmitted through tundra snow at Barrow fell to 37% of that penetrating the surface at 10 cm depth, 13% at 20 cm depth and 1% at the base of the snowpack at 45 cm, giving an extinction coefficient of 0.1 cm^{-1}. However, this simple relationship probably holds good only for freshly fallen snow which is homogeneous in structure.

Transmission is greatest in visible wavelengths but some near-infrared also reaches the ground surface (Curl, Hardy & Ellermeier, 1972). Here,

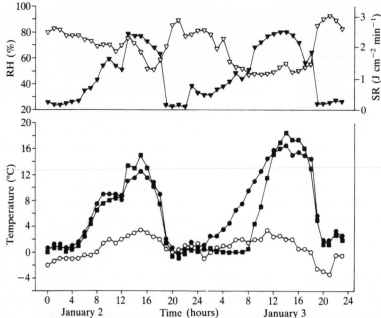

Fig. 4.5. Hourly readings of shortwave radiation (▼), relative humidity (▽), air temperature 2 m above the ground (○), and the temperatures 2–3 mm below the surface of *Bryum argenteum* turfs (■●) on Ross I (frigid-Antarctic) on 2–3 January 1972. After Longton (1974a).

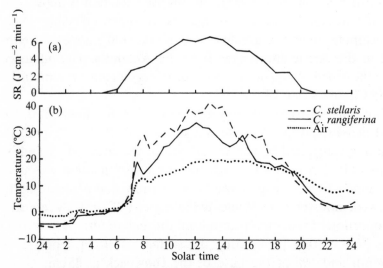

Fig. 4.6. Microclimatic data from Hawley Lake, northern Ontario (mild-Arctic) on 10 July 1978. (a) Hourly readings of shortwave radiation (calculated from data for PAR as indicated on page 162). (b) 15-minute means of air temperature 1 m above the ground and of podetial temperatures in *Cladonia stellaris* at an open site and in *C. rangiferina* under a dwarf shrub canopy. After Tegler & Kershaw (1980).

the energy is effectively trapped, as snow is opaque to longwave radiation emitted by the ground or its plant cover. The heat resulting from this greenhouse effect, combined with stored or geothermal heat conducted upwards to the ground surface, commonly results in a decreasing temperature gradient upwards from the base of the snow cover. A moisture gradient is then created in air between the snow particles as warm air can hold more water vapour than cold air. This results in net movement of water molecules from warm air at the ground surface towards colder snow crystals above, which consequently increase in size. Pruitt (1978) described how this process may create at the base of the snow cover a relatively translucent, lattice-like structure comprising conical air spaces surrounded by a network of ice.

Under relatively warm conditions, thawing and refreezing further increase the density and reduce the extinction coefficient of the snow. Thus light penetration is greater through old summer snow than younger winter snow, and the extinction coefficient decreases in deeper parts of the cover. These effects are likely to result from lowered internal reflection as the ice crystals become fewer, larger and less elaborate in shape in the older, denser snow. Increase in the size of crystals as the snow ages also decreases surface albedo and further enhances penetration (Curl, Hardy & Ellermeier, 1972). Percentage transmission of PAR through snow cover on Signy I is shown in Fig. 4.7. It fell to a minimum close

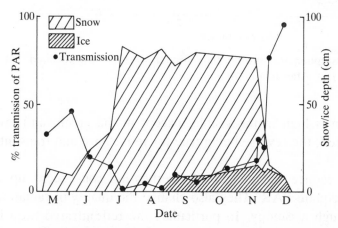

Fig. 4.7. Seasonal variation in percentage transmission of PAR through snow cover at a site on Signy I (cold-Antarctic). After Walton (1984).

to zero when the snow attained its maximum depth of around 80 cm in July, but increased as a layer of ice formed below the snow later in the winter, with transmission through the few centimetres of ice persisting in December reaching almost 100%.

Irradiance of many cryptogams is reduced by an overstorey of flowering plants. When radiation is intercepted by a plant surface varying proportions are absorbed, reflected and transmitted. Mean solar reflectance is commonly 20–30% but it varies widely between species in relation to water content and other features. Mean solar absorptance is commonly 35–60% although reaching 88% in conifer needles (Jones, 1983). Transmittance is generally low, but all three components vary with wavelength (Fig. 4.8). Longwave radiation is predominantly absorbed. Absorptance

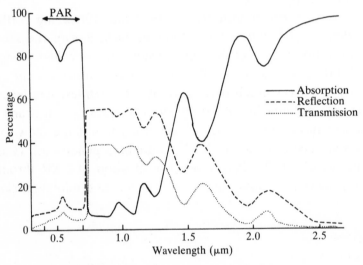

Fig. 4.8. Absorption, transmission and reflection spectra for a generalised angiosperm leaf. After Jones (1983).

of solar radiation is high in the visible, except in the green, and low in the near-infrared, the reverse being true for reflectance and transmittance.

These factors reduce the danger of overheating consequent upon absorption of adequate PAR. They also change the quality of radiation as it passes through a canopy. In particular, the red:infrared ratio is decreased, and this may have morphogenetic effects on plants below through alteration of their phytochrome photoequilibrium, as in mosses from an alpine site in the Yukon Territory (Hoddinott & Bain, 1979). In *Ceratodon purpureus*, a low red:infrared ratio resulted in unusually short plants with small leaves and frequent branching. The responses differed between the moss species, in some cases, as in *C. purpureus*, being the reverse of the typical pattern in angiosperms.

Attenuation of shortwave radiation by a plant canopy varies in relation to such factors as vegetation height and leaf area index, orientation and vertical distribution of the leaves, angle of solar elevation, and proportion of diffuse radiation (Jones, 1983). Irradiance beneath vascular plants shows major spatial and temporal variation, and sunflecks occur under the densest canopies developed by tundra angiosperms, while depletion by deciduous species is low early and late in the growing season. Simulated mean extinction of shortwave radiation by tundra angiosperms at Barrow is indicated in Fig. 4.9 for clear and overcast days. The model suggests

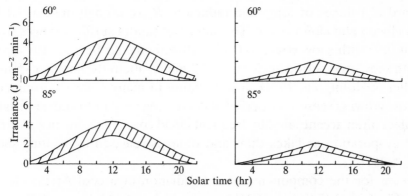

Fig. 4.9. Simulated radiation extinction in tundra vegetation at Barrow (cool-Arctic), for foliage inclination angles of 60° and 85° on a clear day (left) and an overcast day (right). The upper curves represent shortwave radiation incident at the top of the canopy; the lower curves represent radiation penetrating to ground level; the hatched areas represent extinction within the canopy. After Caldwell, Johnson & Fareed (1978).

that approximately 50% of incident solar radiation is intercepted when foliage inclination is assumed to be 60°, with greater penetration when the leaves are almost vertical as in many tundra mires. In contrast, direct measurements by Weller & Holmgren (1974) indicated that only 15% of incident solar radiation penetrated an 8 cm high canopy at Barrow.

Attenuation of solar radiation within bryophyte and lichen colonies is much more intense than in a vascular plant canopy. Skre, Oechel & Miller (1983) found that PAR at a depth of 3 cm was reduced to 17% and 8% respectively of surface values in wefts of *Hylocomium splendens* and *Pleurozium schreberi*, to 12% in a mixed colony of *P. schreberi* and *Polytrichum commune*, and to only 1% in a turf of *Sphagnum subsecundum* in Alaskan spruce forest. Similarly, light irradiance in fruticose lichens fell to 10% of full illumination at depths of 6.5 cm in colonies of *Cladonia rangiferina*, 4 cm in *C. stellaris* and 2.5 cm in *C. arbuscula*, irradiance at the base of the colonies being less than 5% of that at the surface in each case (Kershaw & Harris, 1971). Photosynthetically active tissue is thus concentrated in the uppermost parts of moss and lichen colonies.

Net radiation

Net radiation (page 10) is of particular significance in terms of heat exchange processes, and Equation 4.2 relates net radiation (R_n) to shortwave radiation (S), surface albedo (a), and downward (I_d) and upward (I_u) fluxes of longwave radiation. R_n is strongly influenced by both albedo and cloud cover. The albedo of tundra surfaces ranges from about 0.80 with snow cover to 0.15–0.20 without. At Barrow, the three-month snow-free period begins in mid-June, at about the summer solstice (Weller & Holmgren, 1974). It is a feature of many polar environments that the growing season is a period of decreasing insolation and daylength, an effect often accentuated by frequent cloud cover resulting in part from local evaporation as lakes thaw and water is released by melting snow (Pruitt, 1978).

Means for the components of net radiation at a cool-Arctic site on King Christian I (latitude 77°45′ N) from 15 July to 14 August 1973 were determined by Addison & Bliss (1980) as (J cm^{-2} min^{-1}):

$$R_n = S - Sa + I_d - I_u$$
$$0.62 = 0.87 - 0.10 + 1.85 - 2.00 \tag{4.2}$$

Longwave radiation from clouds (I_d) was the main component of the incoming flux, but it was exceeded by longwave losses (I_u). A feature of these data is the low proportion of shortwave radiation reflected from the ground surface (around 11%).

Fig. 4.10 shows that the Arctic experiences greater seasonal variation in R_n, and a considerably lower total for the year than temperate and tropical regions. As with solar radiation, daily R_n in the northern hemisphere shows comparatively little latitudinal variation for a short period in midsummer, but it falls to negative levels in winter beyond latitude 40° N. On an annual basis, R_n is positive throughout the Arctic tundra but negative over much of the Antarctic continent due to the permanently high albedo of the ice cap.

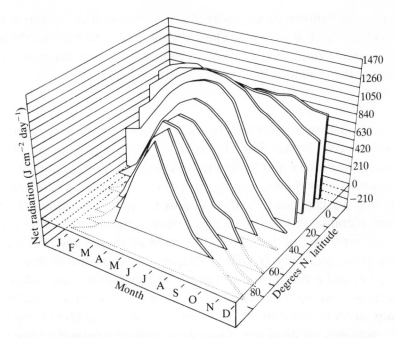

Fig. 4.10. Seasonal variation in net radiation at the earth's surface at various latitudes in the Northern Hemisphere. Dotted lines indicate negative net radiation. After Pruitt (1978).

Energy budgets

Generalised data for tundra

The fate of net radiation at the ground surface may be described by an energy budget formula (Gates, 1962):

$$R_n = LE + H + G \qquad (4.3)$$

where L = latent heat of vaporisation of water
E = rate of evaporation of water from the surface
H = sensible heat flux by convection
G = ground heat flux by conduction

Tundra surfaces differ considerably in their energy budgets. Thus net radiation at a coastal site at Barrow increased tenfold from 155 J cm^{-2} day^{-1} before snow melt to 1591 J cm^{-2} day^{-1} following the thaw, due largely to reduction in albedo. Post-melt R_n was partitioned into sensible heat flux (H, 18%), ground heat flux (G, 9%) and latent heat flux (LE, 73%). In contrast, net radiation after melt, averaging 1578 J cm^{-2} day^{-1}, was lost as H (about 84%) and G (about 15%), with only 1.5% as LE, at an inland site in the Keewatin district, NWT (Bliss *et al.*, 1973). The

incoming radiation is clearly more effective at raising soil and air temperatures at the drier, inland site. These are integrated figures and the magnitude of the various fluxes is subject to local variation depending on the
nature of the ground surface and vegetation cover.

Plant surfaces

Equation 4.3 can also be used to describe the energy budget of
plant surfaces. In this case, G refers to the net conduction of energy
between the surface and the interior. A small proportion of the incoming
energy is fixed as chemical energy by photosynthesis, but G is responsible
primarily for changes in leaf or thallus temperature. Although subject
to considerable short-term change, plant temperatures show little net
change over long periods, and thus most of the energy absorbed (R_n)
is dissipated as LE or H. Where evaporation can occur freely, LE may
exceed H and account for much of the energy release. Restricted evaporation, as a moss or lichen dries or stomata close in a vascular plant,
results in an increase in G and a consequent rise in plant temperature.
This increases the emission of longwave radiation from the plant surface
(Equation 4.4), thereby lowering R_n. It also increases the temperature
difference between the surface and air, and so leads to a rise in H. Surface
temperature continues to rise until gain and loss of energy are in balance.

The preceding discussion concerns a well-insolated surface, and the
direction of the various fluxes can be reversed under other circumstances.
At night, when R_n is negative, surface temperature sometimes falls below
that of the surrounding air, the plant then gaining heat by convection.
Cooling may continue until water vapour condenses on the surface as
dew. Temperature inversions and dew formation are most frequent under
clear skies when little I_u is reradiated to the ground as I_d following absorption in cloud. Evaporation may also lead to the plant temperature falling
rather below air temperature when net radiation is positive but low on
cloudy days.

Vascular plant leaves

Models simulating the energy balance of angiosperm leaves, developed by Gates (1962), Monteith (1973) and others, are useful in consider-
ing the factors involved in the less well documented situation in
cryptogams. A suspended leaf intercepts downward fluxes of shortwave
and longwave radiation, and upward fluxes of shortwave radiation reflected from, and longwave radiation emitted by, the ground. The leaf
emits longwave radiation from both surfaces. According to the Stefan–

Boltzman law, the emission of longwave radiation is strongly dependent on temperature:

$$\varphi = \varepsilon \sigma T^4 \qquad (4.4)$$

where φ = rate of emission
ε = emittance of the surface
σ = a constant
T = absolute temperature

The emittance and absorptance of ground and leaf surfaces for longwave radiation can be regarded as approaching unity. Thus R_n for a single leaf suspended horizontally may be expressed (Jones, 1983) as:

$$R_n = \alpha(S + Sa) + I_d + \sigma(T_{ground})^4 - 2\sigma(T_{leaf})^4 \qquad (4.5)$$

where α = absorptance of the leaf for shortwave radiation (page 114)
a = albedo of ground

R_n is dissipated by energy fluxes due to evaporation and convection, which can conveniently be expressed by relationships analogous to Ohm's law for electrical circuits, i.e. current = potential difference/resistance (Proctor, 1982). Thus evaporation rate is given by:

$$E = \frac{VD_{leaf\,surface} - VD_{air}}{R_{plant} + R_{air}} \qquad (4.6)$$

where VD = water vapour density
R = resistance to water vapour movement

For a fully turgid leaf $VD_{leaf\,surface}$ is approximately equal to the saturated vapour density at the temperature of the surface, but if the tissue has a significantly negative water potential $VD_{leaf\,surface}$ and thus E are reduced. Given that $VD_{air} = 0.217P/T$, where P is the partial pressure of water vapour in the atmosphere (Proctor, 1982), and that relative humidity (RH) $= 100P/P_s$ where P_s is the saturation vapour pressure, it follows that the evaporation rate at the leaf surface is inversely related to the relative humidity of the surrounding air. The value of P_s rises exponentially with increasing temperature, and thus a rise in air temperature leads to a decrease in RH and to more rapid evaporation.

For an angiosperm leaf, R_{plant} may be regarded as stomatal and cuticular resistance operating in parallel. The latter is generally high (2000–10000 s m^{-1}) so that variation in stomatal resistance provides effective control over total leaf resistance. Stomatal resistance in mesophytes commonly ranges from 80–250 s m^{-1} when the stomata are open to 5000 s m^{-1} when closed (Jones, 1983).

R_{air} is determined by the pattern of air movement close to the leaf. When air flows past an object frictional forces create a velocity gradient, wind speed rising from zero in contact with the surface to that of free air some distance away. Thus wind velocity increases upwards from the ground surface for distances of about 10–1000 m depending on topography and other factors. Associated with each object is a shallow boundary layer, comprising an inner laminar sublayer in which the streamlines are smooth and parallel with the surface of the object, a zone of transition, and a region of turbulent air movement. Although measurable in millimetres or fractions of a millimetre, the laminar sublayer is of crucial importance as water vapour moves across it by the slow process of molecular diffusion rather than the more rapid turbulent mixing that occurs in the surrounding air. Laminar sublayer thickness decreases with rising wind velocity. It directly affects the value of R_{air} because R_{air} is equal to the laminar sublayer thickness/diffusion coefficient of water vapour in air (Jones, 1983).

In a similar way, convective heat exchange from an isolated leaf may be expressed (Rosenberg, 1974) as:

$$C = \frac{P_a C_p (T_{surface} - T_{air})}{R_{air}} \tag{4.7}$$

where C = convective heat exchange
P_a = density of air at the prevailing temperature and pressure
C_p = specific heat of air

Here, the most important variables are the temperature difference between the leaf surface and ambient air, and air resistance. As with evaporation, the value of R_{air} is inversely related to wind velocity being determined by laminar sublayer thickness and thermal diffusivity.

Crustose and foliose lichens

The parameters considered above have been quantified for a range of vascular plants (Jones, 1983), but there are fewer data for cryptogams.

Theoretical considerations suggest that details of the energy exchange processes may differ between vascular plants, and mosses and lichens with their low plant resistance to water loss and weakly developed transpiration stream. Moreover, being of low stature, these plants are often exposed to lower wind velocities than angiosperm foliage, and their colony structure may increase air resistance to latent and convective heat exchange.

The simplest case is a crustose or foliose lichen growing closely appressed to its substrate. As the lower surface of the thallus is unaffected by radiation, Equation 4.5 may be simplified to:

$$R_n = \alpha S + I_d - \sigma (T_{thallus})^4 \tag{4.8}$$

Under sunny conditions, R_n is thus influenced strongly by the absorptance of the thallus to shortwave radiation.

Lichen thalli transmit little solar radiation, and thus absorptance varies inversely with reflectance, and this has been shown to differ substantially between species (Gauslaa, 1984). With Norwegian material, reflectance in a series of dark-coloured species such as *Umbilicaria arctica* was negligible in the visible and under 25% throughout much of the near-infrared (Fig. 4.11). Such taxa tend to be chionophobous, and high

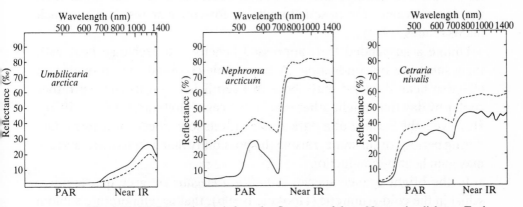

Fig. 4.11. Visible and near infrared reflectance of three Norwegian lichens. Each diagram refers to single thalli when wet (solid line) and when dry (dashed line). After Gauslaa (1984).

absorptance in these wavelengths will favour metabolism and growth during cold weather by increasing both availability of PAR and thallus temperature. In summer, such species are likely to become too hot and therefore too dry to be metabolically active except during cool, cloudy weather or at night. A similar reflectance pattern is shown by some fruticose lichens such as *Pseudephebe pubescens*, and higher temperatures in dark than in light-coloured thalli have been demonstrated among fruticose species (Gauslaa, 1984; Kershaw, 1975a).

A contrasting pattern is seen in chionophytic species such as *Nephroma arcticum* (Fig. 4.11), in which reflectance of solar radiation is higher at all wavelengths than in *U. arctica* or in a generalised flowering plant leaf (Fig. 4.8). Reflectance in *N. arcticum* is particularly high in the near-infrared, with reflectance of PAR lower in wet than in dry thalli. This type of response may delay heating, and therefore drying of the thalli during periods of strong solar radiation, thus favouring photosynthesis during a short summer growing season (Gauslaa, 1984).

Crustose lichens on rock or soil are within the boundary layer of their substrata, and relatively high values for air resistance to latent heat and convective heat transfer (R_{air} in Equations 4.6 and 4.7) may be anticipated, depending on exposure of the substratum to wind. As regards latent heat transfer this must be offset against low plant resistance to water loss. In *Parmelia conspersa* plant resistance of moist thalli was only $39\,s\,m^{-1}$ (Hoffman & Gates, 1970), a value comparable with mesophyll cell-wall resistance in vascular plants. This is roughly half the minimum values for stomatal resistance, and one or two orders of magnitude lower than the normal values for closed stomata or for cuticular resistance in such plants (Jones, 1983).

Unlike a suspended leaf, appressed lichens also exchange heat with their substrata by conduction. A large, pale-coloured rock provides an effective heat sink and may serve to keep lichens on its surface rather cooler by day than might otherwise be the case (Hoffman & Gates, 1970). However, the surface of a dark rock, with its low albedo, warms rapidly during periods of sunshine, raising the possibility that lichens on its surface may gain heat by conduction.

In the bright orange species *Xanthoria elegans* on grey quartz-mica-schist in the cold-Antarctic (Hooker, 1980b), thallus temperature seldom differed by more than one or two degrees from that of the surrounding rock, but reached 18.5 degrees above air temperature. At Churchill, in the mild-Arctic, temperature in black thalli of *Parmelia disjuncta* may be six degrees higher than that of the surrounding grey rock, reaching at least 30 °C under average summer daytime conditions, and probably 45 °C under conditions of maximum radiation and low windspeed (Kershaw & Watson, 1983). The greatest differences between air and thallus temperatures are recorded on rocks receiving direct solar radiation under calm, dry conditions.

Fruticose lichens and mosses

It is impossible to present a realistic equation comparable with 4.5 for net radiation receipt by a moss leaf or a fruticose lichen podetium within a colony. Surfaces are rarely horizontal and S will be reduced to varying degrees by interception above for all but the uppermost parts of the colony. Solar radiation reflected from the ground (Sa) will be negligible, as is the upward flux of longwave radiation from the ground to upper parts of the plants. Thus R_n per unit area of colony may approach the relationship in Equation 4.8, while R_n for surfaces within a colony will vary with angle of orientation, depth in the colony, colony structure and other features.

Mosses and fruticose lichens exchange heat with air by convection and evaporation, and to a limited extent with the ground by conduction. Most moss leaves have their photosynthetic cells exposed directly to the external environment, while in both groups cuticular resistance, and thus R_{plant} in Equation 4.6 is much lower than in angiosperms. Few reliable data for R_{plant} are available for polar mosses and lichens, but it has been confirmed that evaporation rates decrease with decrease in water content (Larson & Kershaw, 1976; Oechel & Sveinbjörnsson, 1978).

For an isolated leaf, R_{air} (Equations 4.6 and 4.7) can be equated with boundary layer resistance, but in dense canopies such as those formed by many mosses and fruticose lichens, the laminar sublayers of adjacent surfaces coalesce at low windspeeds, so that the colony functions as a single object (Proctor, 1982). Molecular diffusion then becomes important in transfer processes in spaces within the colony, increasing the value of R_{air}. The latter may thus be regarded as a canopy resistance plus the boundary layer resistance of the laminar sublayer above the colony, operating in series. This is important since the thickness of the laminar sublayer rises with increasing size of the object, and boundary layer resistance for an isolated moss or lichen plant is likely to be very small.

The depth of the laminar sublayer above a colony, and the extent to which turbulent air movement penetrates into a colony, are strongly influenced by wind speed and colony growth form. Flat moss carpets and short moss turfs having their apices level with the surrounding soil, as in some frigid-Antarctic species (Chapter 2), are likely to support a thicker laminar sublayer than cushions projecting upwards from the surface. Proctor (1980, 1981) has shown that the dense covering of hair points on cushions of *Grimmia* and *Tortula* spp significantly increases R_{air}, thus decreasing evaporation, but that evaporation rates may be increased by other forms of colony roughness.

At low wind speeds he found little variation in rates of evaporation from moss cushions with different degrees of roughness and from wet filter paper. There was little change in evaporation rate from smooth, compact cushions, such as those of *Ceratodon purpureus*, at wind speeds up to $2\,\mathrm{m\ sec^{-1}}$. However, evaporation from mosses with longer leaves and rougher colonies increased sharply above critical wind velocities, in some cases as low as $20\text{--}50\,\mathrm{cm\ sec^{-1}}$, presumably because part of the colony projected through the laminar sublayer allowing turbulent air movement between the shoots. Similarly, R_{air} for colonies of two mosses at Barrow was found to increase sharply at wind speeds below $1\,\mathrm{m\ s^{-1}}$ (Fig. 4.12), and a similar relationship was recorded in mosses from an

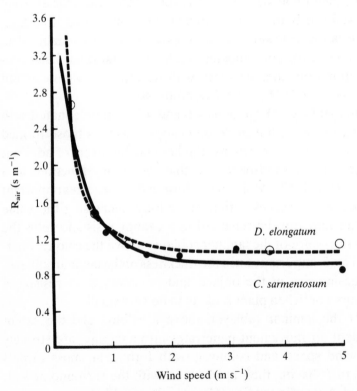

Fig. 4.12. Relationship between wind speed and air resistance to water loss for colonies of *Calliergon sarmentosum* and *Dicranum elongatum* at Barrow (cool-Arctic). Data from Oechel & Sveinbjörnsson (1978).

Alaskan *Eriophorum* tundra (Alpert & Oechel, 1984). However, high wind speeds sometimes reduce evaporation rates because reduction in boundary layer thickness increases sensible heat exchange and thus reduces plant temperature (Monteith in Kershaw, 1985).

It is thus clear that R_{air} for moss and lichen colonies may reach significant levels under calm conditions and that thorough evaluation of wind velocity at the colony surface is essential to an understanding of energy relationships, as wind speeds in the critical range of 0–3 m s^{-1} are probably common at ground level. The effect of canopy resistance in reducing evaporation rates has been demonstrated in Antarctic mosses (Gimingham & Smith, 1971; Chapter 3), and can be seen in Arctic lichens in Fig. 4.13. Gimingham & Smith's experiments were conducted in closed

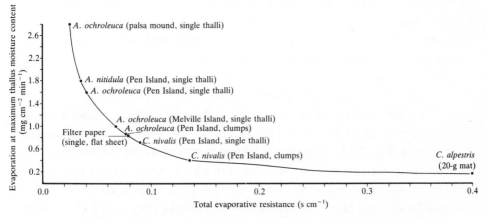

Fig. 4.13. Relationship between total resistance to evaporation and rate of evaporation from water-saturated thalli per unit area of thallus surface in Canadian Arctic lichens. After Larson & Kershaw (1976).

vessels with no air movement. In Larson's (1979) wind tunnel experiments, increase in wind speed in the range 0–3 m s^{-1} increased rates of evaporation from several Arctic lichens, but sensitivity to wind speed varied widely, being much lower in portions of *Cladonia stellaris* (= *C. alpestris*) colonies than in individual podetia. Increased radiation flux also raised evaporation rates, and at low wind speeds the effect was most marked in material that had relatively high resistance to evaporation (Larson, 1979). This apparent anomaly could be due to morphological features which increase evaporative resistance also conferring greater resistance to convective heat exchange, leading to increases in thallus temperature and thus evaporation rates under increased radiant flux.

Complex relationships between resistance, surface area : weight ratios (Chapter 3), evaporation rates and rates of drying were demonstrated in the studies on *Alectoria ochroleuca* summarised in Fig. 4.13, in which water-saturated plants were allowed to dry in a wind tunnel at 23 °C, 50% relative humidity, with a wind velocity of 1.6 m s^{-1} under laboratory lighting. Of the single thalli, those from the palsa mound were small and sparingly branched, thus showing the lowest area : weight ratio (197 cm^2 g^{-1}) and total resistance (2.6 s m^{-1}). The greatest values, i.e. 580 cm^2 g^{-1} and 6.7 s m^{-1}, were recorded for the freely branched Melville I plants. Low resistance resulted in evaporation rates per unit area of thallus surface being highest in the palsa mound plants (2.8 mg cm^{-2} min^{-1} compared with 1.0 mg cm^{-2} min^{-1} in the Melville I plants). However, because of their low surface area : weight ratio, the palsa mound plants nevertheless showed the slowest reduction in water content on a dry weight basis, taking 15 minutes, compared with 11.3 minutes in the Melville I material, to dry to 10% relative water content (RWC: water content as a proportion of that present at saturation).

Energy relations of Cladonia stellaris

Understanding of energy relationships in fruticose lichens owes much to the work of Kershaw (1985) and his associates in the Canadian mild-Arctic. Fig. 4.14 shows the principal components of the energy budget they recorded for *Cladonia stellaris* during a sunny day in July. This pale-coloured species forms dense, almost pure stands 10–15 cm deep formed by robust, erect, freely branched podetia (Fig. 4.15). Rain had resulted in the lichen being water-saturated at dawn. As net radiation increased during the morning energy was lost, at first principally as sensible heat, and later as latent heat when the angle of solar elevation increased and direct solar radiation was better able to penetrate into the colony. Gauslaa (1984) suggested that light penetration into stands of fruticose lichens such as *Cladonia stellaris* and *Cetraria nivalis* is enhanced by relatively high reflectance of PAR in both wet and dry thalli (Fig. 4.11). Although the lichens in Fig. 4.14 were initially water-saturated, latent heat accounted overall for only a slightly greater proportion of energy loss than sensible heat, the Bowen ratio (LE/R_n-G) being 0.65 for the day, and as low as 0.34 in midmorning. The relatively low rate of evaporation was attributed to canopy resistance. Soil heat flux remained a minor component of the energy budget throughout the day.

Measurement of water content at three levels in the *C. stellaris* colony showed that drying occurred from the surface downwards, the basal layer

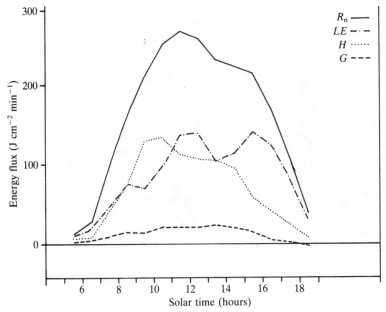

Fig. 4.14. Energy budget for the surface of a *Cladonia stellaris* colony in lichen woodland at Hawley Lake, northern Ontario (mild-Arctic) on 19 July 1970. After Rouse & Kershaw (1971).

maintaining a substantial moisture content throughout the day. Resistance to evaporation created by the lichen canopy was considered an important factor in retaining moisture in the light-textured woodland soil which commonly remains near field capacity throughout the summer (Kershaw 1985). In contrast, fruticose lichen cover in a cushion plant-lichen community on Devon I showed a higher rate of evaporation per unit area than bare soil (Addison, 1977). The dominant lichen at this site was *Pseudephebe pubescens* (Muc & Bliss, 1977), which in common with *Alectoria* and *Bryoria* spp forms a thin cover of narrow, finely divided thalli with a high surface area : weight ratio (Fig. 4.15): it reflects little solar radiation (page 122) and provides a high surface area for evaporation but without the high canopy resistance of *C. stellaris*.

The drying pattern in *C. stellaris* may be related to the humidity profiles recorded by Kershaw & Field (1975). Relative humidity remained close to 100% near the base of the canopy, but by day a vertical gradient became established with RH in the surface layer falling as low as 30–50%, a level considerably below that in the air above (Fig. 4.16). At the same time, a gradient of increasing temperature with height developed in the canopy (Fig. 4.17), the high temperature as well as the dry condition

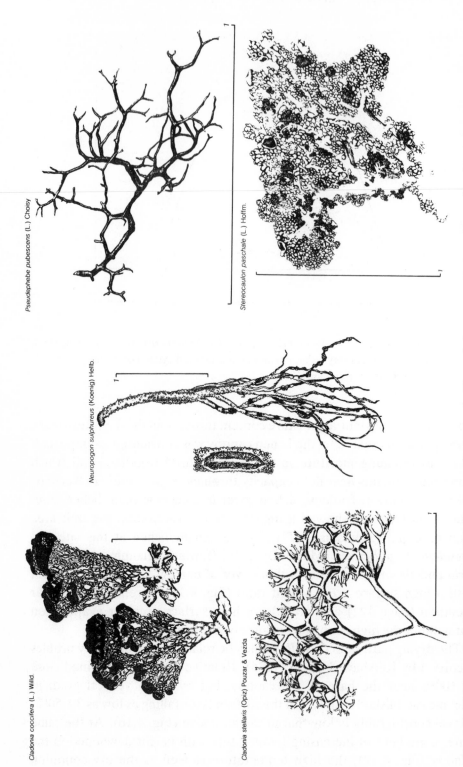

Pseudephebe pubescens (L.) Choisy

Stereocaulon paschale (L.) Hoffm.

Neuropogon sulphureus (Koenig) Hellb.

Cladonia coccifera (L.) Wild

Cladonia stellaris (Opiz) Pouzar & Vezda

Fig. 4.15. Habit sketches of five fruticose lichens common in the Arctic. The scale bars represent 1 cm. Reproduced from Thomson (1984). Copyright © 1984 Columbia University Press. Used by permission.

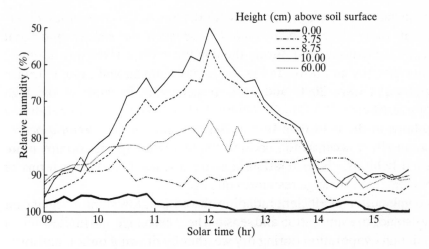

Fig. 4.16. Relative humidity profile within and above a 10-cm deep *Cladonia stellaris* colony at Lake Astray, Labrador on 29 August, 1972. After Kershaw & Field (1975).

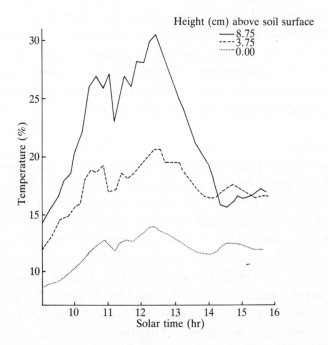

Fig. 4.17. Temperature profile within a 10 cm-deep *Cladonia stellaris* colony at Lake Astray (Labrador) on 29 August, 1972. After Kershaw & Field (1975).

of the surface layer being responsible for the low RH. Temperature differences up to 17 degrees were recorded between the surface and basal layers of the canopy, and during the day the surface layer was normally warmer than the air above. In Fig. 4.17, maximum air and canopy surface temperatures were 20 °C and 31 °C respectively. Because of the high canopy resistance, RWC in *C. stellaris* declines more slowly under drying conditions in the field than that in *Bryoria nitudula* or *Alectoria ochroleuca*, an effect accentuated because the two latter species occupy more exposed habitats: these relationships are discussed further in terms of carbon dioxide exchange responses on page 184.

A combination of efficient radiation absorption, low stature, significant air resistance to evaporation and sensible heat exchange and limited ability to maintain evaporation during dry weather by drawing on soil moisture, even when offset by low internal resistance to water loss and the high degree of exposure characterising tundra sites, suggests that many polar mosses and fruticose lichens should experience pronounced diurnal fluctuations in tissue temperature in response to variation in solar irradiance. The validity of this assumption will become evident in what follows, although the factors responsible have yet to be assessed quantitatively. The rise in moss and lichen temperatures in response to positive R_n accelerates as the colonies dry and, at least in lichens, the highest thallus temperatures seldom coincide with water contents high enough to sustain significant metabolic activity. Energy budgets, temperature and water relations are strongly influenced by wind velocity, and also by plant morphology and colony structure, leading to a diversity of environmental relationships in the species and populations of different habitats.

Plant communities

Energy exchange by a plant community results from the contribution of individual leaves and other plant parts, and the soil beneath the canopy. Based on data for a range of Arctic and alpine vascular plant vegetation types, *LE* appears generally to account for 40–75% of net radiation, *H* for 20–40%, and *G* for 7–30%, although over 85% of R_n was found to be dissipated as convective heat by vegetation on Devon I (Courtin & Mayo, 1975). Unusually low values for latent heat exchange were also recorded on the nearby King Christian I by Addison & Bliss (1980), who determined the energy budget of different components of a lichen–moss–rush community in summer. For the vascular plants, *H* accounted for 74–79% of R_n, *LE* for 17–22% and *G* for only 4% (Table 4.1). The low rate of latent heat exchange was attributed to the shallow

Table 4.1 *Average temperature, net radiation, and its components for various microsites in a lichen–moss–rush plant community on King Christian Island, NWT, for 15 July to 14 August 1973*

Microsite	% cover	R_n (J cm^{-2} min^{-1})	% of R_n			Temperature (°C)
			LE	H	G	
Graminoid	4	0.62	20	76	4	3.7
Broadleaf dicot	3	0.62	22	74	4	3.7
Cushion plant	2	0.62	17	79	4	3.4
Moss	18	0.60	17	75	8	4.1
Lichen	40	0.62	28	58	14	3.4
Bare soil	33	0.63	29	57	14	3.3
Weighted average		0.62	26	61	13	3.5

Data from Addison & Bliss (1980)

vapour pressure gradient between soil and plant surfaces and the air resulting from low temperature and high RH.

The lichen cover was formed principally by black, crustose species. The energy budget for lichen-covered surfaces (Table 4.1) differed little from that of bare soil, but showed higher latent heat and soil heat fluxes, and consequently a lower sensible heat flux, than recorded for flowering plants. The higher latent heat flux may be attributed to the close contact between lichens and the underlying soil, which remained moist throughout the summer, combined with the low plant resistance to water loss. Moss cover and cushion-forming dicots showed the lowest latent heat exchange, possibly as a result of high R_{air} that characterises both types of plant (Courtin & Mayo, 1975).

Annual temperature regimes
Diurnal fluctuation in summer
It is now evident that low-growing cryptogamic vegetation exists in a thermal environment very different from that indicated by standard meteorological recording. Temperature is easier to record than many other microclimatic parameters. For this reason, and because of its perceived importance as a limiting factor at high latitudes, a large body of microclimatic temperature data has accumulated in recent years, ranging from spot readings to data sets extending over several years. Concurrent studies of plant water content have seldom been attempted, which is unfortunate since, given adequate PAR, carbon dioxide exchange and

growth are strongly controlled by the combination of water content and temperature. In the following account emphasis is placed on bryophytes, to which most of the long-term temperature data apply, but short-term observations on lichens suggest that they experience broadly similar patterns of temperature fluctuation.

These studies have demonstrated pronounced diurnal temperature fluctuations in the uppermost layers of moss and lichen colonies during sunny weather in summer, as indicated for *Cladonia stellaris* in Fig. 4.6 and 4.17. A typical summer pattern is illustrated in Fig. 4.18, which compares air temperature with that recorded by thermistors 2–3 mm below the surface of tall moss turfs on Signy I. The data for *Polytrichum alpinum* were recorded in an isolated turf on a sheltered, moderately steep north-facing slope, and those for *P. alpestre* in a more exposed, raised bank on a gentle northeast-facing slope. Air temperature ranged from about 0 to 10 °C, but moss level temperatures reached 35 °C in *P. alpinum* and 28 °C in *P. alpestre*, the difference between the species being attributable to variation in insolation and exposure. The contrast between air and moss level temperature was greatest on sunny days, and the latter remained close to 0 °C during cloudy weather (Fig. 4.18). Slight, nocturnal temperature-inversions (page 118) were also most frequent during the early, clear weather.

Similar patterns are evident in data from short turfs of *Bryum argenteum* on a gentle east-facing slope on Ross I in the frigid-Antarctic. During the relatively warm period illustrated in Fig. 4.5 maximum air and moss level temperatures were 3 °C and 18 °C respectively. In Fig. 4.19, air temperatures ranged only from −8 °C to 1 °C. One of the moss level thermistors was then covered by about 2.5 cm of fresh snow, beneath which the temperature fluctuated between −4 °C and −1 °C. The second moss level probe was not snow-covered, but showed a maximum of only 6 °C. Both moss level probes indicated minima some four degrees above the lowest air temperature. The high water content of the moss turf following recent snowfall, and the insulating effect of the snow cover over the first probe contributed to the reduced diurnal temperature fluctuation in the moss turf. Both Fig. 4.5 and Fig. 4.19 confirm that solar irradiance is correlated positively with moss temperature and negatively with relative humidity in the air above the moss. In more extreme variation at another frigid-Antarctic site, surface temperature in *Schistidium antarctici* and *Usnea sphacelata* rose from −9 °C to over 40 °C within ten hours, while plant water content fell to less than 10%: maximum air temperature was only 1 °C (Smith, 1986).

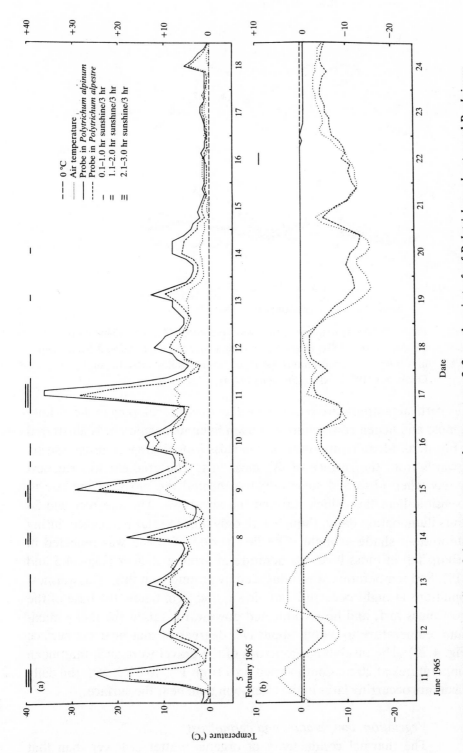

Fig. 4.18. Three-hourly readings of air temperature, and temperatures 2–3 mm deep in turfs of *Polytrichum alpestre* and *P. alpinum*, on Signy I (cold-Antarctic) in relation to duration of sunshine in summer (above) and in winter (below). Data from Longton (1972a).

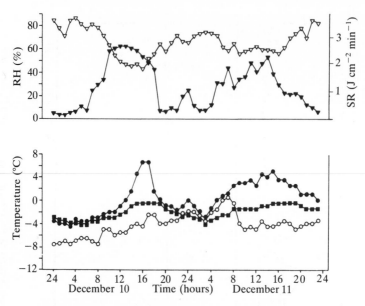

Fig. 4.19. Hourly readings of shortwave radiation (▼), relative humidity (▽), air temperature 200 cm above the ground (○), and temperatures 2–3 mm deep in two *Bryum argenteum* turfs (■ ●) on Ross I (frigid-Antarctic) on 10–11 December (1971). After Longton (1974a).

A vertical temperature profile typical of those developing in the vicinity of moss and lichen communities in open habitats in summer is illustrated in Fig. 4.20. Mean temperature at the Ross I site decreased progressively upwards from the surface of the moss turf. The gradient was steepest by day when absorbed solar radiation heated the moss, giving rise to a sensible heat flux which warmed the air above. The site received 24 hours illumination daily, though with only diffuse solar irradiance during intermittent shade at night. The first period of shade was reflected in a sharp fall in moss level temperature at around 1800 hr (Figs. 4.5 and 4.19). Soil temperature was influenced by ground heat flux. The warmest conditions at night occurred 2 cm down, and 1 cm below the base of the short moss turf, and heat conducted downwards raised the mean maximum temperature to within about two degrees of that near the surface (Fig. 4.20). The insulating properties of the gravel were such that mean temperatures at 20 cm depth varied only from 1 °C to 3 °C, with the daily maximum occurring later in the afternoon than near the surface.

Vegetation, temperature and permafrost

The thermal conductivity of organic matter is lower than that of gravel and other mineral soils. A thick cover of lichens, or particularly

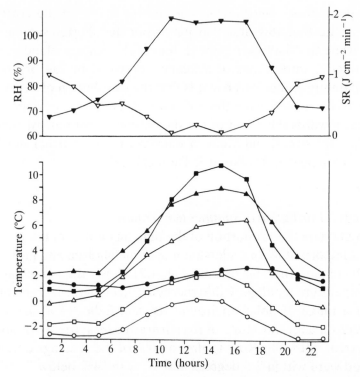

Fig. 4.20. Mean data from Ross I (frigid-Antarctic) for twelve 2-hour time intervals per day during a five-day period from 1600 hr on 28 December 1971 for shortwave radiation (▼), relative humidity (▽), and temperature 2 m (○), 30 cm (□) and 2 cm (△) above the ground, 2–3 mm deep in a *Bryum argenteum* turf (■), and 2 cm (▲) and 20 cm (●) deep in the soil. Based on readings at intervals of 15 minutes (temperature), 7.5 minutes (radiation) and 30 minutes (RH). After Longton (1974a).

of moss or peat, can have a profound effect on soil temperature regimes, by providing insulation beneath which soil temperatures tend to show less diurnal fluctuation, with means lower in summer and higher in winter, than beneath bare ground (Kershaw, 1978; Smith, 1975a). The effect on winter temperatures is limited by the increased thermal conductivity of the moss layer when frozen. In summer, the influence of vegetation cover on soil temperature is enhanced by absorption of water or dew by the vegetation, particularly by *Sphagnum* spp or peat, as subsequent evaporation produces a further cooling of the soil (Benninghoff, 1952).

The effects of moss cover on soil temperature regimes strongly influence the distribution and thickness of permafrost. According to Kudryavtsev (in Ives, 1974), the net effect of moss or peat in decreasing soil temperatures in summer is greater than the converse winter effect in the south of the Arctic permafrost region, while the reverse is true in the north. One may speculate that this arises in part from the shorter summers in the north, while the shallower snow in tundra than in woodland may increase the relative importance of insulation provided by the organic layer. The apparently anomalous result is that the distribution of permafrost is strongly correlated with that of mire vegetation underlain by *Sphagnum* peat towards the south of the discontinuous permafrost zone (page 12), while the occasional areas of unfrozen ground further north also tend to occur in peatlands (Zoltai & Tarnocai, 1975).

Frequency distribution of summer temperatures

Information on the proportion of the time that different temperatures prevail is important to an understanding of metabolism and growth in a rapidly fluctuating thermal regime. Fig. 4.21 shows the percentage of summer temperature readings within successive five-degree increments for mosses in a range of polar environments, and indicates a marked contrast between data for short turfs of *Bryum argenteum* at frigid-Antarctic and mild-Arctic sites. On Ross I, almost 60% of the readings during a 24-day period were within 2.5 degrees of 0 °C, with 84% below 7.5 °C. However, only 3% of readings were below −2.4 °C, with a similar proportion in the range 12.5–17.5 °C. A broader range was recorded at Churchill, where the distribution was skewed upwards from the modal range of 7.5–12.5 °C with maxima up to 40 °C. No temperatures below 2.5 °C were noted during the recording period, although occasional summer frosts may occur. The moss on Ross I received melt water seeping from a persistent snow bank, but that at Churchill was dry, and this difference no doubt influenced the maximum temperatures. Temperatures up to 55 °C were recorded in *B. argenteum* turf on a dry woodland path in the south of the boreal region in central Canada (Longton & MacIver, 1977).

The data for tall turf-forming mosses in the cool- and cold-Antarctic are, predictably, intermediate between those so far considered (Fig. 4.21). Data from Signy I resemble those from Ross I in showing a relatively narrow range with a strong peak in a single five-degree increment, about 55% of the readings being 0–5 °C. Comparison with Ross I is complicated by the different increments to which the data are assigned, but figures in Longton (1974a) show that more than 25% of the Ross I readings were below 0 °C, compared with around 16% on Signy I, and the moss at the latter site clearly experienced temperatures in the range 10–25 °C for longer periods than that on Ross I. The South Georgian data show

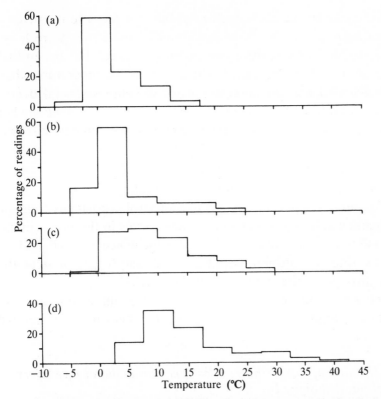

Fig. 4.21. Frequency distributions of temperature readings in the surface layers of moss turfs during periods of 10–24 days in summer at (a) Ross I (frigid-Antarctic); (b) Signy I (cold-Antarctic); (c) South Georgia (cool-Antarctic); (d) Churchill (mild-Arctic). Ross I data for *Bryum argenteum* based on readings at 15-minute intervals for 24 days (Longton & MacIver, 1977); Signy I data for *Polytrichum alpestre* based on data at 60-minute intervals for 10 days (Walton, 1982); South Georgia data for *P. alpestre* based on readings at 3-hour intervals for 14 days (Longton & Greene, 1967); Churchill data for *B. argenteum* based on readings at 30-minute intervals for 17 days (Longton & MacIver, 1977).

a more dispersed distribution about the modal range of 5–10°C, with freezing temperatures rare and maxima 25–30°C.

These figures suggest that mean and modal values, and the diurnal range, of moss level temperatures decrease along a gradient from mild- to frigid-polar regions, but differences in aspect, exposure, colony water content and other factors can lead to considerable local variation in thermal regimes. This is evident in Fig. 4.18, and is illustrated vividly by Kappen's (1985b) data from Birthday Ridge, an arid, frigid-Antarctic site in northern Victoria Land. The highest summer temperatures were recorded on exposed, north-facing rocks, but lichens were restricted to depressions between rocks where maximum temperatures were lower, commonly 10–20°C, but moister conditions prevailed. The lichen thalli were dry, and thus physiologically inactive, at thallus temperatures above 10°C. Rock and lichen temperatures were sometimes several degrees above air temperature during cool, cloudy conditions, a potentially beneficial effect attributed in part to longwave radiation emitted by surrounding rocks.

Frequency and severity of freeze–thaw cycles

Of potential importance in terms of survival in polar environments is the frequency and severity of freeze–thaw cycles in the microclimate. In many mild- and cool-polar regions the incidence of such cycles is greatest in spring and autumn, but in the cold- and frigid-Antarctic they may occur throughout the summer snow-free period.

The annual incidence of freeze–thaw cycles 2–3 mm deep in two moss communities on Signy I is indicated in Table 4.2. They were substantially

Table 4.2 *Number of freeze–thaw cycles* per year in the surface layers of moss colonies on Signy I*

Date	Mesic, tall turf-forming mosses				Wet, carpet-forming mosses		
	1965–66[x]	1972	1973	1974	1972	1973	1974
January	18	17	11	0	1	0	1
February	10	11	8	0	4	0	0
March	19	15	9	1	5	5	0
April	6	8	4	5	3	3	2
May	0	0	0	3	3	2	2
June	0	0	0	0	0	0	0
July	0	0	0	1	0	0	0
August	0	0	0	0	0	0	0
September	1	0	1	0	0	0	0
October	15	2	2	5	0	0	2
November	20	6	1	3	2	2	1
December	21	5	2	5	0	1	5
Total	110	64	38	23	18	13	13

*Cycle criterion: −0.5 to 0.5°C.
[x]April 1965 to March 1966
1965–66 data from Longton (1970, 1972a). Remainder from Walton (1982).

more frequent in mesic turfs of *Polytrichum alpestre* and *Chorisodontium aciphyllum* than in wet carpets of *Calliergon* and *Drepanocladus* spp, the number recorded per year ranging from only 13 in the moss carpet in 1973 to 110 in moss turf in 1965–66. In the latter case, however, the minima were above 2 °C on 90 occasions, and below 5 °C only twice. The most extreme case occurred shortly after snowmelt in October when the temperature in a *P. alpestre* turf fell from 2 °C to −14 °C during 15 hours (Longton, 1970, 1972a). The temperature in damp moss colonies may be expected to change slowly through 0 °C because of latent heat effects. This factor is likely to ameliorate the impact on the mosses of fluctuations through 0 °C in ambient temperature, and probably accounts for the consistently lower frequency of freeze–thaw cycles in the wet moss carpet than in the drier turf.

More severe conditions affect cryptoendolithic lichens in rocks in the dry valley region of southern Victoria Land (Chapter 2). The biotic zone in north-facing rocks experiences freeze–thaw cycles on almost a daily basis in summer, with maxima of up to 10 °C alternating with minima from −5 °C to −10 °C. During sunny conditions with light to moderate wind, a pattern of short-term fluctuations through up to eight degrees is superimposed on the diurnal pattern. Each oscillation lasts a few minutes, and is marked by temperatures passing through 0 °C at the rock surface, but normally remaining above 0 °C in the lichen zone (McKay & Friedemann, 1985).

Temperatures during winter

During the winter, many mosses and lichens are insulated by snow against temperature fluctuations and extreme minima, the extent of the protection depending on the depth and physical characteristics of the snow. Pruitt (1978) draws a contrast between snow cover in the boreal forests, to which he applies the Inuit term api, and that in the tundra (upsik). Api is evenly distributed and of low density, whereas strong winds result in the thickness of upsik varying markedly, though predictably, in relation to microrelief, and in the snow being of greater density than in the forest. He notes that the period of maximum thermal stress, the 'fall critical period', occurs as air temperatures drop in autumn, this period terminating when the snow cover reaches the 'hiemal threshold', a thickness sufficient to stabilise temperatures beneath. For api, the hiemal threshold is 15–20 cm, but for upsik a greater thickness is required as the thermal conductivity of snow increases with density, and the hiemal threshold may never be reached at exposed sites where snow cover remains thin or absent. Snow melt is followed by a 'spring critical period' as air temperatures rise. As previously noted, however, the extent of temperature fluctuations during the fall and spring critical periods is likely to be reduced by the water present within moss and lichen colonies.

The effect of snow thickness on subnivean temperatures on Signy I is shown in Fig. 4.18. During a two-week period in June moss surface temperature at the relatively sheltered *Polytrichum alpinum* site remained almost constant at 0 °C, whereas at the *P. alpestre* site, where the snow was less thick, it fluctuated in line with air temperature. However, *P. alpestre* was sufficiently well insulated to prevent its temperature rising above 0 °C on July 11 and 13, when air temperature reached 5 °C, and Table 4.2 stresses that freeze–thaw cycles are of rare occurrence in some moss communities on Signy I during winter.

The extreme minimum temperature recorded in the *P. alpestre* turf during these observations was −16.5 °C, but minima between −25 °C and −30 °C were recorded in a stand of *P. alpestre* and *Chorisodontium aciphyllum* by Walton (1982). More effective insulation was apparently provided by snow covering short acrocarpous mosses on East Ongul I in the frigid-Antarctic, for here moss temperatures were most frequently between −10 °C and −20 °C, with an extreme minimum of −21 °C, whereas air temperature fell to −40 °C (Matsuda, 1968). The temperature in a *P. alpestre* turf beneath some 40 cm of snow on South Georgia remained within one degree of freezing throughout much of the 1961 winter (Longton & Greene, 1967).

Many mosses and lichens occur in habitats lacking snow cover in winter, and they may be exposed to more severe conditions than those documented above. Kershaw (1985) suggests that *Bryoria nitidula* on exposed raised beach ridges in northern Ontario (page 96) is almost certainly exposed to temperatures as low as −45 °C in winter, when temperatures in *Cladonia stellaris* under snow cover at a more sheltered site are typically around −2 °C. The influence of these contrasting environmental regimes, during both winter and summer, will be considered in Chapter 5 in relation to carbon dioxide exchange and resistance to stress.

5

Physiological processes and response to stress

Environmental control of carbon dioxide exchange
Methods

A plant survives only where the annual environmental regime includes periods favourable for metabolism and growth and where, in the case of perennials, there are no periods of lethal stress. Most polar plants, vascular and non-vascular, are perennials, and must therefore show resistance to both elastic and plastic stress. Levitt (1980) defines a biological stress as 'any environmental factor capable of inducing a potentially injurious strain in living organisms': a plastic stress is one producing irreversible chemical or physical change, e.g. frost damage, while an elastic stress results in a reversible change such as a major reduction in net assimilation rate (NAR) under suboptimal conditions. This chapter begins to examine the features that enable bryophytes and lichens to survive under the apparently severe stress of Arctic and Antarctic environments. Such features include, first, general characteristics of these plants that confer fitness under polar conditions and, second, adaptations specific to polar species or populations which may therefore have evolved in response to local selection pressure.

Carbon dioxide exchange has been widely investigated to determine how polar cryptogams maintain positive carbon and energy budgets under adverse conditions. The results are not fully comparable due to differences in experimental procedure. Most methods have involved infrared gas analysis (IRGA) in the laboratory, either in open systems at ambient CO_2 concentration (Oechel, 1976) or in sealed cuvettes (Larson & Kershaw, 1975a). The latter method allows greater replication but has the disadvantage that CO_2 concentration inevitably fluctuates somewhat during the period of measurement. The relative merits of these systems have been discussed by Lange & Tenhunen (1981) and Kershaw (1985). Other investigators have used differential respirometry at unnaturally high CO_2 concentrations.

The nature of the experimental material is also critical, as thallus size (Larson, 1984), degree of mutual shading, and the proportion of photo-synthetic tissue can all influence rates of CO_2 exchange. Photosynthesis is most active in the green, apical region of moss shoots, in the upper parts of fruticose lichens, and near the thallus margin in some crustose and foliose species, whereas respiration continues in older parts. The lichen material has varied between investigations (Table 5.1). In mosses, only the green region of the shoots has generally been tested (Table 5.2), in part to reduce errors arising from respiration of associated micro-organisms (Smith, 1984a). Pretreatment of material must also be con-sidered in the light of resaturation phenomena (page 193), particularly as lichens are frequently stored dry prior to experimentation, and because of the rapidity with which physiological responses can alter with changes in environment, even in air-dry lichens (page 167). Interpretation of published data is further complicated where water content of the experi-mental materials is not stated precisely. The plants have commonly been described by such terms as 'fully hydrated', but this is not entirely satis-factory as NAR is liable to decrease above, as well as below the optimal level.

Some investigators, notably Gannutz (1971), consider laboratory experiments so artificial as to provide little information about responses to natural environments, and have monitored CO_2 exchange using IRGA systems in the field. A common procedure has been to enclose plants in an assimilation chamber at ambient irradiance, with the temperature of the experimental material controlled to match that in nearby, undis-turbed colonies. However, enclosure inevitably alters the plant environ-ment: temperature control is difficult to achieve at high irradiance, and naturally occurring variation in water content has so far proved impossible to follow.

Field experiments have the disadvantage that the influence of individual factors can seldom be studied in isolation. The most effective approach to investigating plant–environment interaction undoubtedly lies in a com-bination of field and laboratory studies, and where this has been attempted there has commonly been reasonable agreement between the two sets of data (e.g. Hicklenton & Oechel, 1977a). Most results of CO_2 exchange studies with cryptogams have been expressed on a dry weight basis, although expression in terms of chlorophyll content would facilitate com-parison.

Maximum recorded net assimilation rates

Representative data for NAR recorded in polar cryptogams under the most favourable combinations of temperature, light and hydration, and in the most active material employed in the investigations concerned, are shown in Tables 5.1 and 5.2. The data give an indication of the photosynthetic capacity of the various species, but there is no guarantee that such rates are normally sustained for long periods in the field, or that they are never surpassed.

The maximum recorded rates range from $0.1–5.0$ mg CO_2 g^{-1} hr^{-1}. Values for cool-, cold- and frigid-polar lichens are all below 0.4 mg CO_2 g^{-1} hr^{-1}. Elsewhere, the maxima in lichens are variable, but with consistently low rates in xeric species (*Alectoria, Bryoria, Parmelia* spp). Several mosses show intraspecific variation, with lower maximum NAR in polar than in temperate populations (page 208), while the xeric species *Racomitrium lanuginosum* has low maximum NAR in material from contrasting sites. On Signy I, maxima again decrease from hydrophytic (e.g. *Drepanocladus uncinatus*) to mesophytic (*Polytrichum alpestre*) and finally xerophytic species (Smith, 1984a), a trend in line with variation in chlorophyll content (Russell, 1985) and growth rate (Chapter 6) in mosses.

Maximum NAR shows seasonal change, with considerable interspecific variation in the pattern of fluctuation. At Kevo, the maximum in the foliose lichen *Nephroma arcticum* was higher in late summer than in either midsummer or winter (Fig. 5.1). Gannutz (1970) reported minimum metabolic activity during spring in cold-Antarctic lichens. Antarctic mosses were then active, but maximum NAR in *Dicranum* spp at Barrow was also lowest early in the growing season (Fig. 5.2). A similar response in tundra angiosperms has been attributed to high respiration rates accompanying rapid spring growth (Billings, 1974), but respiration rates in *Dicranum* spp were no higher in spring than later in the year. Oechel (1976) suggested that NAR was low because the overwintering green tissue was less photosynthetically active than new growth produced later. In *Calliergon sarmentosum*, little green tissue overwinters at Barrow, and in 1974 the lowest maximum NAR in this species was recorded during an August drought to which the dense turf-forming *Dicranum* spp were less susceptible (Fig. 5.2).

Maximum NAR in *Pleurozium schreberi* at Kevo was higher in plants collected in March and September than under conditions of continuous illumination in June (Kallio & Saarnio, 1986), in line with the results of laboratory experiments (page 203). In tundra mosses, there is little evidence of a major decrease in intrinsic rates of photosynthesis in late

Table 5.1. *Maximum NAR, and optimum temperatures for NAR in polar lichens*

Species	Locality and zone	Max. NAR ($mg\ CO_2\ g^{-1}\ hr^{-1}$)	Optimum Temperature (°C)	Notes
Cetraria nivalis	Kevo, Finland (NB)	0.3	10	Measurements at 300–400% wc. Kallio & Heinonen, 1971.
Cladonia stellaris	Schefferville, Quebec (NB)	2.15	20–25	Uppermost 1.5 cm of podetia at 35–55% rwc. Carstairs & Oechel, 1978.
Cladonia stellaris	Schefferville, Quebec (NB)	0.5	20	Intact colonies cleaned of dead or discoloured podetia, at optimum rwc (70–80%). Lechowicz, 1978.
Nephroma arcticum	Kevo, Finland (NB)	0.55	15	Optimum wc (150–250%). Kallio & Heinonen, 1971.
Parmelia olivacea	Kevo, Finland (NB)	0.3	10	Kallio & Heinonen, 1971.
Sterocaulon paschale	Kevo, Finland (NB)	0.25	0	Uppermost 1.5–2.0 cm of water saturated thalli: increased NAR and T_{opt} of 15–20 °C at thallus water content of 150–215%. Kallio, 1973.
Collema furfuraceum	Moosonee, N Ontario (NB)	5.0	25	Whole thalli at optimum wc (300%). Kershaw & MacFarlane, 1982.

Alectorea ochroleuca	Pen I, NW Ontario (MA)	0.5	14	Whole thalli at optimum wc (100%). Larson & Kershaw, 1975b.
Bryoria nitidula	Pen I, NW Ontario (MA)	0.18	5–10	Whole thalli at optimum wc (100–150%). Irradiance below saturation? Kershaw, 1975b.
Cetraria nivalis	Pen I, NW Ontario (MA)	0.5	14	Material at 150% wc. Larson & Kershaw, 1975c.
Cladonia rangiferina	Hawley Lake, NW Ontario (MA)	3.0	25	Uppermost 2 cm of podetia at optimum wc (150%). Tegler & Kershaw, 1980.
Cladonia stellaris	Pen I, NW Ontario (MA)	0.3	5	50:50 ratio algal and purely fungal parts of podetia at optimum wc (200%). Irradiance below saturation? Kershaw, 1975b.
Cladonia stellaris	Hawley Lake, NW Ontario (MA)	4.3	25	Uppermost 2 cm of podetia from shade form at optimum wc (150%). Kershaw, MacFarlane, Webber & Fovargue, 1983.
Parmelia disjuncta	Churchill, N Manitoba (MA)	1.1	14	Whole thalli at optimum wc (100–150%). Kershaw & Watson, 1983.
Stereocaulon paschale	Abitau–Dunvegan Lake area, NWT (MA)	2.3	22	Whole thalli (?), at optimum wc (150%). Kershaw & Smith, 1978.

(*contd*)

Table 5.1. *Maximum NAR, and optimum temperatures for NAR in polar lichens (contd)*

Species	Locality and zone	Max. NAR (mg CO_2 g^{-1} hr^{-1})	Optimum Temperature (°C)	Notes
Cetraria cucullata	Atkasook, Alaska (CA)	0.4	12	Whole thalli at optimum rwc (60%). Lechowicz, 1981a.
Alectoria ochroleuca	Melville I, NWT (CA)	0.05	—	Whole thalli (?), at optimum wc (100%) NAR little-affected by temperature from 1 to 14 °C. Larson & Kershaw, 1975b.
Buellia frigida	Cape Hallet, Victoria Land (FANT)	—	10	Whole thalli 'completely hydrated'. Lange & Kappen, 1972.
Lecanora melanopthalma	Cape Hallet, Victoria Land (FANT)	0.1	0–5	Whole thalli 'completely hydrated'. Lange & Kappen, 1972.
Neuropogon acromelanus	Cape Hallet, Victoria Land (FANT)	0.25	5	Whole thalli 'completely hydrated'. Lange & Kappen, 1972.
Neuropogon sulphureus	Birthday Ridge, Victoria Land (FANT)	0.2	10–15	Whole thalli within a broad optimum range of wc. Kappen, 1983.

Xanthoria mawsoni	Cape Hallet, Victoria Land (FANT)	—	15–20	Whole thalli 'completely hydrated'. Lange & Kappen, 1972.
Umbilicaria aprina	Mt Larsen, Enderby Land (FANT)	0.8	3–5	Whole thalli 'probably near saturation' with water; 3 °C the lowest temperature tested. Ino, 1985.
Omphalodiseus decussatus	Mt Larsen, Enderby Land (FANT)	0.15	3–5	Whole thalli 'probably near saturation' with water; 3 °C the lowest temperature tested. Ino, 1985.
Neuropogon sulphureus	Mt Larsen, Enerby Land (FANT)	0.06	3–5	Whole thalli 'probably near saturation' with water; 3 °C the lowest temperature tested. Ino, 1985.

NB = northern-boreal
MA = mild-Arctic
CA = cool-Arctic
FANT = frigid-Antarctic
wc = water content
rwc = relative water content

Table 5.2. *Maximum NAR and optimum temperatures for NAR in polar bryophytes*

Species	Locality and zone	Max. NAR (mg CO_2 g^{-1} hr^{-1})	Optimum Temperature (°C)	Notes
Anthelia juratzkana	Abisco, Sweden (NB)	0.7	6–11	Uppermost 5–8 mm of shoots, kept 'moist' by periodic spraying. Lösch, Kappen & Wolf, 1983.
Dicranum elongatum	Kevo, Finland (NB)	0.43	5–10	Uppermost, 5 mm (green) part of shoots at 200–300% wc. Kallio & Heinonen, 1975.
Dicranum fuscescens	Schefferville, Quebec (NB)	2.07	10–20	Uppermost, green part of shoots from an upland tundra population at 400% wc. Hicklenton & Oechel, 1976.
Polytrichum commune	Schefferville, Quebec (NB)	2.99	35	Uppermost, green part of shoots pretreated at 20°C for 3 months. Sveinbjönsson & Oechel, 1983.
Polytrichum sexangulare	Abisco, Sweden (NB)	1.5	6–11	Entire turfs 1–2 cm deep kept 'moist' by periodic spraying. Lösch *et al.*, 1983.
Racomitrium lanuginosum	Kevo, Finland (NB)	0.19	5	Uppermost 1 cm part of shoots at 200–300% wc. Kallio & Heinonen, 1975.
Calliergon sarmentosum	Barrow, Alaska (CA)	3.0	10–15	Uppermost (green) 0.5 to 2.0 cm of shoots, water saturated, in 1972. Oechel & Collins, 1976.
Calliergon sarmentosum	Barrow, Alaska (CA)	0.8	10–20	Uppermost, green parts of shoots, water saturated, in 1974. Oechel, 1976.

Species	Location			Notes
*Dicranum angustum**	Barrow, Alaska (CA)	0.6	10–20	Uppermost, green parts of shoots, water saturated, in 1974. Oechel, 1976.
*Dicranum elongatum**	Barrow, Alaska (CA)	0.5	15–20	Uppermost, green parts of shoots, water saturated, in 1974. Oechel, 1976.
Polytrichum alpinum	Barrow, Alaska (CA)	4.4	10–15	Leafy shoots 2–3 years old, water saturated. Oechel & Collins, 1976.
Polytrichum commune	Barrow, Alaska (CA)	1.59	20	Uppermost, green parts of shoots, pretreated at 20 °C for 3 months. Sveinbjörnsson & Oechel, 1983.
Racomitrium lanuginosum	Spitzbergen (CA)	0.15	5	Uppermost, (green) 1 cm of shoots at 200–300% wc. Kallio & Heinonen, 1975.
Racomitrium lanuginosum	South Georgia (CANT)	0.12	5	Uppermost, (green) 1 cm of shoots at 200–300% wc. Kallio & Heinonen, 1975.
Drepanocladus uncinatus	Signy I (CDANT)	0.95	15–20	Uppermost, green parts of shoots, water saturated. Pretreated at 5/−5 °C day/night. Collins, 1977.
Drepanocladus uncinatus	Signy I (CDANT)	1.4	5–10	Uppermost 8 mm of shoots at 980% wc. Freshly collected in early winter. Davis & Harrison, 1981.

(contd)

Table 5.2. *Maximum NAR and optimum temperatures for NAR in polar bryophytes (contd)*

Species	Locality and zone	Max. NAR $(mg\ CO_2\ g^{-1}\ hr^{-1})$	Optimum Temperature (°C)	Notes
Polytrichum alpestre	Signy I (CDANT)	0.5	10	Uppermost green parts of shoots, water saturated. Pretreated at $5/-5$°C day/night. Collins, 1977.
Polytrichum alpestre	Signy I (CDANT)	0.4	2–3	Uppermost 9.5 mm of shoots at 200% wc. Freshly collected in early winter. Davis & Harrison, 1981.
Bryum antarcticum	Marble Point, Victoria Land (FANT)	—	15–20	Uppermost 5 mm of shoots (40% green:60% brown), freshly washed in water, at 1% CO_2 concentration. Rastorfer, 1970.
Bryum argenteum	Cape Hallet, Victoria Land (FANT)	—	25–30	Uppermost 5 mm of shoots (60% green:40% brown), freshly washed in water, at 1% CO_2 concentration. Rastorfer, 1970.

NB = northern-boreal
CA = cool-Arctic
CANT = cool-Antarctic
CDANT = cold-Antarctic
FANT = frigid-Antarctic
wc = water content

* The identity of these taxa has been revised, here and elsewhere, according to Miller *et al.*, 1978.

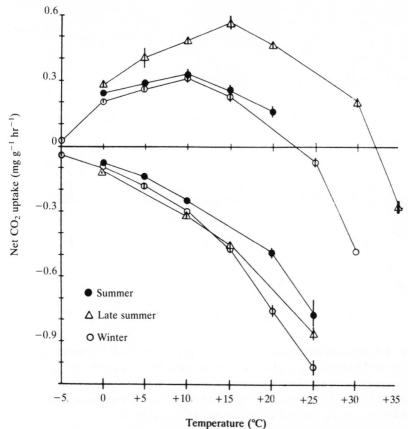

Fig. 5.1. Seasonal variation in NAR and dark respiration in relation to temperature in *Nephroma arcticum* from Kevo at optimum thallus water content (150–250%) and near-saturating irradiance. Data from Kallio & Heinonen (1971).

summer, such as accompanies frost-hardening in tundra angiosperms (Hicklenton & Oechel, 1977a; Oechel, 1976), but more significant decline has been recorded in some boreal forest mosses (Oechel & Van Cleve, 1986).

Several lines of evidence suggest that bryophytes fix CO_2 by the C_3 pathway as in most temperate and polar flowering plants (Rastorfer, 1971a; Valanne, 1984), but maximum NAR in both bryophytes and lichens is considerably lower on a dry weight basis than in angiosperm leaves. Maxima in the foliage of tundra angiosperms discussed by Wielgolaski (1975) were 7–33 mg CO_2 g^{-1} hr^{-1} compared with less than 5 mg CO_2 g^{-1} hr^{-1} in the green parts of associated mosses and lichens. Similarly,

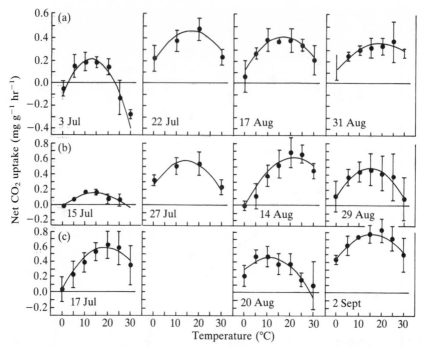

Fig. 5.2. Seasonal variation in the relationship between NAR and temperature in the green portions of water-saturated shoots of *Dicranum elongatum* (a), *D. angustum* (b), and *Calliergon sarmentosum* (c) from Barrow at irradiance of 1.0–1.2 J cm^{-2} min^{-1} PAR. Data are means, usually of six replicates; vertical bars represent ±1 standard deviation. After Oechel (1976).

Oechel & Sveinbjörnsson (1978) considered that NAR in mosses was only 8–9% of that in vascular plant leaves at Barrow. In compensation, growth in bryophytes and lichens is channelled primarily into perennial photosynthetic tissue, whereas flowering plants develop high proportions of non-photosynthetic biomass and, in many species, leaves that function for only a few months.

The proportion of algal to fungal cells is low in lichens, even in young parts of the thallus, resulting in low chlorophyll contents (e.g. 0.2–1.0 mg g^{-1}: Kärenlampi, 1970a; Berg, 1975), and this could be partly responsible for low NAR in lichens on a dry weight basis. The highest NAR in Table 5.1 was recorded in *Collema furfuraceum*, a species in which the photobiont forms an unusually high proportion of the thallus. Chlorophyll contents are rather lower in mosses than in vascular plants from comparable tundra communities (Berg, 1975), but the difference is not sufficient to explain the lower NAR. Valanne (1984) suggested that the latter may be related to features of the photosynthetic apparatus by which mosses resemble sciophytic flowering plants in showing, for example, low chlorophyll *a:b* ratios and low capacity for electron transport.

Carbon dioxide exchange and temperature
Temperature and net assimilation   Response curves of NAR to temperature in cryptogams are typically broad, as in Figs. 5.1, 5.2 and 5.3, a relationship likely to be beneficial to plants experiencing a wide

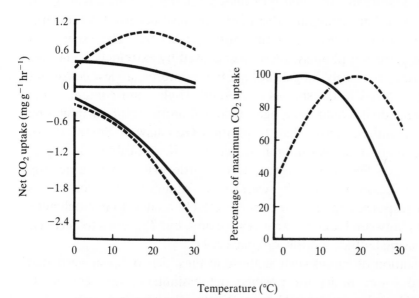

Temperature (°C)

Fig. 5.3. Relationships between temperature, NAR and dark respiration early (solid lines) and late (broken lines) in the growing season in water-saturated shoots of *Polytrichum alpinum* from Barrow at near-saturating irradiance. Data from Oechel & Sveinbjörnsson (1978).

diurnal temperature regime with continuous illumination (Chapter 4). The optimum temperature for net photosynthesis (T_{opt}) is a convenient reference point on such curves, although it conveys only part of the available information and it must be remembered that net assimilation may continue at close to maximum rates over a substantial temperature range around the optimum.

Like maximum NAR, T_{opt} alters both seasonally and in response to variation in other environmental factors, and the data for T_{opt} in Tables 5.1 and 5.2 again relate to the most favourable conditions employed in the various studies. As a further complication, the different relationships of temperature to respiration and to gross photosynthesis imply that an increase in the proportion of non-photosynthetic but respiring tissue will decrease T_{opt} as well as NAR. Experiments on whole plants will thus indicate lower optima than tests involving only the green, photosynthetically active parts.

The optimal temperature for net photosynthesis in polar cryptogams is commonly in the range 5–25 °C, and may vary rather widely between species from a given locality. Optimal temperatures below 5 °C have been reported in several frigid-Antarctic lichens in summer, and in *Polytrichum alpestre* on Signy I in early winter (Tables 5.1 and 5.2). Lange & Kappen (1972) showed that maximum gross photosynthesis occurred at temperatures above 20 °C in several of the Antarctic lichens. They suggested that low T_{opt} for net photosynthesis was caused by extremely high rates of respiration, primarily fungal, at high temperatures, a view consistent with the low NAR in the species concerned. High water content of the experimental thalli could have been a contributory factor by inhibiting gross photosynthesis more than respiration (Kershaw & Smith, 1978), an effect evident in *Stereocaulon paschale* from Kevo (Table 5.1). Apart from the frigid-Antarctic lichens, there is little evidence of a general decrease in T_{opt} in increasingly severe polar climates. Variation in this and other responses appears to be more closely related to microclimatic differences between habitats at a given locality, but T_{opt} tends to decrease with increasing latitude worldwide (page 204).

Extrapolation of curves such as those in Figs. 5.1 to 5.3 suggests that upper and lower limits for positive net assimilation are commonly 30 to 40 °C and −10 to 0 °C respectively. Positive net assimilation below 0 °C has been recorded experimentally in several mosses (Davis & Harrison, 1981; Kallio & Heinonen, 1973), with low but positive rates between −10 and −20 °C in some lichens. These include the frigid-Antarctic species studied by Lange & Kappen (1972), who pointed out that the photobionts must be partially dehydrated following extracellular freezing at such temperatures. This indicates that their photosynthetic enzyme reactions can function under conditions of considerably reduced water potential.

Temperature and respiration   NAR is the balance between gross photosynthesis and respiration. Respiration is essential to liberate energy required for growth, but at sustained high rates it can cause unnecessary loss of carbon and energy. Dark respiration rates in polar mosses and lichens increase with temperature up to 25–30 °C, the maxima commonly investigated (Figs. 5.1 and 5.3). Extrapolation suggests that respiration can continue below 0 °C, and low rates have been confirmed at −5 to

−10 °C in several species (Davis & Harrison, 1981; Kallio & Heinonen, 1973; Lange & Kappen, 1972). In contrast, respiration was found to cease at 0 °C in others, including *Polytrichum sexangulare* and the hepatic *Anthelia juratzkana* (page 201) from Scandinavian late snow beds (Lösch *et al.*, 1983). This could be beneficial in plants that are invariably snow-covered for most of the summer, and perhaps on some occasions continuously for several years.

The shape of the temperature/respiration response curve varies. In some cases increase in respiration rate accelerates as temperature rises (Figs. 5.1 and 5.3). In others Q_{10} decreases with increasing temperature as indicated in Table 5.3 for mosses at Barrow. Changes in Q_{10} during

Table 5.3. *Seasonal and temperature related variation in Q_{10} of respiration rates in mosses from Barrow, Alaska*

Species	Date	Q_{10}		
		0–10 °C	10–20 °C	20–30 °C
Dicranum elongatum	3 July	1.8	1.3	1.2
	22 July	3.2	1.4	1.2
	17 August	3.8	2.2	1.8
	31 August	5.8	2.8	2.0
Dicranum angustum	15 July	3.8	2.7	2.0
	27 July	1.9	2.1	1.8
	14 August	2.5	2.1	1.8
	29 August	3.4	2.4	1.9
Calliergon sarmentosum	17 July	3.6	2.5	1.9
	20 August	4.1	1.9	1.6
	2 September	3.1	2.1	1.7

Data from Oechel (1976).

the growing season were small and showed no generalised pattern in the Barrow mosses, except for a progressive rise in *Dicranum elongatum*.

Seasonal variation in temperature responses of carbon dioxide exchange Relationships between NAR and temperature vary seasonally in ways that often appear to be beneficial in terms of net carbon and energy gain. Both T_{opt} and maximum NAR in *Polytrichum alpinum* from Barrow rose markedly during the growing season in line with field temperatures, with no appreciable change in the temperature response of respiration (Fig. 5.3). This indicates an increase in gross photosynthetic capacity, perhaps caused by changes in CO_2 diffusion resistance or in the activity of photosynthetic enzymes (Oechel & Sveinbjörnsson, 1978). Similarly, T_{opt} rose from 5 °C in June to 15–20 °C in August in *Dicranum fuscescens* from Schefferville (Fig. 5.4), with a comparable rise in the

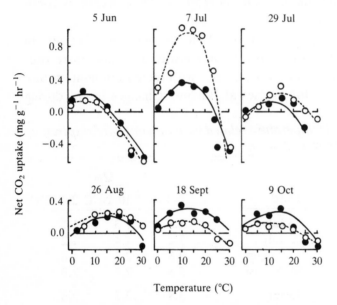

Fig. 5.4. Seasonal variation in relationships between NAR and temperature in the green portions of shoots from upland tundra (O) and lowland woodland (●) populations of *Dicranum fuscescens* near Schefferville kept moist at near-saturating irradiance. Data are means ($n = 3$). After Hicklenton & Oechel (1976).

compensation temperature. Shifts in T_{opt} of ten degrees occurred in *D. fuscescens* in as little as five days in response to increased cultivation temperature, but the response to a temperature decrease was less rapid (Hicklenton & Oechel, 1976). In contrast, *Dicranum elongatum* and several other mosses at Barrow showed only minor seasonal variation in T_{opt} (Fig. 5.2).

Interspecific differences in acclimation potential have also been reported among mosses in Signy I. *Polytrichum alpestre* grown in a light/ dark regime of 5/−5 °C showed a T_{opt} of 5–10 °C, and maintained positive net photosynthesis at about 70% of maximum at 0 °C, whereas plants grown at 10–15/0–5 °C had a slightly higher T_{opt} and considerably increased capacity for net assimilation at supra-optimal temperatures.

The response of *Drepanocladus uncinatus* was quite different, showing a broader curve with little evidence of acclimation (Fig. 5.5). Collins (1977) suggested that acclimation and broad response represent alternative adaptations to seasonally fluctuating temperatures. However, T_{opt} in early winter collections had fallen to 5–10 °C in *D. uncinatus* and 2–3 °C in *P. alpestre* (Table 5.2).

A pattern of temperature acclimation similar to that in *Polytrichum alpinum* has been reported in experiments on *Cladonia stellaris* from lichen woodland at Schefferville. T_{opt} at moderate irradiance was 20 °C in early June, when plants in the field were still snow-covered: it rose to 30 °C in midsummer, but fell to 16 °C in autumn. The high T_{opt} in July and August coincided with increased maximum NAR (Fig. 5.6), and the relationship between temperature and dark respiration again showed little seasonal variation suggesting changes in gross photosynthetic capacity.

Several other patterns of seasonal variation have been recorded in lichens from northwestern Ontario. Maximum NAR in *Alectoria ochroleuca* occurred at 7–14 °C in April, 21 °C in August and 7 °C in October (Larson & Kershaw, 1975b,c). In *Bryoria nitidula*, 20 °C was close to T_{opt} in August but was above the compensation point in plants collected beneath snow in December, T_{opt} then being around 12 °C. NAR in *B. nitidula* was always negative above 25 °C, but respiratory losses were thought to be minimised by dark thallus colour resulting in rapid drying and overall reduction in metabolism under strong irradiance (Kershaw, 1975b). There was again little seasonal change in the response of respiration to temperature in these species, but, in contrast to *Polytrichum alpinum* and *Cladonia stellaris*, maximum NAR also showed little variation.

No change in the response of NAR to temperature was recorded in *Stereocaulon paschale*, *Cladonia rangiferina* or *Collema furfuraceum* from northern Ontario, except for an increase in the compensation point during summer in *S. paschale* (Kershaw & Smith, 1978; Tegler & Kershaw, 1980). *Cladonia rangiferina* normally grows in the shade of dwarf shrubs and *Collema furfuraceum* occurs as an epiphyte on the northern side of trees, so that both species are shielded from high summer temperatures. Seasonal respiratory changes also vary between species. The rates showed little seasonal variation in *C. furfuraceum*, but decreased markedly at all temperatures with the onset of winter in *C. rangiferina* and increased in winter in *Parmelia disjuncta* from Churchill (Kershaw & Watson, 1983).

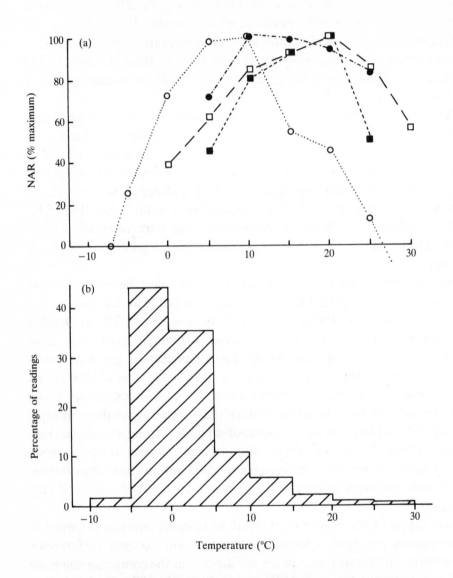

Fig. 5.5. (a) Relationship between NAR and temperature in moist, green portions of *Polytrichum alpestre* (circles) and *Drepanocladus uncinatus* (squares) shoots from Signy I at an irradiance of $0.75\,\mathrm{J\,cm^{-2}\,min^{-1}}$ PAR: plants previously maintained in light/dark temperature regimes of $5/-5\,^{\circ}\mathrm{C}$ (open symbols) and 10–$15/5\,^{\circ}\mathrm{C}$ (closed symbols). (b) Frequency distribution of temperature readings recorded hourly 5 mm below the surface of a *P. alpestre* turf on Signy I during 120 days in summer. Data from Collins (1977). Data in Walton (1982) suggest that the summer temperature regime in *D. uncinatus* carpets shows less fluctuation, with a higher proportion of readings close to $0\,^{\circ}\mathrm{C}$ and fewer above $10\,^{\circ}\mathrm{C}$ than in the *P. alpestre* turf.

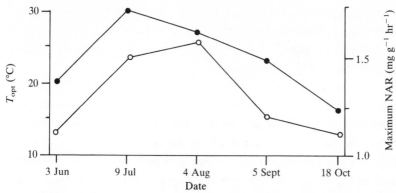

Fig. 5.6. Seasonal variation in T_{opt} (●) and maximum NAR (○) in water-saturated thalli of *Cladonia stellaris* from Schefferville at irradiance of $0.6 \, J \, cm^{-2} \, min^{-1}$ PAR. After Carstairs & Oechel (1978).

The significance of seasonal variation in CO_2 exchange patterns in lichens has been discussed by Kershaw & Watson (1983). Like Collins (page 157), these authors considered acclimation and broad but stable response to represent alternative strategies. They pointed out that acclimation appears to be characteristic of species occupying habitats with pronounced seasonal variation in microclimate, such as *Alectoria* and *Bryoria* spp of open, well-insolated terrain. This contrasts with the position in *Cladonia rangiferina* and *Collema furfuraceum* growing in more shaded situations. Similarly, *Drepanocladus uncinatus* occupies wetter habitats on Signy I, with less temperature variation than those favoured by *Polytrichum alpestre* (Walton, 1982). Under these circumstances expenditure of energy in maintaining a mechanism permitting pronounced acclimation could be counterproductive.

Kershaw & Watson (1983) interpreted seasonal variation in CO_2 exchange relationships as representing several distinct sets of physiological processes. One involves changes in respiration rate, either decreasing in winter as in *Cladonia rangiferina* or increasing as in *Parmelia disjuncta*. Here, by analogy with flowering plants (Levitt, 1980), the contrasting responses were viewed as reflecting different mechanisms of winter hardening, but it should be noted that there is little evidence that frost resistance in polar cryptogams decreases during summer (page 198). Second, there are seasonal changes in T_{opt} in line with ambient temperature without major variation in maximum NAR, as in *Alectoria* and *Bryoria* spp. This was seen as a homeostatic mechanism enabling similar NAR to be maintained throughout the growing season despite changes in climate. Different again is the situation in *Cladonia stellaris*, and in *Polytrichum alpinum*, where there is an increase in both T_{opt} and maximum NAR leading to enhanced photosynthetic capacity in midsummer. Physiological aspects of seasonal changes in the CO_2 exchange responses of lichens are discussed further by Kershaw (1985).

Relationships of carbon dioxide exchange to field temperature regimes   The degree to which net photosynthesis/temperature relationships maximise assimilation under field conditions varies widely, despite seasonal changes. In *Dicranum fuscescens* at Schefferville, the optimum temperatures in Fig. 5.4 were within a few degrees of the most frequently prevailing field temperatures throughout June, July and August (Hicklenton & Oechel, 1976, 1977a), with maxima on warm days sufficient to depress NAR at midday (page 186). Conversely, rapidly declining temperatures in September and October were not accompanied by a comparable shift in T_{opt}, in line with the results of laboratory experiments. *Cladonia stellaris* at Schefferville and mosses at Barrow seem less in harmony with field temperatures: T_{opt} was substantially above the most frequently recorded plant temperatures for most of the summer, although the latter remained above the minimum for positive net assimilation except early in the growing season (Carstairs & Oechel, 1978; Oechel, 1976).

On Signy I, *Polytrichum alpestre* seems better fitted than *Drepanocladus uncinatus* to the summer temperature regime (Fig. 5.5), but the most frequently occurring daytime temperatures are suboptimal for both species, and *D. uncinatus* is less likely than *P. alpestre* to experience heat-induced depression in NAR at midday (Collins, 1977). At Cape Hallett, daily maximum and minimum rock surface temperatures during two summer months were commonly 10 to 30°C and −10 to −2°C respectively (Rupolph, 1966). Optimal temperatures at high irradiance in four epilithic lichens lay between these extreme (Table 5.1): *Xanthoria mawsoni*, with a T_{opt} of 15–20°C, appears to be the species least well fitted to the local temperature regime as the most frequently occurring microclimatic temperatures are usually towards the lower end of the daily range. Optimal temperature was close to 0°C at reduced irradiance in all four species, but maximum NAR was then low. Kappen (1985b) noted that frigid-Antarctic lichens were always at less than their optimal temperature when in a wet, physiologically active condition.

Even less correspondence between T_{opt} and field temperatures is seen in Fig. 5.7, where physiological data for *Bryum argenteum* from Cape Hallett are related to summer microclimatic temperature data from a comparable frigid-Antarctic site. The response of NAR to temperature

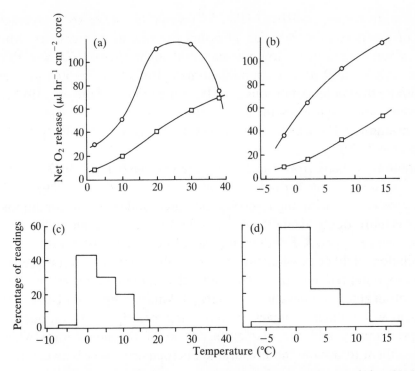

Fig. 5.7. Relationships between temperature, dark respiration ($\square$) and NAR ($\bigcirc$) at irradiance of $0.5\,\mathrm{J\,cm^{-2}\,min^{-1}}$ PAR (a) and $0.28\,\mathrm{J\,cm^{-2}\,min^{-1}}$ PAR (b) in 1 cm-deep cores of *Bryum argenteum* from, Cape Hallett, with frequency distributions of temperature readings from 0400–1859 h (c) and from 1900–0359 h (d) in *B. argenteum* on Ross I during 27 days in summer. Data from Rastorfer (1970) and Longton (1974a).

is presented at high and low irradiance as the best available comparison with day and night field irradiance. Field temperatures were adequate to support positive net assimilation most of the day, but were most frequently within 2.5 degrees of 0 °C, and almost always below 17.5 °C. Thus the experimentally determined T_{opt} of around 25 °C was seldom approached during the 27-day recording period.

Relationships between field temperatures and CO_2 exchange may be more favourable than the previous discussion suggests. The physiological data for several of the species considered were based on the most photosynthetically active parts of the plants and T_{opt} for whole plants would be lower than the values indicated. Also, the typically broad response curves imply that NAR would be substantial at significantly suboptimal temperatures. Thus for frigid-Antarctic *Bryum argenteum*, apparently the most extreme case considered, calculations suggest that the microclimatic temperature regime on Ross I (Fig. 5.7) would enable NAR to average 25% of the maximum rate for 24 hours per day in midsummer, with only minor limitation by low irradiance at night (page 167), and that mean NAR in the *B. argenteum* microclimate would be some 50% greater than under the cooler conditions 2 m above the ground (Longton, 1974a). Moreover, photosynthesis in *B. argenteum* was investigated at a CO_2 concentration of 1% which probably resulted in an unnaturally high T_{opt} (Rastorfer, 1970).

Endolithic lichens from the frigid-Antarctic have T_{opt} 1 to 7°C, and upper and lower limits for net assimilation of 5 to 16°C and −8 to −6°C respectively, and with high respiration rates under warm conditions. Water is more freely available in the endolithic habitat than on the rock surface, and Kappen & Friedmann (1983a,b) calculated that positive net assimilation might be possible for 13 hours daily in summer, despite generally suboptimal temperatures. The endoliths seem no better adapted for metabolism at low temperatures than frigid-Antarctic epiliths (Table 5.1). Kappen & Friedmann suggested that their principal adaptations are low productivity and the filamentous form of the mycobionts which enables them to occupy the favourable microenvironment between rock particles.

The most frequently occurring microclimatic temperatures thus seem generally suboptimal for photosynthesis in polar cryptogams, but adequate to support net assimilation at substantial rates during summer. However, the response of NAR to temperature interacts strongly with responses to other factors, which must therefore be considered before the overall pattern can be understood.

Carbon dioxide exchange and light

Comparison of relationships between photosynthesis and light is complicated by the variety of radiation units in common use. Photon flux density is generally preferred in physiological studies as photosynthesis is controlled more directly by the number of light quanta than by energy receipt, but energy units are used here to facilitate comparison with data for irradiance in natural environments. Data expressed in other forms have been converted as follows: $2 J\ cm^{-2}\ min^{-1}$ solar radiation $= 1 J\ cm^{-2}\ min^{-1}\ PAR = 33.5\ k\ lux = 670\ \mu E\ m^{-2} s^{-1}$. The conversions are approximations as the relationships vary with spectral composition.

Polar environments are often considered light-limited because maximum irradiance is lower than elsewhere, and NAR in flowering plants rises with increasing irradiance to levels approaching full sunlight (Tieszen, 1978b). This is less true of cryptogams, particularly bryophytes which reach saturation at low irradiance in line with their other shade plant characteristics. At favourable temperature and hydration, light compensation in most mosses and lichens is 0.01–0.06 J cm^{-2} min^{-1} PAR. Saturation is commonly 0.3–0.6 J cm^{-2} min^{-1} PAR, but reaches 1.5 J cm^{-2} min^{-1} PAR in some species (Kallio & Kärenlampi, 1975; Lechowicz, 1982a; Proctor, 1981).

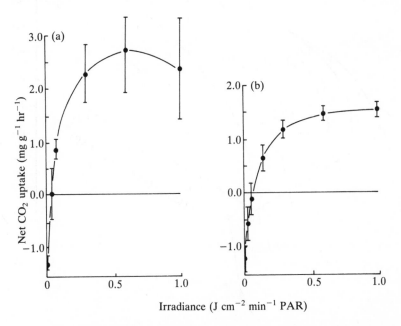

Fig. 5.8. Relationships between NAR and irradiance at 15 °C in water-saturated plants of *Polytrichum alpinum* (a) and *Calliergon sarmentosum* (b) from Barrow. After Oechel & Collins (1976).

In polar species, light compensation points for both mosses and vascular plants in open tundra at Barrow are commonly 0.03–0.07 J cm^{-2} min^{-1} PAR, but mosses saturate at 0.5–0.7 J cm^{-2} min^{-2} PAR, approximately 30% of the level typical for local angiosperms. Some of the mosses show reduced NAR at supra-optimal irradiance (Fig. 5.8) due to photo-inhibition or photo-oxidation of the photosynthetic apparatus (Oechel & Sveinbjörnsson, 1978). Similar values are known for other polar mosses, e.g. *Drepanocladus uncinatus* from Signy I (Fig. 5.9). *Stereocaulon paschale* from northern Ontario has a higher saturation point of around 1.5 J cm^{-2} min^{-1} PAR, in keeping with its typical habitat in open woodland

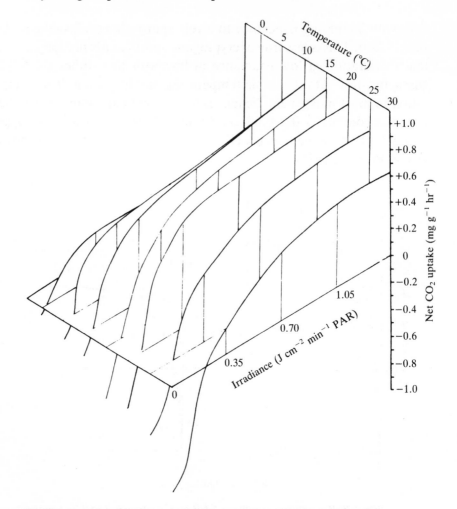

Fig. 5.9. Relationships between NAR, irradiance and temperature in green parts of *Drepanocladus uncinatus* shoots from Signy I, at a water content of 700%; plants previously maintained at a light/dark temperature regime of 5/−5 °C. After Collins (1977).

and its disappearance with closure of the canopy during succession (Kershaw & Smith, 1978). Relatively high saturation points have also been recorded in frigid-Antarctic lichens (Lange & Kappen, 1972) and in *Polytrichum alpestre* from Signy I (Fig. 5.10), also from open habitats. Lower saturation levels characterise several mild-Antarctic Cladonias (Lechowicz, 1978).

Net assimilation represents a balance between rates of gross photosynthesis and of respiration, with the latter the more temperature-dependent. Thus relationships between NAR, temperature and irradiance interact

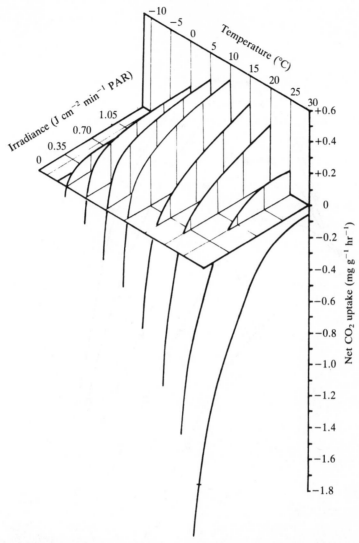

Fig. 5.10. Relationships between NAR, irradiance and temperature in green parts of *Polytrichum alpestre* shoots from Signy I, at a water content of 400%; plants previously maintained in a light/dark temperature regime of 5/−5 °C. After Collins (1977).

strongly: light compensation and saturation points decline with decreasing temperature while T_{opt} falls with decreasing irradiance. Such relationships have been confirmed in many polar cryptogams and imply that net assimilation will continue at low but positive rates at low temperature and irradiance, e.g. during summer nights. Thus in Fig. 5.11, T_{opt} in *Buellia frigida* fell from 12 °C to 2 °C as irradiance decreased, with a

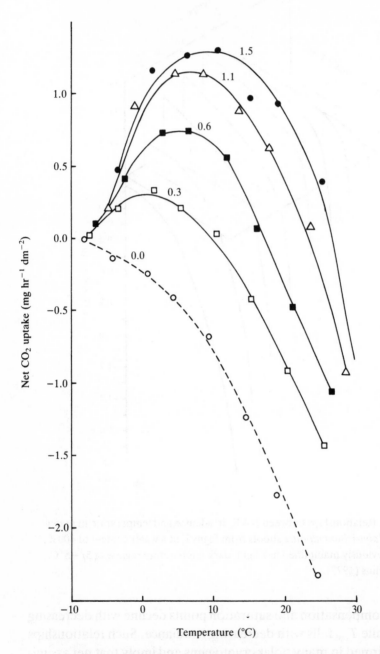

Fig. 5.11. Relationships between NAR and temperature at different irradiance in fully hydrated thalli of *Buellia frigida* from Cape Hallett; plants previously maintained at 4 °C. Figures by the curves indicate irradiance in $J\,cm^{-2}\,min^{-1}$ PAR. After Lange & Kappen (1972).

corresponding fall in NAR. Similarly in *Polytrichum alpestre* from Signy I, positive net assimilation was maintained under low irradiance at temperatures close to freezing, whereas at 30 °C compensation was not reached at $1.4 \text{ J cm}^{-2} \text{ min}^{-1}$ PAR. *Drepanocladus uncinatus* again behaved differently, showing a broader relationship with little temperature-related change in saturating irradiance (Figs. 5.9 and 5.10).

Relationships of NAR to irradiance can show seasonal variation comparable with those to temperature. Saturating irradiance in *Dicranum fuscescens* at Schefferville increased from early June to July, but fell again during August and September, and the extent of inhibition by irradiance above saturation also declined in midsummer (Table 5.4). As with temperature, only *Polytrichum* spp of the mosses tested at Barrow showed substantial acclimation in the response of NAR to light, and here the principal effect was a reduction in compensation points between June and September (Oechel & Sveinbjörnsson, 1978). Seasonal acclimation in light saturation has been recorded in mild-Arctic lichens (page 178), and acclimation to both light and temperature occurs rapidly in some lichens when air-dry, possibly enabling them to respond positively to occasional, brief periods of hydration (Kershaw & MacFarlane, 1980).

Thus NAR in polar cryptogams may seldom be light-limited at midday during summer, and positive net assimilation can continue 24 hours per day at the highest latitudes if other conditions are favourable. The frequency distribution of solar irradiance for *Bryum argenteum* on Ross I (latitude 78° S: Table 5.5), when compared with Rastorfer's (1970) physiological data for this species from Cape Hallett, suggests that irradiance during midsummer may have been above the saturation point at 15 °C for 90% of the time during the day (Fig. 5.12). At night, when moss level temperatures were almost invariably near or below 5 °C (Fig. 5.7), irradiance was close to or above saturation for that temperature for 85% of the time. Solar irradiance thus appears to have only a minor limiting effect, even at night, if the experimental results accurately reflect field behaviour (Longton, 1974a).

At Barrow (latitude 71° N), solar irradiance is considered to be above saturation for mosses for about 35% of the time during midsummer, and for about 80% of the time between 1200 and 1300 hr. The mosses are not light-saturated between 2300 and 0200 hr, although midsummer irradiance remains continuously above compensation. Day to day variation in irradiance is thought to have little effect on daily photosynthetic gain in mosses at Barrow (Oechel & Sveinbjörnsson, 1978), and even

Table 5.4. *Seasonal changes in light compensation and saturation ranges, and in inhibition of NAR by high irradiance, in* Dicranum fuscescens *from upland tundra and lowland woodland sites near Schefferville, Quebec*

Sampling date	Light compensation range ($J\ cm^{-2}\ min^{-1}$ PAR)		Light saturation range ($J\ cm^{-2}\ min^{-1}$ PAR)		Percentage of maximum NAR recorded at $2.2\ J\ cm^{-2}\ min^{-1}$ PAR	
	Tundra	Woodland	Tundra	Woodland	Tundra	Woodland
5 June	0.04–0.09	0.02–0.04	0.09–0.19	0.09–0.19	67	29
7 July	0.04–0.09	0.02–0.04	0.19–0.37	0.19–0.37	93*	65
29 July	0.02–0.04	0.19–0.37	0.19–0.37	0.37–0.75	97*	100*
26 August	0.02–0.04	0.09	0.19–0.37	0.09–0.19	81*	100*
18 September	0.02–0.04	0.02–0.04	0.09–0.19	0.19–0.37	29	26
9 October	0.02–0.04	0.04–0.09	0.04–0.09	0.09–0.19	52	69

* Percentages not significantly different at 95% probability level, by *t*-test.
Data from Hicklenton & Oechel (1976).

Table 5.5. *Percentage of total solar radiation readings within the ranges shown for a stand of* Bryum argenteum *on Ross I, southern Victoria Land, from 8 December 1971 to 4 January 1972*

Range (J cm $^{-2}$ min^{-1})	Percentage of readings*	
	Day[x]	Night
0.04–0.25	2.2	14.2
0.26–0.54	8.6	19.9
0.55–0.80	9.9	14.2
0.81–1.22	20.7	15.0
1.23–1.64	20.2	12.8
1.65–2.06	16.5	10.3
2.07–2.48	14.5	9.1
2.49–2.89	6.5	4.1
2.90–3.31	0.7	0.4
3.32–3.73	0.3	0.2

* Based on readings at 7.5 minute intervals.
[x] Day starts at 0400 hr solar time and night at 1900 hr.
Data from Longton (1974a).

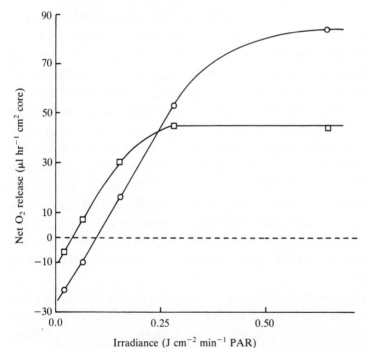

Fig. 5.12. Relationships between NAR and irradiance at 15 °C (○) and at 5 °C (□) in 1 cm-deep cores of *Bryum argenteum* from Cape Hallett. Data are means ($n = 3$). After Rastorfer (1970).

at Schefferville (latitude 55° N), where there is a period of darkness during summer nights, daily photosynthetic gain in *Dicranum fuscescens* appears to be influenced more by variation in temperature than in irradiance (Hicklenton & Oechel, 1977a). Diurnal variation in NAR as recorded in field experiments is discussed on page 185.

Carbon dioxide exchange and water relations

Water uptake and loss   Hydric mosses grow with the base of the colonies almost permanently submerged in water. Other mosses and most lichens depend on intermittent supplies of ground water, often as seepage through soil and over rock surfaces, and of atmospheric moisture in the form of precipitation, fog, dew or water vapour. Lichens and most mosses are primarily ectohydric (Chapter 3). Upward movement of ground water occurs largely by external capillarity, and water is absorbed through most plant surfaces. Moisture below the uppermost few millimetres of the substratum is unavailable. Control of water uptake and loss is primarily morphological, and uptake by dry plants is rapid, particularly in species with a high surface area to volume ratio.

In lichens, the free space system of air-dry thalli commonly becomes filled within five minutes after the thalli are immersed in water or within ten minutes when they are subjected to a stream of fine water droplets, due to lateral movement in internal capillary spaces between the fungal hyphae. Further uptake occurs over a longer period as a result of absorption of water by colloidal materials. Only limited longitudinal movement occurs, even in species with longitudinally orientated cortical hyphae. Some species of *Umbilicaria* and a few other genera rehydrate relatively slowly. Many lichens are dependent on precipitation, snow melt or dew absorbed directly into actively growing parts of the thalli. Within the saturated thallus water is held largely in inter-hyphal spaces, in swollen hyphal walls, and in gelatinous sheaths around cyanobacterial photobionts in some species. Water is also present in cell cytoplasm and vacuoles, with some retained by capillarity on the outer surface of the thallus (Blum, 1973; Larson, 1981).

External water movement in bryophytes occurs in spaces among adjacent shoots and between stem and leaf surfaces. Features such as papillae create narrow channels through which particularly extensive capillary movement may be anticipated (Proctor, 1979). Internal water movement, either apoplastic in the free space system of the cell walls or symplastic through the cytoplasm and across the walls between adjacent cells, is probably of limited extent. These various pathways appear to be inadequate to raise water from the substratum to the shoot apices in some of the larger ectohydric mosses which, like many lichens, depend largely on precipitation and other forms of atmospheric moisture (Longton, 1980). Water uptake into photosynthetic cells can be particularly rapid as the leaf laminae are unistratose. A fully hydrated bryophyte colony contains water within the cells, in cell walls, and in external capillary spaces. The latter are commonly better developed in mosses than in lichens. Aggregation of shoots into colonies is likely to extend periods of metabolic activity by increasing both water-holding capacity and canopy resistance to evaporation. Shoot arrangement is strongly correlated with habitat, the most compact growth forms generally occupying the driest habitats (Chapter 3).

An inevitable consequence of ectohydric water relations is restricted cuticular development and low internal resistance to water movement both into and out of the plant (Chapter 4). Lack of an effective cuticle permits many lichens and ectohydric bryophytes to absorb water from fog and dew, and water vapour from air at high RH. It is notable that lichen distribution is correlated with high frequency of fog in arid Antarctic regions (Chapter 3). Under laboratory conditions, absorption of water vapour has been demonstrated in mosses from Signy I (Gimingham & Smith, 1971), while air-dry thalli of some frigid-Antarctic lichens were shown to resume respiration and positive net assimilation following one to two days' exposure to air are 98% RH (Kappen, 1983; Lange & Kappen, 1972). They did so without a burst of rapid respiration such as occurs when dry mosses and lichens are rehydrated with liquid water (Fig. 5.13).

The significance of water vapour absorption is not yet clear. Lange & Kappen (1972) considered it important for epilithic lichens in regions of low precipitation but generally high RH, as in coastal regions of the frigid-Antarctic. Even here, however, most mosses and many lichens grow in habitats providing at least an intermittent supply of liquid water, and field studies at Cape Hallett suggested that lichen photosynthesis after rehydration solely by water vapour reaches at most 1% of maximum rates (Gannutz, 1969). Resumption of net assimilation by lichens following absorption of water vapour in the field has been confirmed in Alaska, the rates varying between species (page 189).

Some acrocarpous mosses, notably *Polytrichum* spp, are partially endohydric, having a central conducting strand of hydroids in the stem which permits upward, internal water movement from soil to leaves (Hébant,

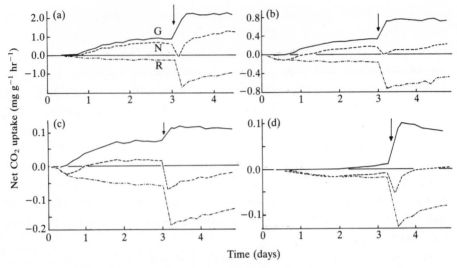

Fig. 5.13. Rates of gross photosynthesis (G), net photosynthesis (N) and dark respiration (R) at 4 °C in initially desiccated thalli of *Xanthoria mawsoni* (a), *Buellia frigida* (b), *Neuropogon acromelanus* (c) and *Lecanora melanopthalma* (d) from Cape Hallett, placed in an atmosphere with 98% RH at time 0, and sprayed with water after 3 days (arrows). After Lange & Kappen (1972).

1977). Partially endohydric mosses possesses a relatively thick cuticle, although they are reliant to some extent on absorption of water by upper parts of the shoots (Bayfield, 1973). *Polytrichum* spp also seem able to restrict water loss to a limited degree by bending leaves against the stem, folding the leaves longitudinally, or arching the leaf lamina over the photosynthetic lamellae when stressed (Bazzaz, Paolillo & Jagels, 1970; Oechel & Sveinbjörnsson, 1978). Mode of water uptake and retention can influence photosynthesis. Thus water content in *Polytrichum commune* at a boreal forest site in Alaska was above the compensation point on all but two days in July 1976. In the ectohydric species *Hylocomium splendens* and *Pleurozium schreberi* it was below compensation on 14 and 10 days respectively (Skre, Oechel & Miller, 1983a).

Water content and net assimilation rate   In compensation for their primarily ectohydric water relations, with the corollary of low internal resistance to water loss, most lichens and many bryophytes are poikilohydrous. Metabolism is most active when the plants are hydrated, but unlike flowering plants many mosses and lichens can tolerate periods of desiccation at water contents as low as 5–10% dry weight. The relation-

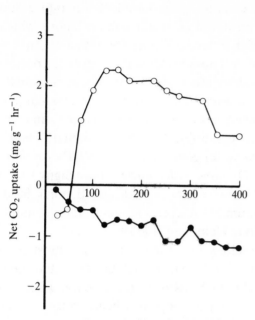

Water content (% dry weight)

Fig. 5.14. Relationships at 22 °C between dark respiration (●), NAR at irradiance of 1.05 J cm^{-2} min^{-1} (○), and thallus water content in *Stereocaulon paschale* collected from the Abitau-Dunvegan Lake area, NWT, during summer. Data from Kershaw & Smith (1978).

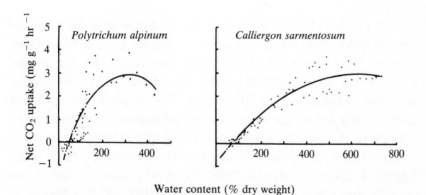

Water content (% dry weight)

Fig. 5.15. Relationships between NAR and water content in two mosses from Barrow at 10 °C and irradiance of 0.63 J cm^{-2} mmin^{-1} PAR. Data from Oechel & Collins (1976).

ship between NAR and water content at favourable temperature and illumination commonly shows an optimum water content above which NAR falls slowly and below which it declines more sharply, as in *Stereocaulon paschale* (Fig. 5.14). The optimum varies widely between species. In *Polytrichum alpinum*, an endohydric moss of mesic and seasonally wet habitats, maximum NAR at Barrow occurred at a water content of 300–400%, compared with 600% in *Calliergon sarmentosum*, an ectohydric species of permanently wet conditions (Fig. 5.15). Similarly, maximum NAR was attained at 200% water content in the mesic *Polytrichum alpestre* and 500% in the hydric *Drepanocladus uncinatus* on Signy Island (Collins, 1977). *Racomitrium lanuginosum*, a widespread polar lithophyte, maintains close to maximum NAR over a range of water contents from only 150% to 400% (Kallio & Heinonen, 1973).

The optima in lichens are generally lower than in mosses, most commonly 100–200% (Kallio & Heinonen, 1975), and habitat correlated variation is again evident. Thus maximum NAR occurred at a water content of 85% in *Alectoria ochroleuca* from the top of a beach ridge on Pen I (Chapter 3), compared with 160% in *Cetraria nivalis* from a moister habitat near the ridge bottoms (Larson & Kershaw, 1975b,c). Net assimilation rate in lichens has also been related to relative water content (page 126), although several definitions of RWC have been employed with respect to the inclusion or otherwise of externally held water as part of the water content of a saturated thallus. Worldwide data assembled by Lechowicz (1982a) show that maximum NAR occurs at RWC ranging from 20% to 90%, but most commonly from 40% to 70%: optimum RWC shows little relationship to latitude, but is positively correlated with precipitation.

Carstairs & Oechel (1978) reported no seasonal change in the relationship between NAR and RWC in *Cladonia stellaris* from Schefferville, with optimum RWC stable at 35–55%, but they noted that the water-holding capacity of the podetia fell progressively from June to September and rose again during October (Fig. 5.16). This implies that optimum water content on a dry weight basis also shifted seasonally, with maxima early and late in the growing season when moisture is most freely available. A decrease in the rate of water loss from saturated thalli (Fig. 5.16) suggested increased internal resistance in midsummer, when it was observed that the lichen became noticeably tougher and less pliable in texture than in spring. The potential for net assimilation under drying conditions was enhanced by the fact that the lower compensation point was not reached until 5–10% RWC, when the lichen was almost air-dry

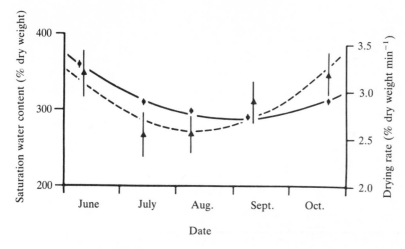

Fig. 5.16. Season variation in saturation water content (solid line) and drying rate (broken line) of photosynthetically active portions of *Cladonia stellaris* podetia from Schefferville. Data are means; vertical lines represent ±1 standard error. Data from Carstairs & Oechel (1978).

(3–4% RWC). Lechowicz (1978) reported a higher optimum RWC for *C. stellaris* from Schefferville (70–80%), and the difference has not been satisfactorily explained.

The relationship between NAR and water potential varies less than that between NAR and water content, at least in mosses (Proctor, 1982). Maximum NAR occurs at water potential close to zero, the rate falling as the cells lose water and their water potential becomes increasingly negative. Maximum NAR is thus likely to occur at the water content required to maintain full turgor. Xeric mosses generally have smaller cells with thicker walls than hydrophytes, perhaps giving greater capacity for water conduction in the free space system of the walls (Proctor, 1979). An increase in the proportion by volume of cell wall to cell lumen is likely to decrease water content at full turgor on a dry weight basis, and this may explain the relatively low optimum water content in xeric mosses. Larson & Kershaw (1975c) likewise suggested that the difference between the optima recorded for *Alectoria ochroleuca* and *Cetraria nivalis* had a morphological basis.

The relationship between water content and water potential is also influenced by the proportion of the water content at saturation that is held within the cells, as shown by the contrasting patterns in Fig. 5.17. In *Racomitrium lanuginosum*, apoplastic and externally held water form

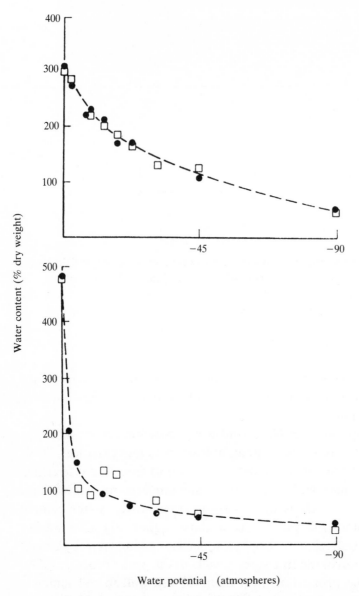

Fig. 5.17. Relationship between water potential and water content in British material of *Polytrichum commune* (above) and *Racomitrium lanuginosum* (below). The symbols (● and □) represent two separate determinations. After Bayfield (1973).

the major components of water content at field capacity. Extensive evaporation can then occur with little change in water potential, but as the water content falls below 150% loss of symplastic water leads to a sharp decline in water potential, in line with a fall in NAR below 150% water content (p. 174). In *Polytrichum commune*, the widely spreading leaves permit less storage of external water, so that loss of symplastic water and a consequent decline in water potential begin early in the drying process.

Ectohydric mosses of dry habitats generally resemble *R. lanuginosum* in having a high capacity to store external water (Proctor, 1982). It is thus somewhat surprising, in view of the humid environment, that limited measurement on Signy I suggested that such species commonly have field water contents below 200%, with little held externally (Gimingham & Smith, 1971). Mosses of wet habitats are variable in this respect. In some, water is retained primarily in the vacuoles of large leaf cells so that a minor loss of water leads to a reduction in water potential (Proctor, 1982). In others, much water is held outside the cytoplasm, as among the concave, imbricate leaves of *Calliergon* spp or in the hyaline cells of *Sphagnum* spp. Hydric mosses on Signy I commonly have water contents of 500–1200%, with 20–65% held externally (Gimingham & Smith, 1971).

Water content and respiration   Dark respiration rates in lichens rise with increasing thallus water content, with low but measurable rates maintained down to water contents of 10–15% and no decline above an optimum (Fig. 5.14). In many species, including *Alectoria ochroleuca*, *Bryoria nitidula* and *Cladonia stellaris*, the rate declines linearly as the thallus dries from saturation (Kershaw, 1975b; Larson & Kershaw, 1975b). In others, such as *Peltigera aphthosa*, the respiration rate remains constant over a wide range of thallus hydration, declining linearly only at low water contents (Kershaw & MacFarlane, 1980). The relationships of respiration to water content and to temperature commonly interact. The respiration rate is low and stable over a range of water contents at low temperature, while both maximum respiration rate and sensitivity to thallus water content increase with rising temperature (Fig. 5.18). This response presumably arises because limitation by water content is most significant at temperatures that permit relatively high respiration rates.

There are few comparable data for polar bryophytes. In temperate mosses, respiration is commonly stimulated by slight water stress at water potentials of −2 Mpa, its rate then declining progressively but slowly down to water potentials of around −20 Mpa (Dilks & Proctor, 1979). Proctor (1981) pointed out that where most water is held in the cells, as in some hydric mosses, photosynthesis begins to decline early as the plants lose water whereas respiration remains active until later, leading to the possibility of a net carbon loss should the plants become desiccated.

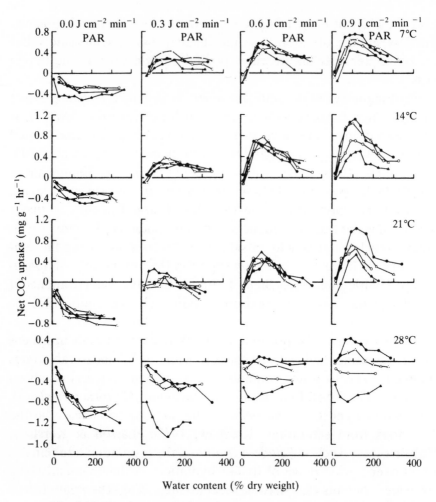

Fig. 5.18. Net photosynthetic and respiratory matrices for thalli of *Parmelia disjuncta* collected at Churchill in February (▲), May (○), June (△) and August (●). After Kershaw & Watson (1983).

Conversely, where substantial amounts of water at field capacity are held outside the cells, as in xeric mosses, photosynthesis and respiration rates fall in quick succession later in the drying process.

Supra-optimal water content and carbon dioxide exchange   Net assimilation rate declines at water contents above an optimum in some bryophytes and lichens (Figs. 5.14 and 5.15), but not in others (Coxson, Brown & Kershaw, 1983; Kappen, 1983). The reduction appears to result principally from resistance to CO_2 diffusion through water to the photosynthetic cells, the diffusion coefficient of CO_2 in water being less than that in air by a factor of 10^4. In some lichens it has been shown that NAR is reduced by high water content at ambient CO_2 concentration but not at higher concentrations due to the increased concentration gradient (Lange & Tenhunen, 1981). An explanation of the differing responses of NAR to high water content must therefore be sought in features that determine how water content influences resistance to CO_2 uptake.

Carbon dioxide entry in mosses is restricted by three variable resistances in series. First is boundary layer resistance (Chapter 4); second is the resistance caused by any water on the plant surface or in the free space system of the cell walls; third is resistance within the photosynthetic cell. The latter has several components including resistance to CO_2 movement in the liquid phase from the inner wall of the cell to the chloroplasts, and a carboxylation resistance reflecting the demand for CO_2 molecules by the uptake mechanism. Although the distance involved is short, the third resistance will often exceed the first (Clayton-Greene, Collins, Green & Proctor, 1985), while the second resistance varies with water content in relation to the detailed architecture of the plants.

In *Polytrichum* spp, photosynthesis occurs principally in cells of lamellae on the adaxial surface of the broad leaf midrib, and marginal cells of the lamellae carry plates or granules of waxy, water-repellent material. There is thus an approach towards internalisation of the photosynthetic tissue as shown by lichens and vascular plants, but generally not by bryophytes. When the surface of the lamellae is dry the second CO_2 resistance is low and, given adequate soil moisture, water lost by evaporation can be replaced by internal transport. However, if the spaces between the lamellae become flooded, CO_2 must diffuse through water for significant distances to reach the photosynthetic cells thus reducing NAR (Clayton-Greene *et al.*, 1985), as seen in *P. alpinum* in Fig. 5.15.

By contrast, photosynthesis in *Calliergon sarmentosum* takes place in the unistratose laminae of concave, imbricate leaves. Here, distal parts of the abaxial leaf surface may be dry, even while water is retained between the adaxial surface and other parts of the shoot, and near-maximum NAR is maintained over a substantial range of water content (Fig. 5.15). One advantage of the loose growth forms characterising *C. sarmentosum* and other hydric mosses may be the provision of air spaces between shoots facilitating CO_2 uptake when the base of the colony is submerged. Similarly, the leaves of some South Georgian *Tortula* spp bear hollow papillae (Lightowlers, 1985), the apices of which provide a possible pathway for CO_2 when much of the leaf surface is wet. In *Schistidium antarctici*, water rapidly rises up the outside of the shoot by capillarity when the shoot base is placed in water, thus soaking most exterior surfaces, but the relatively few, bright green leaves in the shoot apex do not readily become wetted, suggesting that the actively photosynthetic leaves may have a water-repellent surface.

Mosses thus display several solutions to the problem of maintaining both freedom of gaseous exchange and the level of hydration necessary to maintain active metabolism. The problem may be particularly intense in lichens as algae, which form the photobiont in most species, are normally aquatic. The paraplechtenchymatous cortex is thus viewed as a structure protecting algal cells from desiccation (Kershaw, 1985), but it represents a further resistance to CO_2 diffusion. This becomes large when the interhyphal spaces in the cortex are flooded or constricted by cell expansion, although the experiments of Snelgar, Green & Wilkins (1981) suggest that diffusion in the looser-textured medulla takes place almost entirely in the gas phase. Indeed it seems reasonable to suggest that the diffusion resistance presented by the cortex is one of the reasons why lichens are most abundant in dry and mesic habitats. Some correlation exists between habitat and the pattern of response to thallus hydration in lichens, with species showing the greatest reduction in NAR at high water content tending to occupy drier habitats than less sensitive species (Kershaw & Smith, 1978).

Snelgar, Green & Wilkins (1981) confirmed that reduction in NAR in water-saturated lichens results from CO_2 diffusion resistance. Total resistance was shown to be high in dry thalli, a finding attributed to carboxylation resistance. It falls to low levels at moderate water contents, and rises again at high water content in those species showing a corresponding decline in NAR. These authors also showed that resistance at high water content is related to thallus structure. It is increased by tomentum that retains surface moisture and reduced by pseudocyphellae, which comprise pores up to 2 mm wide through the upper cortex, as Hale (1981) noted for some Arctic species.

Further examples of relationships between lichen structure and CO_2 diffusion resistance have been detected in coniferous woodland in northern Ontario (Coxson, Brown & Kershaw, 1983). In *Peltigera aphthosa* of shaded, mesic habitats, the lower part of the thallus comprises a thick medulla covered below by an extensive tomentum allowing efficient water storage. The upper cortex is thin and provides the major pathway for gaseous diffusion, but in consequence it is relatively transparent and woodland populations show a stress response at only moderate levels of illumination (Kershaw & MacFarlane, 1980).

Quite different is the fruticose species *Bryoria trichodes*, which occupies a xeric habitat on upper tree branches. Its thalli are richly branched, thus reducing resistance to entry of both CO_2 and water by providing a high surface area to volume ratio and ensuring that the internal CO_2 pathway is short. Photosynthesis is maximised during rain, fog or dew formation, but as resistance to water loss is also low the thalli are likely to spend considerable periods dry and metabolically inactive. The darkly pigmented cortex is thought to give effective light screening, with excessive heating prevented by efficient sensible heat transfer resulting from the high surface area to volume ratio. Both these species show relatively low resistance to CO_2 entry when wet and thus maintain maximum NAR over a considerable range of water contents. In *Neuropogon sulphureus*, NAR declined at supra-optimal water contents only at relatively high temperatures, and Kappen (1985c) suggested that this response was of little significance in frigid-Antarctic environments.

Field water content   Measurement of field water content in the hydric *Drepanocladus uncinatus* and the mesic *Polytrichum alpestre* suggested that water availability is unlikely to be limiting for photosynthesis at any stage of the summer at some sites on Signy I (Collins, 1977). Oechel & Sveinbjörnsson (1978) reached a similar conclusion regarding mosses at Barrow. Of the species considered, the highest water content was maintained in *Calliergon sarmentosum* and the lowest in *Polytrichum alpinum* with intermediate values for *Dicranum* spp. Comparison of Figs. 5.15 and 5.19 suggest that water content was adequate to maintain near maximum NAR throughout the growing season in the first two species, although *C. sarmentosum* occasionally dried out in some sites. Differences in water relations seem likely to underlie the preference of *C. sarmentosum* for wet depressions and of *P. alpinum* for drier polygon rims. Even at Cape Hallett in the frigid-Antarctic, water content in *Bryum argenteum* was commonly 60–100% RWC during summer due to melt water seepage (Rudolph, 1971).

Significant diurnal variation in water content may be superimposed on these broad seasonal patterns as demonstrated by Hicklenton & Oechel (1977a) in *Dicranum fuscescens* at Schefferville (Figs. 5.20 and 5.21). Water content remained continuously above 200% on two of the days considered, but on the other two it fell to about 100% in the upper, green part of the moss turf and was then likely to limit net assimilation. July 12 was warm and sunny with temperatures up to 25 °C in the moss turf. The high temperature was considered responsible for the decline

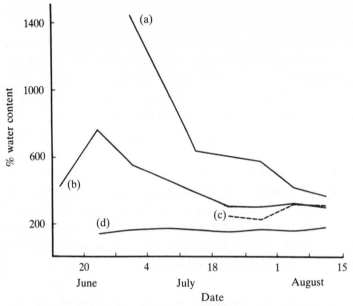

Fig. 5.19. Seasonal variation in water content in *Calliergon sarmentosum* (a), *Dicranum angustum* (b), *Dicranum elongatum* (c) and *Polytrichum alpinum* (d) at Barrow. After Oechel & Sveinbjörnsson (1978).

in water content in the early morning, but the increase from 1000 hr onwards is less easy to explain as moss temperature continued to rise until 1300 hr. The fall in water content on 14 August occurred despite light rainfall and cool conditions: it was attributed to interception of precipitation by a vascular plant canopy and indicates that ectohydric mosses may lose water even under low saturation deficits.

The mosses that have so far been intensively studied thus occupy habitats where they are sufficiently hydrated to remain metabolically active for long periods in summer. Some cold-Antarctic lichens are also subjected to only short dry periods in summer (Kappen, 1985c). Under these circumstances, NAR is probably controlled largely by interaction between temperature and irradiance. Less favourable water regimes may be anticipated in other cases, with epilithic mosses in arid regions, and lichens more generally, likely to be so dry as to be metabolically inactive for long periods during summer. In the frigid-Antarctic, Rudolph (1971) found that three epilithic lichens were most frequently below 20% RWC, a level severely limiting for photosynthesis, and RWC in *Neuropogon sulphureus* may never exceed 60% (Kappen, 1985c). In these cases, water availability assumes critical importance in determining growth rates and

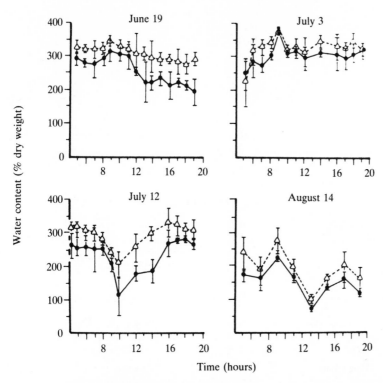

Fig. 5.20. Diurnal variation in water content in green (solid line) and brown (broken line) portions of *Dicranum fuscescens* shoots at woodland (19 June, 22 July) and upland tundra (3 July, 14 August) sites near Schefferville. Data are means ($n = 6$); bars represent ± 1 standard deviation. Data from Hicklenton & Oechel (1977a).

abundance. Restriction of net assimilation by low water content in Arctic lichens is discussed on pages 189 and 195.

Seasonal response matrices

It is now clear that responses of NAR to the major environmental variables both interact and change seasonally. Thus a full understanding of relationships between CO_2 exchange and environment for a given bio-type requires factorial experiments in which CO_2 exchange is investigated under different combinations of water content and environmental conditions in plants collected at different times of the year. The nearest approach to such seasonal response matrices relates to Canadian mild-Arctic lichens, e.g. the epilithic species *Parmelia disjuncta* from Churchill (Fig. 5.18). Seasonally varying patterns in the response of CO_2 exchange to temperature, light and thallus hydration are evident. Particularly

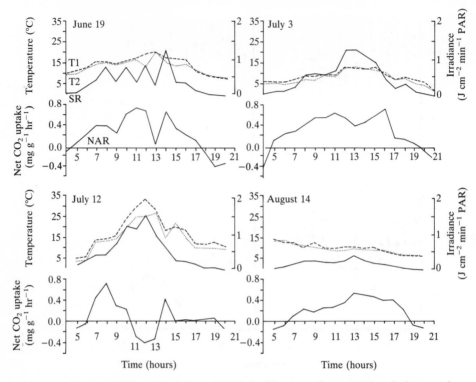

Fig. 5.21. Diurnal variation of NAR in *Dicranum fuscescens* in relation to solar radiation (SR), and to moss leaf temperature in the assimilation chamber (T1) and in undisturbed colonies (T2) at woodland (19 June, 12 July) and upland tundra (3 July, 14 August) sites near Schefferville. After Hicklenton & Oechel (1977a).

noticeable are the high respiration rates in plants collected from beneath thin snow cover in February, especially at 28 °C where NAR was negative even at high irradiance, and the enhanced NAR at high irradiance recorded during the summer.

Comparison of response matrices can provide many clues as to factors determining habitat preferences, for example the distribution of *Cladonia stellaris* on slopes and *Bryoria nitidula* on the dry summits of raised beach ridges on Pen I (Fig. 5.22). Respiration rate in *B. nitidula* increases sharply at high temperature but field respiration is limited under warm conditions in summer by low thallus water content, the latter resulting from the exposed habitat, dark thallus colour and low resistance to evaporation arising from the high surface area to weight ratio (page 127). During summer, *B. nitidula* is thought to undergo most of its net assimilation during cool periods with cloud cover, conditions that favour a high water content but low thallus temperature.

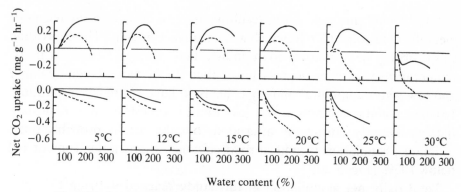

Fig. 5.22. Relationship between NAR (above), dark respiration (below) and thallus water content in *Bryoria nitidula* (broken lines) and *Cladonia stellaris* (solid lines) collected from Pen I in August; NAR recorded at irradiance of 0.22 J cm^{-2} min^{-1} PAR. After Kershaw (1975b).

C. stellaris proved better able to utilise a combination of high temperature and high thallus water content, suggesting that during summer it maintains a greater NAR under most conditions than *B. nitidula*. In compensation, *B. nitidula* may have the capacity for positive net assimilation under cold conditions in spring and autumn when *C. stellaris* is snow-covered (page 190). Somewhat similar considerations apply to *Alectoria ochroleuca* on the crests and *Cetraria nivalis* on the slopes of other ridges. In some species physiological differences were also detected between populations from different habitats within the ridge system (Larson & Kershaw, 1975c).

Net assimilation rate under field conditions
Diurnal variation in summer
Dicranum fuscescens at Schefferville   The value and limitations of the approach outlined on page 142 for monitoring CO_2 exchange in the field by IRGA are illustrated in Fig. 5.21, which shows NAR in *Dicranum fuscescens* during climatically contrasting summer days near Schefferville. On 19 June, intermittent cloud cover led to considerable fluctuation in irradiance; NAR showed comparable variation, apparently controlled primarily by light as temperature in the assimilation chamber was relatively steady during the midday period. Irradiance was more consistent on 3 July: conditions were cool, but NAR was moderate to high, with a minor depression at midday coincident with maximum temperature and irradiance. Field water content in the moss remained high on both these days (Fig. 5.20), and thus NAR in the fully hydrated experimental plants (page 156) may have been close to that in undisturbed colonies.

July 12 was unusually warm, resulting in a severe depression in experimentally determined NAR, which was negative at midday. The depression was probably less pronounced in undisturbed colonies, where maximum temperature was eight degrees lower than in the assimilation chamber and low water content at midday is likely to have restricted respiration. Irradiance and temperature were lower on 14 August: NAR was substantial, particularly during the afternoon, but the results probably overestimate assimilation in undisturbed colonies where water content fell below 100% (Fig. 5.20).

Total daily net assimilation in this study reached 9–10 mg CO_2 g^{-1}. Maxima were recorded on cloudy days with mean moss temperatures of around 10 °C, a few degrees below the experimentally determined optimum (Fig. 5.4). Temperature was considered the major limiting factor (Hicklenton & Oechel, 1977a), and daily totals of net assimilation showed a strong negative correlation with mean moss leaf temperature in midsummer (Fig. 5.23). However, the effect of the continuously high water content on respiration rates in the experimental plants must again be considered when evaluating this result.

Polytrichum alpinum *at Barrow*   Field studies at Barrow have confirmed that positive NAR in *Polytrichum alpinum* is maintained for 24 hours per day under continuous summer illumination. Maximum irradiance occurred during June when the moss had recently emerged from snow cover and all the photosynthetic tissue had overwintered. On 15–16 June (Fig. 5.24), irradiance was reduced by cloud cover but NAR was positive throughout, and the rate at midnight was 20% of the maximum for the day despite moss leaf temperature then being below 0 °C.

Maximum NAR was higher on 19 July, when some current season growth was present: daylight was still continuous, but irradiance fell to the compensation point at midnight. Maximum NAR of 1.3 mg CO_2 g^{-1} hr^{-1} was recorded at 0800 hr, the rate falling slightly at midday in line with increasing leaf temperature and irradiance. The latter (Fig. 5.8) was considered the more significant as temperature remained below the laboratory-determined optimum, and there was no comparable reduction of NAR in other plants subjected to a 50% reduction in irradiance to simulate the shade of a vascular plant canopy. The NAR in the shaded *P. alpinum* reached 1.6 mg CO_2 g^{-1} hr^{-1} during the day but fell to negative levels at night. Other experiments showed that daily totals of net assimilation are similar in shaded and unshaded mosses at Barrow in early summer when a reduction in NAR by high irradiance at midday is compensated

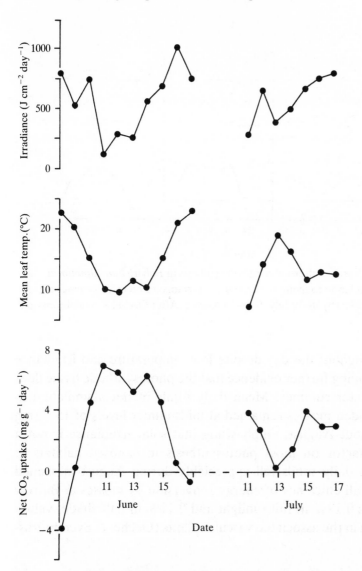

Fig. 5.23. Variation in daily net assimilation in *Dicranum fuscescens* at a woodland site near Schefferville in relation to mean daily moss leaf temperature and daily totals of solar radiation. After Hicklenton & Oechel (1977a).

for by higher rates at night in exposed than in shaded habitats (Oechel & Sveinbjörnsson, 1978).

By 8 August, near the end of the snow-free period at Barrow, low solar irradiance resulted in a seven-hour period of negative net assimilation at night in unshaded *P. alpinum*. High rates of fixation were still

188   *Physiological processes and response to stress*

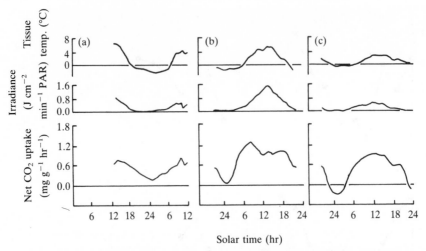

Fig. 5.24. Diurnal pattern of net photosynthesis in *Polytrichum alpinum* at
Barrow in relation to temperature and solar irradiance on three summer days:
(a) 15–16 June; (b) 18–19 July; (c) 9–10 August. After Oechel & Sveinbjörnsson
(1978).

maintained throughout the day despite low temperature and irradiance
(Fig. 5.24), providing further evidence that the photosynthetic tissue does
not senesce in later summer. Mean daily totals of net assimilation in
a range of unshaded mosses remained at midsummer levels of 15–20 mg
$CO_2 g^{-1}$ throughout August, emphasising that solar irradiance is not a
major limiting factor on moss photosynthesis in exposed habitats at
Barrow, but the daily totals fell appreciably during August in shaded
plants. The overall efficiency of energy conversion by mosses at Barrow
was estimated as 0.18% in full sunlight and 0.11% in half shade, values
much lower than in the associated vascular plants (Oechel & Sveinbjörnsson, 1978).

   Arctic lichens   Field studies on lichens have utilised the technique
in which the rate of $^{14}CO_2$ uptake is monitored, this rate approaching
that of gross photosynthesis. In the experiments of Baraskova (1971)
on *Cladonia* species in cool-Arctic tundra on the Taimyr Peninsula, daily
uptake showed a bimodal pattern comparable with that in Fig. 5.24. Maxima were recorded during the morning and evening, with the rate commonly depressed at midday as well as at night. As in the bryophytes,
it was concluded that photosynthesis is favoured by relatively cool, moist
conditions, and substantial fixation was recorded in twilight during September. Interpretation of these data is complicated by the above ambient

experimental CO_2 concentration (0.3–1.0%) and the high rates of dark CO_2 fixation that were also recorded. Non-photosynthetic CO_2 uptake has been reported in other lichens (Lechowicz, 1978; Richardson & Finegan, 1977): Lechowicz speculated that lichen acids take up CO_2 by a physical process in dry lichens and that its release upon wetting increases photosynthesis during the subsequent phase of activity.

Different patterns were revealed by studies of $^{14}CO_2$ uptake in *Cladonia* spp during summer in relatively dry tundra in northern Alaska (Moser & Nash, 1978; Moser, Nash & Link, 1983). The sites were shaded at night and rates of gross photosynthesis then fell to levels that were minimal, but detectable when the thalli were wet. Water availability proved the main limiting factor during the day, with gross photosynthesis recorded only after the thalli had been wetted by rain, fog or absorption of water vapour from air at high RH. Absorption of water vapour resulted in minimal rates of $^{14}CO_2$ uptake in *C. rangiferina* and *C. stellaris*. In *C. cucullata* it gave rates of up to $1.3 \, \text{mg} \, CO_2 \, g^{-1} \, hr^{-1}$ in apical parts of the thalli compared with maxima up to $2.0 \, \text{mg} \, CO_2 \, g^{-1} \, hr^{-1}$ following rain or fog. Gross photosynthesis showed much lower maxima in central regions of the thalli, but it was sometimes maintained after the tips had dried and become inactive, reflecting retention of moisture within the lichen mat (page 126). The $^{14}CO_2$ uptake technique gives no information on respiration, and Lechowicz (1981a) concluded that resaturation respiration results in a net loss of carbon following moistening of *C. cucullata* by dew or atmospheric humidity, following an IRGA study at another Alaskan locality (page 195).

Photosynthesis and snow cover

Capacity for net assimilation in tundra angiosperms declines in late summer coincident with the development of frost-hardiness, and little photosynthesis occurs until the spring melt (Tieszen, 1974). The mosses and lichens are continuously frost-hardy (page 198), and at least in mosses there is no evidence of late season withdrawal of nutrients from actively growing regions, with photosynthesis thought to continue as long as favourable conditions prevail (Oechel & Sveinbjörnsson, 1978). Photosynthesis could thus continue while much of the tundra is under snow, either in exposed habitats or beneath the snow.

Epilithic mosses and lichens are commonly snow-free almost throughout the year and wind reduces the depth and duration of snow cover in other exposed habitats. Continuous darkness prevents photosynthesis in midwinter at the highest latitudes, but this restriction does not apply in spring or autumn. Kershaw (1983) considered thallus colour to be particularly important in this context, as exemplified by the dark brown or black thalli of *Bryoria nitidula*. The resulting low albedo appears to accelerate clearance of the thin snow cover from exposed beach ridges, while the high temperatures attained by virtue of the dark thallus colour are thought to permit photosynthesis when melt water is available but low ambient temperatures are unfavourable for paler forms. This species is also likely to assimilate under snow, where conditions are at times favourable for both lichens and bryophytes, with their ability to maintain positive NAR under low irradiance at temperatures close to freezing, a feature shown by *B. nitidula* particularly in spring and autumn (Kershaw, 1975b; Larson & Kershaw, 1975d).

Photosynthesis under snow is most likely in spring when both solar irradiance above the snow and light penetration through it are at their maxima (Chapter 4). Water is then available from snow melt, a high RH is maintained, and subnivean CO_2 concentration may be enhanced by respiration of organisms beneath the snow (Aaltonen, Pasanen & Aaltonen, 1985). Temperatures beneath snow in spring are relatively steady: they are commonly 0–5 °C (e.g. Longton, 1974a), but vary between habitats. Temperatures up to 15 °C have been recorded in mosses and lichens under thin sheets of ice (Lange & Kappen, 1972; R. I. Lewis Smith, 1982b), but soil surface temperature at a Barrow site remained at −7 °C until five days before snow clearance (Tieszen, 1974).

Laboratory experiments suggest that some lichens could undergo positive net assimilation at substantial rates under these conditions (Carstairs & Oechel, 1978; Lange & Kappen, 1972), and Gannutz (1971) demonstrated that lichen cover at Cape Hallett is greatest on rocks that are snow-covered for much of the summer (but see also page 75). However, there is evidence that photosynthetic capability in some lichens is low in spring. Barashkova (1971), while suggesting that Arctic lichens may assimilate under spring snow, reported that maximum rates of light CO_2 fixation in plants taken from beneath snow in March and May were lower than in August collections. Cold-Antarctic lichens also showed low photosynthetic activity in early spring and Gannutz (1970) suggested that parasitism of algal cells by the mycobiont during winter might be responsible. From Devon I comes evidence that the concentration of algal cells in the thalli of some lichens is lower immediately before snow melt than during summer (Richardson & Finegan, 1977). However, Kershaw (1983) attributed a reduced photosynthetic capacity in *Cladonia stellaris* in winter to uncoupling of energy transduction in some photosynthetic units in response to short days and low temperature.

The condition of bryophytes in early spring appears to vary. Near Kevo, *Dicranum elongatum* occurs in tall turfs on rock ledges where snow is seldom deep and melts as early as mid-April, with photosynthesis thought to commence almost immediately at temperatures of 0–3 °C. Plants in spring are actively photosynthetic under laboratory conditions, with T_{opt} only 5–10 °C: NAR is 70% of maximum at 0 °C, even in August (Kallio & Heinonen, 1975). Gannutz (1970) reported cold-Antarctic mosses to be photosynthetically active in spring, but low maximum NAR was recorded in early season collections of several species from Barrow (Fig. 5.2). Nevertheless, subnivean net photosynthesis at 0.34–0.61 mg CO_2 $g^{-1} hr^{-1}$ (8–14% of the maximum summer rate) was demonstrated during spring in field experiments on *Polytrichum alpinum* at Barrow (Oechel & Sveinbjörnsson, 1978). Simulation models predict that cold-Antarctic mosses may also photosynthesise beneath fresh snow in autumn (Collins & Callaghan, 1980; Davis & Harrison, 1981).

Prediction of carbon dioxide exchange

Diurnal and seasonal patterns of CO_2 exchange and net production in polar cryptogams have been predicted by simulation models based on microclimatic data combined with laboratory-determined physiological responses (Collins & Callaghan, 1980; Kallio & Kärenlampi, 1975; Lechowicz, 1981a; Longton, 1974a). In some cases there has been reasonable agreement between net annual production as predicted and as determined by harvest techniques. This may be to some extent fortuitous as comprehensive response matrices (Fig. 5.18) have seldom been available, the mathematical techniques are still undergoing refinement (Davis, 1983) and data on field conditions, particularly degree of plant hydration, are generally incomplete.

Miller and his associates tried to overcome the latter problem by predicting moss water content from climatic data using an energy budget approach. Their comprehensive models also simulate the influence on net assimilation of such factors as nutrient availability and resaturation respiration (Miller, Oechel, Stoner & Sveinbjörnsson, 1978; Miller *et al.*, 1984). Conclusions concerning the influence of a vascular plant canopy on net annual production in mosses along a gradient of soil moisture availability at Barrow are summarised in Fig. 5.25. Simulation models will undoubtedly become increasingly important as a means of investigating plant/environment relationships as their precision improves.

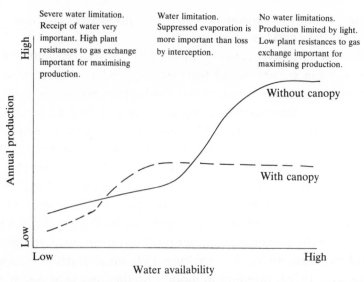

Fig. 5.25. Simulated relationships between bryophyte productivity, water availability and the presence or absence of a vascular plant canopy at Barrow, indicating factors important in maximising production. Data from Stoner *et al.* (1978).

Response to plastic stress
Desiccation

Drought resistance Apart from the difficulty of maintaining adequate rates of net assimilation and growth in a short, cool and often dry growing season, polar cryptogams are subjected to conditions that would represent severe levels of plastic stress in temperate angiosperms. Frost, and in many areas desiccation, are the most obvious factors, but continuous illumination and high tissue temperatures in summer are also potentially damaging.

Problems of desiccation exist at two levels. First, shortage of liquid water at most times of the year characterises some polar regions. Mosses are most abundant in habitats where water is available at least intermittently, but some mosses, and particularly lichens of drier habitats, are likely to be subjected to extended periods of desiccation. Second, many polar plants are likely to experience considerable diurnal fluctuation in water content in response to fluctuations in irradiance, tissue temperature and atmospheric RH in summer (Fig. 5.20). There have been surprisingly few studies of desiccation tolerance in Arctic or Antarctic cryptogams, but work in temperate regions, reviewed by Kershaw (1985), Longton (1980) and Proctor (1981), indicates the responses likely to be involved.

Many mosses and lichens can survive months or even years of water contents as low as 5–10%, with water potentials down to −150 MPa, and resume metabolism and growth on rehydration. There is some correlation between desiccation tolerance and habitat, with many hydric mosses and liverworts killed by air-drying for a few hours. There is also evidence of seasonal variation, with tolerance tending to increase during dry periods. Xeric bryophytes are characterised by more strongly negative water potentials when turgid, particularly near shoot apices, than are mesic species. This could accelerate water movement by creating steeper water potential gradients within the plant, or increase absorption of atmospheric water vapour, but osmotic potentials are thought to remain too close to zero to cause significant reduction in evaporation (Proctor, 1981). Biochemical and ultrastructural changes associated with desiccation in cryptogams have been discussed by Ascaso, Brown & Rapsch (1985), Moore, Luff & Hallam (1982) and Oliver & Bewley (1984).

Resaturation phenomena   When air-dry mosses or lichens are immersed in water, three main features characterise subsequent CO_2 exchange (Farrar & Smith, 1976; Proctor, 1981). These are: first, an immediate and substantial release of absorbed gas, mostly CO_2 of non-metabolic origin; second, a period of up to several hours with respiration at above normal rates, this resaturation respiration differing from basal respiration in being cyanide-sensitive; and finally, recovery of photosynthesis and other processes. Resaturation respiration could perhaps result from transient uncoupling of mitochondria (Brown, MacFarlane & Kershaw, 1983). Rewetting may also be followed by rapid loss of organic and inorganic solutes for a brief period (1–2 minutes), probably due to increased permeability resulting from withdrawal of water from cell membranes during the dry period. Metabolic recovery is progressively slower and less complete as desiccation time increases, even in resistant species, and eventually rewetting is followed by a period of rapid respiration culminating in death.

Resaturation respiration under laboratory conditions often results in a phase of negative net assimilation (Fig. 5.13), particularly after long periods of drying. In view of this, and solute losses on rewetting, alternating wet and dry periods might be regarded as particularly damaging, especially as resaturation respiration occurs in some lichens after drying to only 40% RWC and may thus be a frequent occurrence (Farrar & Smith, 1976). However, a hardening effect leading to increased desiccation tolerance in successive dry periods has been demonstrated in some species, and there is limited evidence that both resaturation respiration and solute leakage are less intense in xeric than in mesic and hydric species. The extent of resaturation respiration increases in some species following rapid drying and with the length of time during which the material had previously been wet (Brown & Buck, 1979; Dilks & Proctor, 1976; Smith & Molesworth, 1973). These factors complicate interpretation of the ecological consequences of fluctuating levels of hydration.

Wetting and drying cycles   Cycles of wetting and drying are less deleterious to some lichens than continuous saturation in terms of polyol (soluble carbohydrate) content and capacity for both photosynthesis and potassium uptake, a result that led Farrar (1976, 1978) to formulate his concept of physiological buffering. According to this theory, which has so far been applied only to lichens, a substantial pool of polyols is present in the thallus and is subject to depletion or turnover during periods of stress, so protecting proteins and structural material. Utilisation of photosynthate for polyol synthesis could thus limit the amount available for growth. Support for this view was provided by the observation that radioactivity remained principally in the ethanol-soluble fraction after 12 months in thalli of *Umbilicaria lyngei* exposed to sodium ^{14}C-bicarbonate and then returned to a beach ridge on Devon I: turnover was low, as total radioactivity remained at 32% of the original level and growth within the year was not measurable (Richardson & Finegan, 1977).

Farrar's experimental plants were maintained in either continuous light or continuous darkness, and Kershaw & MacFarlane (1980) pointed out that this might have been responsible for some of the damage attributed to continuous hydration. However, there is some evidence that the proportion of assimilate transferred from algal to fungal cells declines with decreasing water content, suggesting that alternating wet and dry periods might be necessary to satisfy the metabolic requirements of both bionts (MacFarlane & Kershaw, 1982).

The effect of intermittent moisture has been investigated by Lechowicz (1981a) for *Cetraria cucullata* in two inland tundra habitats in northern Alaska. Irradiance, lichen temperature and RWC were recorded over 16 days in July. These data were related to experimental response matrices for, first, dark respiration as a function of temperature and RWC following rewetting of dried plants and, second, NAR as a function of temperature, irradiance and RWC in plants that had been soaked for 24 hours so that resaturation respiration was complete. Simulation of field NAR was then undertaken. The weather was mainly dry with only a single spell of rain, and most wetting was by fog or dew.

The response differed in detail between populations (Fig. 5.26), but

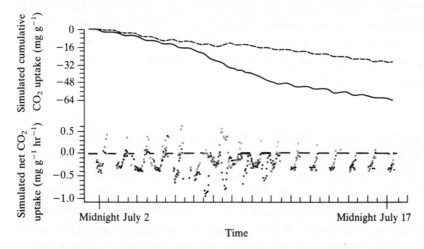

Fig. 5.26. Simulations of hourly net CO_2 exchange rates and cumulative net CO_2 balance for *Cetraria cucullata* on ridge tundra (open circles, broken line) and upland tundra (solid circles, solid line) near Atkasook, Alaska in 1977. After Lechowicz (1981a).

both simulations suggested that NAR was zero for long periods when the lichens were dry, and that negative NAR resulting from resaturation respiration was of longer duration and often of greater rate than the succeeding periods of positive net assimilation. Thus mean NAR was predicted as being negative, a result in agreement with the demonstration of negative growth rate over the 16 days by dry-weight determination. It was suggested that loss of assimilate was from the polyol reserve which would be replenished during climatically more favourable periods. Lechowicz (1981a) also noted that lichen biomass in Alaska is greater in relatively dry inland tundra than on the coastal plain where prolonged wet periods occur, supporting the view that periodic drying is beneficial to lichens despite the apparently unfavourable gas exchange relationships predicted in Fig. 5.26. How far this distribution pattern reflects problems of CO_2 diffusion as opposed to other effects of sustained high water contents remains to be determined.

Polar bryophytes and lichens are clearly able to occupy habitats where they are subject to varying degrees of periodic and long-term desiccation. Some of the detrimental effects, e.g. loss of assimilate due to resaturation respiration, may be reduced by the prevailing low temperatures. Aridity severely restricts the development of vegetation in some regions, but it is by no means clear whether this should be attributed to lethal effects of desiccation or to water availability restricting net assimilation, and thus propagule establishment and the growth of mature plants. Physiological buffering by extensive polyol synthesis could be an important factor leading to the abundance of lichens in relatively dry habitats where competition from other types of plants is low, and high growth rates are in consequence of limited benefit.

Frost

Frost resistance   As with desiccation, polar plants face the problem of both continuous and intermittent frost, but the annual regimes vary widely. Winters are periods of continuous, severe frost in the cold- and frigid-Antarctic and much of the Arctic, whereas winter air temperatures commonly fluctuate through 0 °C on cool-Antarctic islands and in parts of the mild-Antarctic. Freeze–thaw cycles in air temperature occur throughout the summer in the cold-polar regions and coastal parts of the frigid-Antarctic. Other parts of the Arctic experience similar conditions in spring and autumn, with only occasional frost at night during summer.

While some mosses and lichens are exposed to severe ambient temperatures in winter, much of the polar cryptogamic vegetation is insulated by snow cover against extreme low temperature and fluctuations through 0 °C, while high external water content may reduce the frequency and severity of freezing and thawing, especially in spring and autumn when maximum temperatures, and thus evaporation rates, are low. High internal water content increases susceptibility to frost damage (Dilks & Proctor, 1975), but the freezing point of cell sap is always depressed a few degrees below 0 °C, thus reducing the danger of the regime of short-term freeze–thaw cycles on, for example Signy I (Table 4.2).

Development of ice crystals within plant cells is usually, if not invariably, lethal due to physical damage to membranes and other organelles, but the occurrence of intracellular freezing has yet to be confirmed in nature. With slow cooling, at rates normally experienced in natural environments, ice forms first in intercellular spaces or on the surface of plant tissues, and this helps to prevent intracellular freezing because extracellular ice attracts water from surrounding cells, thereby raising the concentration of the remaining cell sap and lowering its freezing point. Formation of extracellular ice is potentially damaging to neighbouring cells as a result of mechanical stress and increased concentrations of solutes as the protoplasts lose water, but many plants can tolerate extracellular ice formation by mechanisms involving avoidance or tolerance of freeze-dehydration. Repeated freezing and thawing is potentially more dangerous than continuous frost, and rapid freezing is damaging under laboratory conditions because it increases the likelihood of intracellular ice formation. Rapid thawing may also induce injury in some plants for reasons not fully understood, as may very slow thawing which leads to recrystallisation with the formation of larger ice crystals. For many plants, rates of freezing and thawing influence survival only within a critical range of minimum temperatures, which varies from species to species (Fig. 5.27). These responses, and the molecular basis of frost injury and resistance in higher plants, are discussed in detail by Levitt (1980).

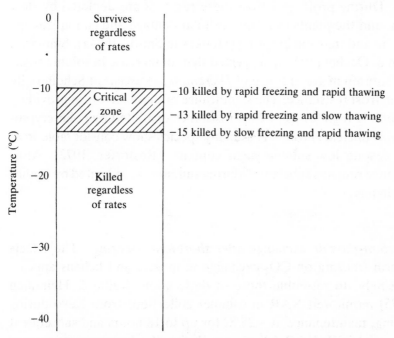

Fig. 5.27. Effects of freezing and thawing on plants of moderate hardiness. Reproduced from Levitt (1980) by courtesy of Academic Press.

Mosses and lichens might be expected to survive frost more readily than angiosperms because of such features as lower internal water content when turgid, low resistance to water loss and the considerable tolerance of desiccation in many species. Correlation between resistance to frost and to desiccation has been detected among temperate bryophytes, as among angiosperms (Clausen, 1964; Proctor, 1981). Kappen (1973) suggested that cold tolerance in lichens is related to the presence of small vacuoles and elastic cell walls, and that it is indicative of high membrane permeability allowing water to leave the cells quickly. The freezing point depression of lichen protoplasts is likely to be considerable, perhaps as low as −10 °C, even without removal of water by extracellular freezing, due to high concentrations of lichen acids and other substances (Gannutz, 1970).

Increase in soluble sugar content is thought to be a mechanism increasing frost resistance in vascular plants (Billings, 1974), and it has been suggested that changes in lichen respiration rates in winter are also associated with frost hardening (Kershaw, 1985). Lichens on Signy I develop high concentrations of glycosides and soluble carbohydrates which are thought to confer frost resistance during periods of freezing and thawing in autumn. During prolonged frost these reserves are depleted by slow respiration, and the plants can suffer cellular damage and cation leakage from freezing and thawing in spring (Hooker in Smith, 1984a). Similarly, Hicklenton & Oechel (1977b) suggested that an increase in soluble sugar content in autumn in green tissue of *Dicranum fuscescens* at Schefferville had a role in frost resistance. The significance of these factors in providing the undoubtedly high levels of frost resistance exhibited by polar cryptogams remains unclear, however, as many species show considerable frost resistance despite low soluble sugar contents (Rastorfer, 1972). Also, frost resistance remains substantial during summer, as indicated by studies of CO_2 exchange.

Carbon dioxide exchange after short-term freezing The effects of short-term freezing on CO_2 exchange in mosses and lichens appear, not surprisingly, to resemble those of desiccation. Kallio & Heinonen (1973, 1975) monitored NAR in summer collections from Kevo during slow freezing, maintenance at −25 °C for up to 18 hours and subsequent slow thawing at 0 °C. NAR fell to zero during the freezing process and its recovery on thawing was preceded by a burst of rapid respiration giving a short period of net carbon loss. Net assimilation rate at 0 °C reached its original value ten hours after thawing in *Dicranum elongatum*, but in *Pleurozium schreberi* and *Racomitrium lanuginosum* recovery was only 70% during the same period. The behaviour of lichens was more variable. Some species, e.g. *Nephroma arcticum* (Fig. 5.28), showed a brief respiration surge on thawing, whereas in *Parmelia olivacea* no period of negative NAR was evident. Recovery was slowest in *Umbilicaria vellea*, although this is a species of habitats lacking thick winter snow cover:

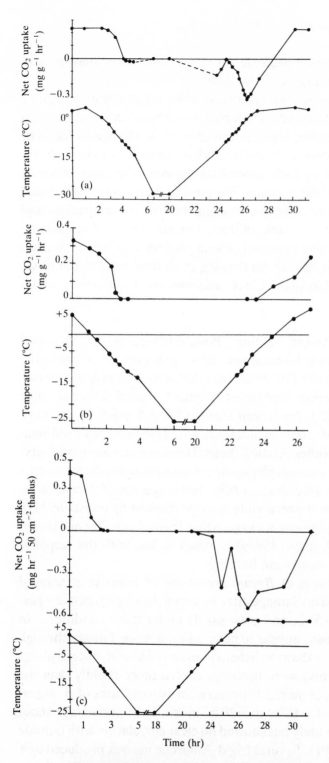

Fig. 5.28. Fluctuations in NAR in *Nephroma arcticum* (a), *Parmelia olivacea* (b) and *Umbilicaria vellea* (c) from Kevo during the course of slow freezing, maintenance at about −25 °C for several hours and slow warming to 0 °C. After Kallio & Heinonen (1971).

NAR in *U. vellea* remained negative for seven hours after thawing and total recovery was not achieved in two days.

Variation in response was also recorded when fully hydrated thalli of four frigid-Antarctic lichens were immersed for six hours in liquid nitrogen at an ecologically unrealistic temperature of $-196\,°C$ (Lange & Kappen, 1972). With gradual cooling, almost immediate recovery of gross photosynthesis was recorded in each species on thawing, the rates reaching 100–140% of those before freezing. Respiration rates were stimulated to more than three times the pre-freezing rate in some species, and remained above normal for several days, but no enhanced respiration was recorded in *Xanthoria mawsoni*. Rapid cooling to $-196\,°C$ resulted in reduced gross photosynthesis on thawing in all four species, but NAR returned to normal within 20 days in *X. mawsoni* and *Lecanora melanopthalma*.

Survival of prolonged freezing   Polar cryptogams thus seem well able to survive short-term freezing and thawing in summer. Concerning long-term freezing, Larson (1978) showed that *Alectoria ochroleuca* collected from Pen I in early September resumed normal CO_2 exchange within a few hours after dry-frozen storage for 3.5 years in the dark (20% water content and $-60\,°C$), and Lechowicz (1981a,b) reported similar findings for several other Arctic lichens. However, lichens stored dry-frozen in the dark may undergo physiological changes normally associated with winter conditions (Kershaw, 1985). Some species of *Umbilicaria*, notably *U. vellea*, showed irreversible damage marked by unusually high respiration rates and pigment leakage when thawed after being frozen when fully hydrated (Larson, 1983a), a result in line with the response of NAR in *U. vellea* to short-term freezing.

In mosses, agar-cultures of *Bryum argenteum* of temperate, tropical and polar origin showed no damage after 10 days with a light/dark temperature regime of $5\,°C/-5\,°C$, and grew slowly under these conditions. In other experiments, mosses appear to have suffered more damage through prolonged frozen storage than the lichens considered above, but the plants were generally moist, and were probably cooled more rapidly than the lichens. Substantial shoot mortality occurred in agar-cultures of *B. argenteum* kept for 11 days at $-15\,°C$ to $-20\,°C$ in the dark: percentage shoot survival varied rather widely but showed no clear correlation with latitude of origin (Longton, 1981). Several frigid-Antarctic mosses produced new shoots and secondary protonema after being frozen for 5 years (Horikawa & Ando, 1967). Mosses from Signy I also resumed growth in a cool greenhouse after storage for 3 years at −15 °C in the dark by producing new, bright green shoots as branches below the original apices which did not resume growth (Longton & Holdgate, 1967). This behaviour may reflect a relatively high susceptibility to frost in young moss tissue (Hudson & Brustkern, 1963), and parallels observations that some shoot apices of *Polytrichum alpestre* and other mosses on Signy I die during winter (Chapter 6), a fact attributed by Collins (1976b) to frost damage.

In the Signy I material, most shoots of *Chorisodontium aciphyllum* were yellow-green in colour after thawing and new shoots were produced freely throughout the experimental turf. In contrast, carpets of *Drepano-cladus uncinatus* became predominantly brown in colour on thawing, with only local areas of yellow-green shoots and branch production restricted to the latter. It is not clear whether this difference was caused by a high water content in the *Drepanocladus* carpet when collected or to inherently lower frost resistance in *D. uncinatus* than in *C. aciphyllum*, perhaps related to the thicker snow cover and higher winter temperatures normally experienced by the former.

Indeed the significance of insulation by winter snow is uncertain, given that many cryptogams survive in exposed habitats. An experimental investigation of this point, involving transplant experiments, comparison of frost resistance in species normally protected by different depths of snow, and investigation of the effects of artificial snow removal, would be of great interest. The results might not always support the conventional view that winter snow cover is beneficial, since the photosynthetic capacity of *Umbilicaria deusta* was not affected by transplantation from a snow-covered to a snow-free site in central Ontario, whereas *U. vellea* was adversely affected by the reciprocal treatment (Scott & Larson, 1985), despite its apparent susceptibility to frost.

Heat

Bryophytes and lichens are well known to be susceptible to permanent damage by even short periods of moderate heat when moist. The hepatic *Anthelia juratzkana*, a species largely confined to Arctic and alpine late snow beds, is one of the most sensitive species, 30 minutes at 39 °C causing plasmolysis and severe electrolyte leakage (Lösch *et al.*, 1983). Even in *Racomitrium lanuginosum*, a species of exposed, sunny habitats, NAR shows a permanent decline in moist plants kept at 30 °C for a few hours (Kallio & Heinonen, 1973). Similarly, a light/dark regime of 35 °C/30 °C resulted in slow growth with extensive protonema and shoot mortality in agar-cultures of *Bryum argenteum*. Rather curiously, these effects were less pronounced in polar than in temperate and tropical provenances (Longton, 1981).

Considerably higher tissue temperatures, sometimes reaching 60 °C (Kershaw, 1978), are experienced by mosses and lichens under full irradiance in the mild-Arctic, and rock surface temperatures at times exceed 30 °C even in the frigid-Antarctic. Unlike desiccation and frost, these high temperatures are maintained continuously for at most a few hours, but they may recur on a diurnal basis for several weeks during fine, summer weather, sometimes alternating with frost at night (Chapter 4).

High tissue temperature in polar cryptogams is usually associated with low water content, and mosses and lichens are generally thought to be remarkably resistant to heat when dry (Lange, 1953; Nörr, 1974). However, this conclusion is based in part on the results of single, short exposures to ecologically unrealistic temperatures, often 70–105 °C, using reduction in respiration rate as the sole criterion for damage. More recently, it has been shown that NAR in temperate and Arctic lichens may be severely depressed after exposure of dry thalli to much lower temperatures (25–45 °C) for 12 hours daily. Interspecific differences in response were recorded, with species of open habitats which normally experience the highest tissue temperatures being the least susceptible to the experimental high temperature regimes (MacFarlane & Kershaw, 1980; Tegler & Kershaw, 1981). However, Larson (1982) found that the effects of high temperature stress on lichens increase with the length of exposure, noting that the 12 hour daily periods of stress in Kershaw's experiments were unnaturally long (Figs. 4.18 and 4.19). For this and other reasons he questioned the ecological significance of the findings, and the matter remains unresolved.

Light

Maximum solar irradiance in polar regions is relatively low, but NAR in some mosses and lichens is reduced at irradiance above saturation (Fig. 5.8). There is also evidence of damage to the photosynthetic apparatus by light levels above saturation and by continuous illumination in summer. A striking example is provided by Kershaw & MacFarlane's (1980) work on *Peltigera aphthosa* from deep shade in spruce woodland in northern Ontario. Plants were maintained air-dry at 15 °C with a 12 hour daily period of irradiance at either $0.06\,\mathrm{J\,cm^{-2}\,min^{-1}}$ PAR or $0.52\,\mathrm{J}$ $\mathrm{cm^{-2}\,min^{-1}}$ PAR, the former designed to simulate woodland conditions and the latter to give a degree of light stress. The NAR under standard conditions was reduced within seven days in plants kept at the higher irradiance, and was negative after 14 days. *P. aphthosa* from a nearby open habitat was unaffected by the higher irradiance. The cause of the reduced photosynthetic capacity of the woodland plants is unclear, but could be related to inefficient light screening by a thin upper cortex. It was caused by irradiance lower than that reported to cause chlorophyll destruction and other damage to woodland angiosperms.

Continuous illumination causes a rapid and continuous decline in NAR in some temperature lichens (Kershaw & MacFarlane, 1980), and this response has been studied in detail in Arctic mosses (Aro & Valanne, 1979; Kallio & Valanne, 1975). Plants from temperate and boreal forest populations of several species, including *Pleurozium schreberi* and *Racomitrium lanuginosum*, showed depressed NAR after cultivation in continuous light, and this was related to reduced chlorophyll content and ultrastructural modification of the chloroplasts. Arctic populations of *P. schreberi* showed a similar response, but continuous illumination had no adverse effect on Arctic provenances of *R. lanuginosum*, or on the primarily Arctic species, *Dicranum elongatum*. Inability to adapt to continuous illumination could be one reason for the failure of *P. schreberi* to extend far into the Arctic, as its photosynthetically active season at Kevo (latitude 70° N) coincides with periods of alternating light and darkness in spring and autumn.

Adaptation and selection
Basic fitness of bryophytes and lichens

Polar mosses and lichens appear to be physiologically well equipped to function in their exacting environments. Particularly important features include: first, the broad response of net photosynthesis to temperature, with relatively low optima and continuation of both photosynthesis and respiration below 0 °C; second, the lower irradiance required for compensation and saturation; third, interactions between irradiance and temperature that result in positive net assimilation, albeit at low rates, under cold, low-light conditions; and fourth, the seasonal variation in response patterns shown by many species. Also crucial are the poikilohydrous character of the plants, their considerable resistance to both desiccation and frost maintained throughout the growing season, and their ability to switch rapidly between states of metabolic rest and activity as dictated by fluctuations in external environment.

Among lichens, black, brown or other dark-coloured forms predominate in the cool-Arctic and frigid-Antarctic regions, with dark rocks also favoured (Ahmadjian, 1970; Siple, 1938; Thomson, 1982). The significance of this probably lies principally in its effect on thallus temperature, and thus on CO_2 exchange, at times when water is most freely available, often under relatively cool ambient conditions coinciding with fog, dew deposition or spring snow melt. High thallus temperatures could also accelerate melting of light snow accumulation, minimising losses through ablation or wind. Metabolism is likely to be curtailed by desiccation at the high temperatures reached by such thalli under strong summer irradiance, the lower water content then conferring heat resistance.

None of these features can be regarded as specific adaptations to polar environments, since all are shown to a greater or lesser extent by bryophytes and lichens growing under apparently more favourable conditions. There is little evidence that polar species are better able than those from elsewhere to photosynthesise at low irradiance (Kallio & Kärenlampi, 1975). The degree of drought and frost tolerance shows habitat-correlated variation in both groups, but extreme tolerance is a feature of many species outside the polar regions, some lichens from central Europe and the Negev Desert surviving gradual cooling of water-saturated thalli to $-196\,°C$ (Kappen, 1973).

Using data from 42 lichen species, Lechowicz (1982a) showed that there is no significant relationship between latitude and maximum NAR on a dry weight basis, saturating irradiance, or optimum thallus RWC (Fig. 5.29). The data did demonstrate a significant tendency for T_{opt} to decrease at high latitudes, a conclusion reached also by Gannutz (1969). In some polar bryophytes (Table 5.2) T_{opt} is also below the values of 15–20°C typical of temperate species (Proctor, 1981). It is not clear whether latitudinal variation in T_{opt} represents genetic differences or acclimation, and some species from relatively warm climates are known to be capable of photosynthesis at low temperature. Positive net assimilation continues below $-5\,°C$ in several central European mosses, at least in winter (Atanasiu, 1971), and gross photosynthesis has been recorded below 0°C in some tropical lichens (Kappen, 1973).

The low T_{opt} for net assimilation in frigid-Antarctic lichens has been attributed to rapid respiration at high temperatures (page 154), but it will be noted that this response, while lowering T_{opt}, does not increase NAR under cool conditions. The relationship between respiration and temperature was found to be similar in a range of Arctic and tropical species, except in the Peltigeraceae and Stictaceae where Arctic species

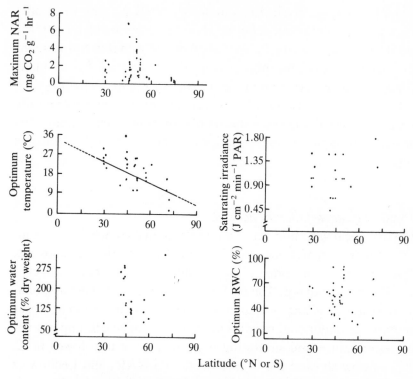

Fig. 5.29. Relationship between latitude of origin, maximum NAR and environmental conditions favouring maximum NAR in a range of lichen species. Line fitted by least squares regression for the only significant relationship. After Lechowicz (1982a).

again showed higher rates at all temperatures than related tropical taxa (Scholander, Flagg, Walters & Irving, 1952). High respiration rates are a feature of Arctic flowering plants; their importance may lie in permitting adequate respiration to support tissue maintenance and growth in short cool summers, but the response is often one of acclimation rather than genetic adaptation (Billings, 1974). The significance of the high respiration rates under warm conditions in frigid-Antarctic lichens could conceivably be that the Antarctic species expend less energy in controlling respiration than do species that normally experience high tissue temperatures when hydrated.

There is little evidence of latitudinal variation in maximum NAR on a dry weight basis in either lichens (Fig. 5.29) or mosses (Kallio & Kären-lampi, 1975). This conclusion is supported by Ino's (1985) comparison of three frigid-Antarctic lichens and seven species from subalpine habitats in Japan. However, both chlorophyll content and NAR per unit weight of chlorophyll varied widely in Ino's material, NAR ranging from only 0.2–0.8 mg CO_2 mg^{-1} chlorophyll per hour in the Antarctic material compared with 0.9–6.0 mg CO_2 mg^{-1} chlorophyll per hour in that from Japan. Moreover, comparison of morphologically similar species from similar habitats in different climatic regions, or of species occupying contrasting habitats at the same locality, may be expected to reveal environmentally significant physiological variation that is obscured in more general surveys. Numerous examples of such variation are known among both mosses and lichens.

Physiological variation in lichens

Physiological variation was demonstrated in a comparison of *Cladonia stellaris* from Schefferville and *C. evansii* from Florida (Lechowicz, 1978). The species are morphologically similar and both were collected in open coniferous woodland. The optimum temperature for net assimilation was lower in *C. stellaris*, which maintained net assimilation at closer to maximum rates under cold conditions than did *C. evansii*; upper compensation temperature and saturating irradiance were also lower in *C. stellaris*. *C. evansii* showed the higher maximum NAR, and Lechowicz thought that this might reflect a restriction of metabolically active periods by the rainfall and evaporation regimes in Florida, and the need to compensate for resaturation respiration.

Intraspecific variation on a geographical basis was recorded in *Alectoria ochroleuca* from a mild-Arctic beach ridge on Pen I, a treeless palsa mound in forest 65 km inland from Pen I, and an arid, cool-Arctic heath on Melville I. The populations differed in morphology and thus in resistance to evaporation, with the highest resistance and lowest evaporation rate in the Melville I plants (Fig. 4.13). The low NAR under all conditions tested in the Melville I plants was also striking (Fig. 5.30). These maintained positive NAR over the range 1–14 °C, but 21 °C was above the compensation temperature except at low water content. The Pen I material predictably operated more effectively at higher temperatures, with a T_{opt} of 14 °C and a substantial, positive NAR at 21 °C. Irradiance of 0.22 J cm^{-2} min^{-1} was shown to be saturating for the cool-Arctic plants, and surprisingly for those from the palsa mound, but not for the Pen I material.

An example of intraspecific physiological variation between plants from contrasting habitats in one locality is provided by the greater tolerance to strong illumination shown by plants of *Peltigera aphthosa* from open

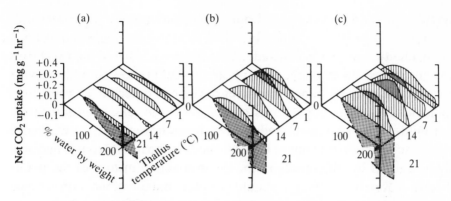

Fig. 5.30. Relationships between NAR, thallus temperature and water content at irradiance of $0.2\,J\,cm^{-2}\,min^{-1}$ PAR in thalli of *Alectoria ochroleuca* from Melville I (a), a palsa mound 65 km within the tree line south of Pen I (b) and a beach ridge crest at Pen I (c). Dark respiration rates at 21 °C are also shown. Data from Larson & Kershaw (1975b).

as opposed to shaded habitats (page 203). Similarly, *Neuropogon sulphureus* exists in southern Victoria Land as distinct sun and shade forms, the first occurring on rock surface and the second under reduced irradiance and temperature beneath pebbles (Kappen, 1983). The shade form was found to have a chlorophyll content more than double that in the sun form, but showed a lower NAR per unit weight of chlorophyll under a range of experimental conditions. The differences in NAR were smaller on a dry weight basis, and indeed the shade form had a slightly higher NAR than the sun form under cool conditions, showing overall fitness under conditions of low temperature and irradiance. In *Cladonia stellaris* from northwestern Ontario, the relationships of NAR to temperature, irradiance and thallus water content were similar in morphologically distinct sun and shade forms, but the latter had a higher NAR on a dry weight basis due to a greater chlorophyll content, the rates expressed in terms of chlorophyll content being similar (Kershaw *et al.*, 1983).

Prolonged growth of lichens in culture has so far presented intractable difficulties and therefore it has not been feasible to test the genetic basis of intraspecific variation by the method of collateral cultivation traditionally applied to other groups. In an alternative approach, Kershaw *et al.* (1983) investigated isozyme and general protein distribution in the sun and shade forms of *C. stellaris*. They found qualitative differences between the upper and lower, purely fungal parts of the podetia, and differences between the strength of various bands in the zymograms of the two forms. However, apart from a single difference in esterase band-ing, the results from comparable thallus segments suggested genetic homogeneity, and it was concluded that the occurrence of the two forms was an expression of phenotypic plasticity. Differences in both enzyme patterns and CO_2 exchange responses have been detected between mor-phologically similar populations of *Peltigera rufescens* from temperate and mild-Arctic sites in Canada. Even here, however, the occurrence of genetic variation remains to be confirmed, as the results could conceiva-bly be caused by differences in the environmental history of the plants (Brown & Kershaw, 1985). It should be noted that use of the term ecotype to describe variants between which no genetic differences have been detected, as is common in recent literature on lichens, is misleading because the term has otherwise been consistently applied only to geneti-cally differentiated variants since its original definition by Turesson.

Contrasting relationships of CO_2 exchange to environment have also been recorded between lichen species occupying different habitats. In these cases some, at least, of the variation is likely to be genetically determined and partially responsible for the different habitat preferences of the taxa concerned, as in the examples considered on pages 184–5. Adaptations in the photosynthetic responses of lichens are discussed in detail by Kershaw (1985).

Physiological variation in mosses

The genetic basis of intraspecific physiological variation has been investigated in several mosses by collateral cultivation or reciprocal field transplants. Particularly detailed studies have been carried out on *Racomi-trium lanuginosum*, a species widespread in oceanic north-temperate and Arctic regions and occurring more locally in the Antarctic. Striking simi-larities were recorded in T_{opt} (Table 5.2) and in response to freezing (page 198) in freshly collected material from a range of temperate and polar localities, and it was originally concluded that little genecological differentiation had occurred (Kallio & Heinonen, 1973). However, lower maximum NAR was recorded in this species, and also in *Hylocomium splendens* and *Pleurozium schreberi*, in plants from Kevo than from south-ern Finland. These differences were still evident, although to a reduced degree, after reciprocal transplants lasting several years (Kallio & Saar-nio, 1986).

Maximum NAR in these mosses was generally lowest in provenances originating in the coldest natural environments, and a similar trend was reported by Sveinbjörnsson & Oechel (1983) in *Polytrichum commune*.

Carbon dioxide exchange was investigated in material from five populations along a transect extending from Barrow south to Florida, after pretreatment in two temperature regimes. Both environmentally and genetically induced variation was detected (Fig. 5.31). Each provenance

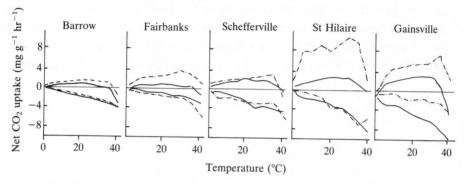

Fig. 5.31. Relationhips between temperature, NAR (upper lines) and respiration rate (lower lines) in *Polytrichum commune* from a cool-Arctic site at Barrow (Alaska), boreal forest sites at Fairbanks (Alaska) and Schefferville (Quebec), and temperate sites at St Hilaire (Quebec) and Gainsville (Florida) after acclimation at 5 °C (solid lines) and 20 °C (broken line). After Sveinbjörnsson & Oechel (1983).

showed a higher maximum NAR after cultivation at 20 °C than at 5 °C. This difference was greatest in the southern provenances, which experience wider seasonal temperature variation in their field environments than do the Arctic plants. Maximum NAR in the northern provenances was lower than in the southern strains after either treatment, NAR in the latter being among the highest so far recorded in bryophytes. Similarly, cool growing conditions led to a decrease in T_{opt} in most strains, while T_{opt} was lower in Arctic than in temperate strains after either treatment. Unlike results for flowering plants, the lowest respiration rates were recorded in the northern provenances. The lower growth temperature resulted in a substantial decrease in respiration rate only in the material from Florida. The significance of the lower maximum NAR in Arctic populations of these mosses is not clear, and contrasts with the position in *Dicranum fuscescens* at Schefferville, where plants from an upland tundra population showed a higher maximum NAR than those from a lowland forest after collateral cultivation under warm conditions (Hicklenton & Oechel, 1976).

The limited information so far available thus suggests that many of the features that permit bryophytes and lichens to maintain favourable carbon and energy budgets in severe Arctic and Antarctic environments are characteristic of these groups as a whole, rather than specific adaptations to polar conditions. Nevertheless, detailed investigation of individual species or groups of related species growing under different environmental regimes indicates that both acclimation and, at least in mosses, genetic differentiation are important in fitting populations to their particular habitats. This topic is explored further in relation to growth as a function of plant size in Chapter 6, and to reproductive biology and other factors likely to control evolutionary processes in Chapter 8.

6

Vegetative growth

Patterns of growth in relation to assimilation and translocation
Bryophytes

Growth and net assimilation Positive net photosynthesis tends to increase dry weight due to accumulation of assimilate. In this chapter we are concerned with the translation of this process into growth, in the sense of increase in plant size. Adaptations that permit positive net assimilation at substantial rates under severe environmental conditions are commonly viewed as the key to plant success in polar regions (e.g. Mooney, 1976), and considerable effort has been directed towards investigating environmental relationships of CO_2 exchange in mosses and lichens. As discussed in Chapter 5, the results confirm that polar species *in situ* are able to photosynthesise at reasonable rates, but the assumption that assimilation would be maximised by a close correspondence between optimum conditions for net photosynthesis and the most frequently prevailing environmental conditions has not been fully substantiated (page 160; Lechowicz, 1981b). However, the parallel assumption, that survival is favoured by maximum rates of photosynthesis and growth, is not necessarily valid for plants with essentially opportunistic growth responses in environments where competition is not everywhere intense, and low stature may be advantageous.

Moreover, conditions promoting maximum NAR are often very different from those favouring growth in size. This is particularly true in polar species in which maximum NAR may occur at temperatures low enough to cause a severe depression of respiration, and probably other processes essential to growth. The latter may then be restricted more by direct limitation than by availability of assimilate, and adaptation further increasing net assimilation would be superfluous.

As an example, the optimum temperature for net photosysthesis in *Dicranum elongatum* at Kevo is 5–10 °C throughout the year, with the

rate at 0 °C depressed by only 30% compared with the maximum. However, no growth was recorded in experiments at 1 °C, while shoot elongation during four weeks was only 1.5 mm at 6 °C compared with 6.5 mm
at 23 °C. The latter temperature was above compensation for net assimilation, and growth in length was accompanied by a reduction in weight,
particularly in brown, subapical parts of the plants, and could not have
been sustained indefinitely. The restriction on growth imposed by the
local environment is evident from the fact that extension of *D. elongatum*
in the field at Kevo is normally only 2–3 mm per year (Karunen, 1981;
Karunen & Kallio, 1976; Karunen & Liljenberg, 1981). A rather different
response is seen in *Bryum argenteum* from McMurdo. Relatively warm
conditions are optimal for both NAR and growth, but growth is more
severely depressed than net assimilation by low temperatures comparable
with those in the field (Figs. 5.7 and 6.1).

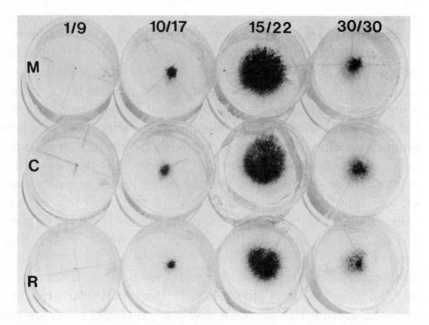

Fig. 6.1. Five-week old cultures of *Bryum argenteum* from clones originating
at a frigid-Antarctic site on Ross I, McMurdo Sound (M), a mild-Arctic site
near Churchill, Manitoba (C), and a site near the southern edge of the boreal
forest zone at Richer, Manitoba (R), grown at the day/night temperature
regimes (°C) indicated. The cultures were initiated by placing single shoot apices
on nutrient agar in the centre of the 5 cm diameter petri dishes. After Longton
& MacIver (1977).

Field environments must thus allow for a balance between net assimilation and growth. In this context it may be noted that the short daily period of relatively high temperature commonly experienced by polar cryptogams in summer (Chapter 4), while often leading to a midday depression in NAR (Chapter 5), may nevertheless be beneficial in providing brief periods favourable for respiration and growth. This point awaits clarification because studies on the environmental regulation of growth have lagged far behind those on control of CO_2 exchange.

Patterns of growth   Growth in bryophytes results from the division of a single apical cell in each shoot or branch axis. The apical cell functions as an initial. It divides infrequently, each derivative giving rise, following further division and differentiation, to one leaf and an adjacent portion of stem. Extension growth thus occurs principally at the shoot apex. Branches arise where cells behind the apex begin to function as apical initials. Tundra mosses are typically erect-growing, either sparingly branched or with erect, indeterminate branches, and form cushions, turfs or carpets (Chapter 2). Even in *Drepanocladus* spp and other pleurocarps branching is less frequent than in temperate populations. More frequent, determinate lateral branches are seen in some mat-forming species such as *Racomitrium lanuginosum*, and pinnate or bipinnate branching patterns are characteristic of *Pleurozium schreberi*, *Hylocomium splendens* and other weft-forming species in mild-Arctic woodland.

Most polar mosses are perennial, but the duration of active growth in individual shoots varies widely. In *Sphagnum* spp shoots apparently continue growing indefinitely. In contrast, those of *Pohlia wahlenbergii* by streams in South Georgia grow during only one summer. They arise early in the growing season from the apical region of decumbent shoots formed during the previous year: they are erect-growing during the summer, but themselves become decumbent as a result of winter snow cover and the force of meltwater during the following spring (Clarke, Greene & Greene, 1971). Most mosses lie between these extremes. In some, including carpet- and hummock-forming species of *Brachythecium*, *Calliergon* and *Drepanocladus*, many shoot apices are killed in winter, being replaced, in the absence of apical inhibition, by branches arising in subapical positions, but other apices survive so that the shoots may grow for two or more years (Fig. 6.2).

In *Polytrichum* spp seasonal variation in the length of newly formed leaves gives the shoots a segmented appearance so that annual increments can readily be detected, and it can be seen that each shoot continues

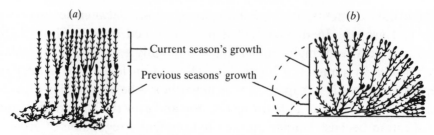

Fig. 6.2. Patterns of growth in carpet-forming species of *Calliergidium* and *Drepanocladus* (a) and hummock-forming species of *Brachythecium* (b) in the cold-Antarctic, showing horizons of dead shoot apices resulting from winter kill. After Callaghan & Collins (1981).

to grow for several years. In *P. alpestre*, replacement occurs through the formation, some distance behind the apex, of erect-growing branches bearing scale leaves. Their growth rate exceeds that of the older shoots until their apices reach the surface of the turf, after which foliage leaves develop on further new growth and dense tomentum on the lower region (Fig. 6.3). In some other species of *Polytrichum* young shoots arise at the apex of horizontal rhizomes extending several centimetres from the parent shoots.

Bipinnate branching is a conspicuous feature of *Hylocomium splendens*, but it is superimposed on a sympodial pattern in the principal axes, which gives rise to clearly defined segments. Each arises early in the growing season as a bud on the upper surface of the axis of the previous segment, and its development continues throughout that growing season and the next (Callaghan, Collins & Callaghan, 1978). Sometimes two new segments develop from a single parent, giving rise to a third order of branching (Fig. 6.4).

The extent of green, photosynthetically active tissue also varies widely. It extends only 1–2 mm into short turfs of *Bryum* spp and other mosses in the frigid-Antarctic. Growth here is very slow and it is not clear whether the green region comprises more than one year's growth (Longton, 1974a). In *Pohlia wahlenbergii* the shoots are green only during their single year of active growth, but in many other mosses chlorophyll is retained in the uppermost two or three annual growth increments. Studies of gross photosynthesis as indicated by $^{14}CO_2$ uptake have shown that photosynthetic capacity in second and third year segments of *Polytrichum alpinum* at Barrow averaged 75% and 42% respectively of that in the current year's growth, whereas at Abisco the assimilation rate in second year segments of both *P. commune* and *Hylocomium splendens* was below

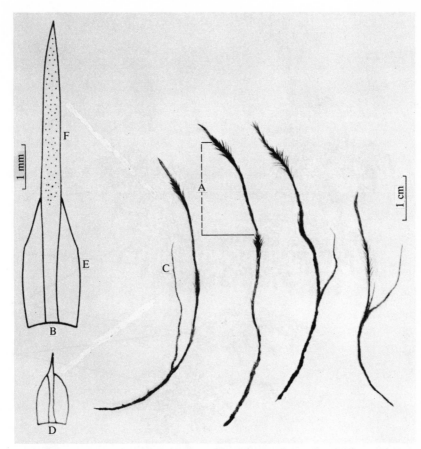

Fig. 6.3. Typical shoots of *Polytrichum alpestre* from a boreal forest site at Pinawa, Manitoba, showing annual growth increments (A) resulting from variation in the length of foliage leaves (B) produced at different times of the year. Also shown are young lateral shoots (C), which bear tomentum and scale leaves (D) until their apices reach the surface of the turf. Foliage leaves comprise a sheathing base (E) and a spreading limb (F). Stippling indicates the distribution of chlorophyll. After Longton (1979b).

60% of that in first year growth (Table 6.1), even though branch development in *H. splendens* segments is not completed until their second growing season. Low rates of photosynthesis were detected in segments of *H. splendens* up to seven years old, whereas few shoots of the Polytricha were more than four years old.

Movement and utilisation of assimilate As a result of apical growth, colonies of perennial bryophytes comprise an upper green, photo-

Fig. 6.4. Pattern of growth in *Hylocomium splendens* at Abisco, Sweden. The figures 1–4 indicate regions of the shoot in their first, second, third and fourth growing seasons respectively. After Callaghan, Collins & Callaghan (1978).

synthetically active region overlying a considerable amount of brown stem and leaf material that may gradually become incorporated into the substratum or, in mires, play a major role in peat formation (Chapter 7). It is clear, however, that living tissue extends well below the green/brown interface. On Ross I, *Bryum antarcticum* and *B. argenteum* form turfs up to 4 cm deep, and although only the uppermost 1–2 mm is green, secondary protonema develops freely from stems leaves and rhizoids to the base of the profiles when portions of moist turf are placed on their side in a warm, light environment (Longton, 1974a). Similarly, new shoots

Table 6.1 *Relative rates of gross photosynthesis per unit dry weight in annual increments of different age in three moss species*

Species	*Hylocomium splendens*	*Polytrichum commune*	*Polytrichum alpinum*
Locality	Abisco, Sweden	Abisco, Sweden	Barrow, Alaska
Habitat	Birch woodland	Birch woodland	Open tundra
Sampling date	July	July	August
Year 1	100.0	100.0	100.0
Year 2	54.9	57.9	74.6
Year 3	9.3	—	42.3
Year 4	1.7	—	—
Year 5	2.0	—	—
Year 6	0.2	—	—
Year 7	0.3	—	—

Year 1 increments are those initiated during the current growing season.
Data from Callaghan, Collins & Callaghan (1978) and Collins & Oechel (1974).

arise from old stems up to 25 cm below the surface when turfs of *Polytrichum alpestre* are treated in a similar way (Collins, 1977; Longton, 1972b).

The extent of physiological interaction between brown and actively photosynthetic biomass is of considerable interest, but a topic about which little is known. Upward movement of water and mineral nutrients is presumed to occur internally through the brown region in partially endohydric species. Respiration and initiation of branch shoots may also occur actively in the brown region. The lowermost, non-photosynthetic parts of the young shoots (Fig. 6.3) were shown to account for 15–20% of annual growth on a dry weight basis in *P. alpestre* at Churchill (Longton, 1979b). At Barrow, below ground biomass in *P. alpinum* and *P. commune* constitutes 30% of total biomass, and its respiration can exceed 30% of the daily photosynthetic gain by the green component (Sveinbjörnsson & Oechel, 1981). These observations raise questions concerning the translocation of assimilate between the green and brown regions for storage, tissue maintenance, and to support the development of new shoots.

Anatomical studies suggest that the potential for translocation is greatest in the Polytrichaceae. The central strand in the stem of these plants commonly includes, in addition to water-conducting hydroids, leptoids which show many structural similarities to the sieve elements of higher plants (Hébant, 1977). The leaf midrib of *Polytrichum commune* contains cells with wall ingrowths and other features of transfer cells (Scheirer,

1983). Upward movement of photosynthate at velocities up to 32 cm hr^{-1} has been demonstrated in the leptoids of temperate *P. commune* (Eschrich & Steiner, 1967), and translocation has been demonstrated in several Arctic mosses in studies involving exposure of shoot apices to $^{14}CO_2$ followed by autoradiography and liquid scintillation counting in different parts of the plants. Limited translocation was recorded in *P. commune* at Abisco, both downwards from the shoot apices, and upwards into developing sporophytes (Callaghan, Collins & Callaghan, 1978). In *P. alpinum* at Barrow, preferential translocation occurred via rhizomes into new underground shoots, in some cases with accumulation of labelled carbon in the young shoot apices (Collins & Oechel, 1974).

Translocation is not restricted to the Polytrichaceae. Redistribution of labelled photosynthate into brown portions of the shoots has been reported in *Dicranum fuscescens* at Schefferville (Hicklenton & Oechel, 1977b) and comparable movement apparently occurs in *D. elongatum* (page 219). Callaghan, Collins & Callaghan (1978) were unable to detect significant movement of assimilate between annual segments of *Hylocomium splendens*, and they concluded that each one acts as an independent physiological unit. In contrast, Skre, Oechel & Miller (1983b) showed that translocation from green to brown tissue does occur in *H. splendens*, and also in *Pleurozium schreberi*, at several periods during the growing season in Alaska. However, translocation was less extensive and consistent in these species than in *Polytrichum commune*, and also in *Sphagnum subsecundum* where, rather surprisingly in view of the lack of anatomically specialised conducting tissue, labelled assimilate was detected 15 cm below the green/brown interface. This study also demonstrated an accumulation of assimilate at the apex of mature shoots, particularly in *P. commune*, where it is presumably used to support current shoot growth. Rates of $^{14}CO_2$ fixation and loss of labelled assimilate during a 35 day period of measurement were also greater in *P. commune* than in the other species investigated, a result attributed to the high respiration rates necessary to maintain the relatively complex stem structure in this species.

Some of the assimilate translocated downwards into the brown region of bryophyte colonies has a storage function, possibly assisting in survival during years when summer conditions are unfavourable for net assimilation. As well as carbohydrates, lipids commonly act as storage products in polar plants, and their significance has been widely discussed (Bliss, 1962; Tieszen, 1978c). Mosses, especially Arctic species, are unusual in that their lipid component includes high levels of unsaturated triglycerides, steryl esters and wax esters (Gellerman, Anderson, Richardson

& Schlenk, 1975). These compounds may be of adaptive value in harsh environments as the high degree of unsaturation results in the triglycerides being fluid, thus facilitating their metabolic utilisation at low temperature and possibly providing protection for membranes (Karunen & Mikola, 1980; Prins, 1982). Karunen (1981) also viewed unsaturated lipids as an effective form of energy storage under conditions favouring high rates of energy fixation but low levels of utilisation in respiration and growth.

Such conditions are experienced by *Dicranum elongatum* during spring at Kevo, when irradiance is high and water is freely available following early melting of thin winter snow, but temperatures are too low to permit rapid growth (page 212). Photosynthesis is then active, with triglycerides and other lipids the principal products. Lipid content of the plants declines during warmer conditions later in summer as energy is liberated for respiration and growth. There is evidence that lipids produced following photosynthesis in the uppermost, green region of the *D. elongatum* turf during spring move downwards and are stored in the still-frozen, brown tissues below. It has also been suggested that they act as an energy source during regeneration processes similar to those described on page 214 (Karunen & Mikola, 1980). Seasonal variation in lipid content has been recorded in *Sphagnum fuscum*, but here storage is principally in the capitulum (Karunen & Salin, 1982). In contrast, lipid content in *Dicranum fuscescens* at Schefferville showed only minor seasonal variation with a maximum in July, perhaps because deep snow restricts photosynthesis during cold weather in early spring (Hicklenton & Oechel, 1977b). Triglycerides are also known in some lichens, including Antarctic species of *Umbilicaria* (Huneck, Sainsbury, Rickard & Smith, 1984).

Lichens

Fruticose species Growth of a lichen thallus depends on the elongation of numerous component fungal hyphae. Cell division occurs principally at the thallus margin or apex, behind which, as in bryophytes, occurs a region of cell enlargement and differentiation (Hawksworth & Hill, 1984). Studies on growth in fruticose lichens were pioneered in the Soviet Union, as reviewed by Andreev (1954), being stimulated by the importance of these plants as reindeer forage (Chapter 7).

Detailed work centred on species of *Cladonia*, in which growth is initiated as small squamules which quickly give way to colonies of erect, hollow, freely branched podetia (Fig. 4.15). Andreev recognised three phases in the life of a colony. During the first, the accumulation phase, podetial length increases steadily for several years. In some species, a single whorl of branches is produced annually in an apical position, one branch becoming erect and continuing the upward growth of the podetium. The period of accumulation may last up to 25 years, but 10–15 years is more typical. It is followed by a renewal phase, lasting several decades, during which the podetia decompose from the base at approximately the rate at which the upper parts are extending. Finally during the degeneration phase, which is of similar duration to that of accumulation, the rate of decomposition exceeds the rate of growth and eventually the podetia die.

In a detailed study of *Cladonia stellaris* (Kärenlampi, 1970a, b), growth was found to be concentrated in the uppermost 6 mm of the podetia, but a given internode continued to increase in both length and breadth by intercalary growth for up to 8 years (Fig. 6.5). Absolute annual growth

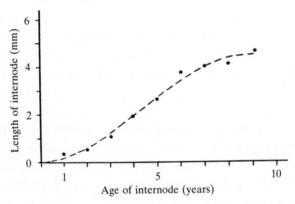

Fig. 6.5. Relationship between length and age in internodes of *Cladonia stellaris* from Kevo, Finland. Each point is the mean of 4–26 measurements. Data from Kärenlampi (1970b).

rates in length and dry weight increased logarithmically during the first 12 years in the life of a podetium in line with a rise in the amount of actively growing tissue, but relative growth rate (RGR) declined with age (Fig. 6.6). A similar pattern has been reported in Antarctic fruticose lichens (Hooker, 1980c, d).

The reduction in RGR with age presumably results from an accumulation of non-productive biomass in older parts of the plant. This interpretation is supported by the fact that chlorophyll, and thus photosynthetic

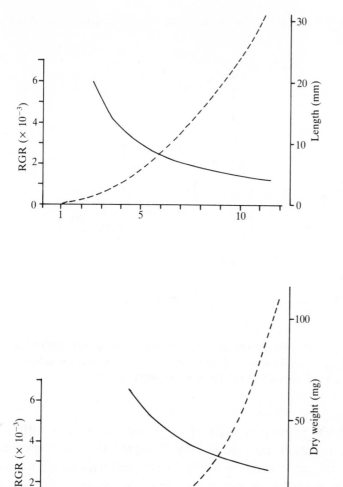

Fig. 6.6. Relationships between podetial age, length, and RGR for length (above), and between podetial age, dry weight and RGR for dry weight (below) in *Cladonia stellaris* from Kevo, Finland. RGR is indicated by a solid line in each case. After Kärenlampi (1970b).

activity, are concentrated in the youngest regions of the podetia (Kärenlampi, 1970b; Lechowicz, 1983; Nash, Moser & Link, 1980). At Schefferville, Lechowicz found that branch whorls of *C. stellaris* continued to increase in weight for up to nine years, a figure in broad agreement with Kärenlampi's data for internode length. Whorls six years old and younger accounted for only 18% of podetial biomass but 50% of the photosynthetic activity (Fig. 6.7).

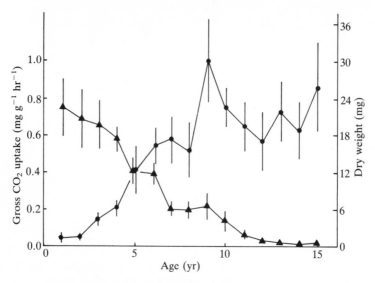

Fig. 6.7. Dry weight (●) and rate of gross photosynthesis (▲) in branch whorls of different age in *Cladonia stellaris* from Schefferville, Quebec. The values are means ± 1 s.e. ($n = 10$). After Lechowicz (1983).

The decline in photosynthetic capacity with age corresponds with decreasing light penetration into the colony (Chapter 4), a factor which is likely to affect the concentration of algal cells, and with an associated colour change from yellowish-green to white. Respiration rates also decline in older parts of the podetia, but to a lesser degree than photosynthesis. In *C. stellaris* from Anaktuvuk Pass, Alaska, almost 100% of gross photosynthesis and 80% of respiration were shown to occur in the uppermost 4 cm of the podetia, corresponding figures for *C. rangiferina* being 100% and 67% (Nash, Moser & Link, 1980).

Continuing respiratory demand by older, non-photosynthetic parts could conceivably contribute to the eventual degeneration of the podetia noted by Andreev (1954). Information on the extent to which this demand is met by assimilate translocated from the podetial apices is necessary to establish this point, and the possibility that older, non-photosynthetic parts of the thallus have a storage function has yet to be investigated. *C. stellaris* is particularly abundant during succession following fire in northern woodlands (Chapter 3), and it is possible that eventual colony degeneration results from a reduction in irradiance by the development of a vascular plant canopy or other successional processes, rather than intrinsic factors. Another possible explanation, suggested by Andreev, is that the phytobiont becomes weakened by repeated asexual reproduction within the podetia.

Branching patterns vary widely between different fruticose lichens (Fig. 4.15), and there may also be considerable intraspecific variation as in Finnish populations of *Cladonia uncialis* (Kärenlampi & Pelkonen, 1971). Branching is predominantly dichotomous in Lapland and coastal areas in the southwest, but more complex patterns characterise specimens from the southern interior (Fig. 6.8). The Lapland populations have short, relatively thick internodes, and a lower proportion of algal cells in cross-sectional area than the southern plants. Although clearly related to differences in environment, the variation is clinal and it is unclear how far it was genetically determined.

Crustose and foliose species   The most obvious manifestation of growth in many crustose and foliose lichens is centrifugal expansion leading to the development of roughly circular thalli. Circular form may be advantageous in habitats where competition is significant since it maximises increase in surface area for marginal expansion of a given distance (Hill, 1984). Expansion is thought to be essentially marginal, with lateral intercalary growth restricted to parts of the thallus up to two years old, although older areas may increase in thickness and develop reproductive structures (Hale, 1973). Theoretical as well as empirical approaches have been adopted in studying this relatively simple system, growth being expressed as increase in thallus radius or area per unit time, or as relative growth rate according to the following formula (Armstrong, 1975):

$$\text{RGR} = \frac{\log A_2 - \log A_1}{t_2 - t_1} \tag{6.1}$$

where   RGR = relative growth rate
A_1 = thallus area at time t_1
A_2 = thallus area at time t_2

As with fruticose species, growth rates in crustose lichens vary during the life of a thallus. In one model, young thalli pass through a prelinear phase, marked by an increase in radial growth rate with time (1 in Fig. 6.9) and a more or less logarithmic increase in thallus radius (6). This is followed by a linear phase during which radial growth rate is constant (2) and the radius increases at a steady rate (7). However, the growth

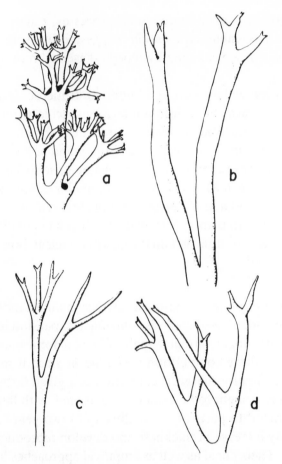

Fig. 6.8. Finnish specimens of *Cladonia uncialis* from the southern interior (a), the southwestern archipelago (b, c), and the Arctic coast (d). Reproduced from Kärenlampi & Pelkonen (1971) by courtesy of the Editor, *Reports from the Kevo Subarctic Research Station.*

patterns differ between species. The linear phase may theoretically continue until the thalli begin to fragment at the centre (Armstrong, 1975), but many species show a post-linear phase during which the radial growth rate of older, senescent thalli declines (Fig. 6.10). Some Arctic and alpine crustose lichens eventually enter a stationary phase when older thalli show no appreciable change in size over many years (Beschel, 1961). Growth rates in polar crustose lichens are often slow: some thalli of *Rhizocarpon geographicum* are thought to be as much as 9000 years old and yet have a diameter of only 28 cm (Andrews & Barnett, 1979).

Proctor (1977) found that the prelinear phase in *Buellia canescens* was

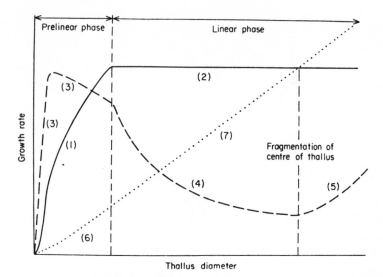

Fig. 6.9. Changes in growth rates during the life of a crustose lichen thallus showing: (1) increasing radial growth rate in prelinear phase; (2) constant radial growth rate in linear phase; (3) changes in RGR in prelinear phase; (4) exponential decline in RGR in linear phase; (5) increasing RGR after fragmentation of thallus; (6) logarithmic increase in radius in prelinear phase; (7) constant increase in radius in linear phase. Reproduced from Armstrong (1975) by courtesy of Academic Press.

marked by increasing radial growth rate as in Fig. 6.9, the growth curve approximating to a simple model in which relative growth rate is proportional to the area of a band of constant width in the expanding margin. Aplin & Hill (1979) presented a rather more complex model for growth in a thallus of radius r, where:

$$\frac{\mathrm{d}r}{\mathrm{d}t} = \frac{\alpha r}{\beta(r + 2\alpha)} \tag{6.2}$$

The value of the constants α and β can be determined by measuring growth rates in thalli of known radius. β is regarded as referring broadly to the rate of algal photosynthesis and α to the extent to which organic nutrients produced in inner parts of the thallus contribute to marginal growth.

The model was shown to fit growth curves for *Buellia canescens*, but not those for *Rhizocarpon geographicum* in later stages of growth. Beschel (1961) indicated that the radial growth rate in young thalli of *R. geographicum* at first increases slowly but then rises to a maximum during a 'great

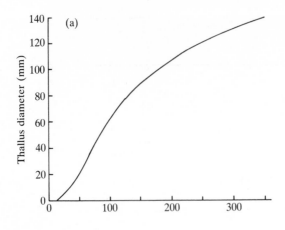

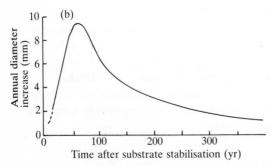

Fig. 6.10. Thallus diameter (a) and yearly increase in thallus diameter (b) for *Pseudephebe miniscula* on dated substrata on eastern Baffin I. After Miller (1973).

period' when the radial growth rate is higher than that during the subsequent linear phase (Fig. 6.11), a pattern contrasting with that in Fig. 6.9. Aplin & Hill (1979) suggested that the pattern in *R. geographicum* is related to thallus morphology. In this species and a number of other crustose lichens the thallus cracks, possibly as a result of repeated wetting and drying, to form discrete, algae-containing aereolae (Fig. 6.12), attached to a purely fungal, endolithic hypothallus that pioneers radial growth. The cracks could reduce growth rates in older thalli by restricting radial translocation.

In Armstrong's model, relative growth rate (RGR) increases sharply during the early part of the prelinear phase (3 in Fig. 6.9), but it subsequently declines, particularly during the linear phase (4), increasing again after the thallus begins to fragment. The decline during the linear phase is attributable to an increase in thallus area leading to a steady rise in

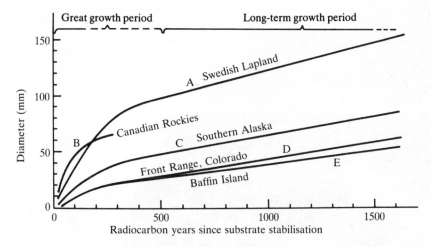

Fig. 6.11. Growth curves for *Rhizocarpon geographicum* on dated substrata in climatically different regions. Data from Calkin & Ellis (1980).

A_1 (Equation 6.1), the later increase in RGR resulting in a reduction in A_1 as the thallus fragments. However, it is not always clear how far photosynthesis in the older part of the thallus contributes to peripheral growth, and Equation 6.1 takes no account of resources utilised in increasing thallus thickness or in reproductive development.

Interpretation of lichen growth curves thus demands more information than is currently available concerning translocation, the relative contribution of different parts of the thallus to overall net photosynthesis, and other factors such as rates of turnover of protein and structural material. Rapid movement of assimilate from algal cells to the mycobiont has been shown in several lichens (Smith & Drew, 1965; MacFarlane & Kershaw, 1982, 1985), but little is known about lateral movement in fungal hyphae. Hale (1973) suspected it to be of minor importance, but Aplin & Hill (1979) assume that it occurs freely. They suggest that some extracellular movement of carbohydrate released upon rewetting of dry thalli may even occur in areolate lichens.

In fruticose species, Barashkova (1971) has reported transport of ^{14}C at rates up to 57 mm hr^{-1} through both living and dead portions of *Cladonia rangiferina* podetia, movement being more rapid and extensive in wet than in dry material. Regarding photosynthetic capacity, NAR was found to be substantially higher in young, marginal lobes than in older material of *Peltigera polydactyla* (Kershaw, 1977b), and Armstrong (1979) showed that removal of the central region of *Parmelia conspersa* thalli

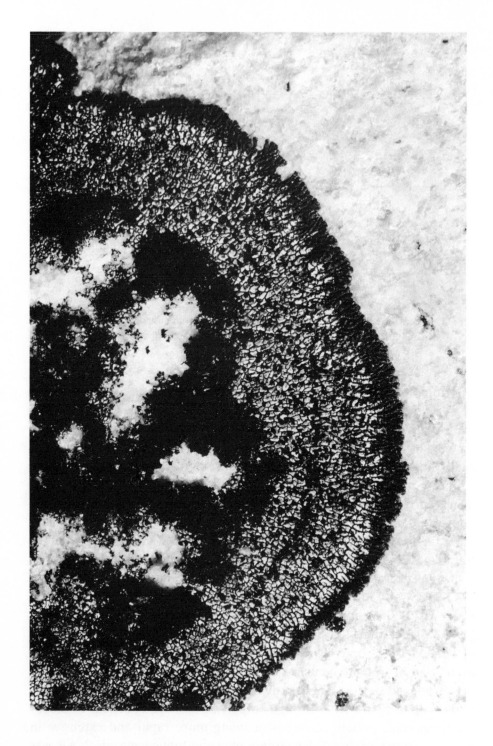

Fig. 6.12. *Buellia frigida* on a rock in Wilkes Land, frigid-Antarctica. The thallus
shows marginal lobing, and much of it is divided into aereolae. The black dots
in the aereolate region are apothecia. The central, older part of the thallus
has been colonised by *Alectoria minicscula*, and is partly eroded.

had no significant effect on radial expansion which thus appeared to be independent of assimilate from the centre. In contrast, no gradients in NAR were detected across thalli of *Umbilicaria* spp (Larson, 1983b), but here each thallus remains permanently attached to the substratum by a central stipe so that maintenance of older parts of the thallus is essential.

The thallus margin is concentrically zoned in some crustose lichens (Fig. 2.1). In Antarctic species of *Lecanora* and *Ochrholechia* this results from summer growth being paler in colour than the narrower zone developing during the rest of the year (Hooker, 1980a). Many foliose lichens and placodioid crustose forms (Table 2.3) have lobed margins (Fig. 6.12), a feature that may assist in restricting tangential cracking and thus favour radial translocation (Aplin & Hill, 1979). For *Xanthoria elegans* on Signy I, Hooker (1980a) describes how bifurcation at the lobe apices as the thallus expands produces new lobes in excess of the number required to maintain the circularity of the thallus. However, some lobes succumb to competition from their neighbours, a process Hooker calls engulfment. Armstrong (1984) showed that chemical exchange via hyphal connections is not essential for lobe growth in *Parmelia conspersa*, but that adjacent lobes do interact in some way, because thalli reconstructed by gluing excised lobes to the substratum in their original configuration grew similarly to intact controls, whereas slower growth was recorded in lobes kept separate from each other. Theoretical aspects of the maintenance of circularity in placodioid thalli are discussed by Hill (1984) and Hooker (1980d).

Vegetative phenology and annual growth increments
Methods of analysis
How fast does a moss or a lichen grow, and when? At first sight these seem difficult questions to answer for such small and often densely tufted plants, but a variety of approaches is possible once the pattern of growth in a species is understood. These are discussed for lichens by Hale (1973) and for bryophytes, with particular reference to polar studies, by Longton (1980) and Russell (1984). Several of the methods can be adapted to estimate increases in both size and dry weight and, depending on the frequency of measurement, to indicate either annual increments or seasonal patterns of growth.

Methods based on innate markers of annual growth increments have been the most widely used in studies of bryophytes and fruticose lichens.

Growth increments are clearly demarcated in some mosses by such features as seasonal variation in the length of newly formed leaves (Fig. 6.3), in branching pattern (Fig. 6.4), or by horizons with abundant dead shoot apices resulting from winter kill (Fig. 6.2).

Andreev (1954) reviewed several methods of assessing growth rates in fruticose lichens based on the annual production of a single whorl of branches by each podetium. Western authors have generally reported a parameter termed the annual linear growth rate (Lindsay, 1975; Pegau, 1968; Scotter, 1963), which is derived by dividing the mean height of the living portion of the podetia by the mean number of nodes on the living portion in individuals drawn from stands in the renewal phase. Andreev considered that this value corresponds to the mean annual increment during the accumulation phase of the colony, but that it underestimates annual elongation during the renewal phase. The latter can be determined as the mean length of the lowest living internodes, i.e. those which have completed intercalary growth, and the values were found to exceed the average annual linear growth rates by factors of about 1.7 in the Soviet forest-tundra and 1.5 in the tundra. Andreev (1954) further considered that the mean number of internodes in the living parts of the podetia from a colony in the renewal phase corresponds with the length in years of the accumulation phase.

Innate markers provide an accurate indication of growth once it has been confirmed that they are annually produced, as in *Polytrichum alpestre* (Longton & Greene, 1967), and not subject to environmental modification as with the branching pattern in *Racomitrium lanuginosum* (Russell, 1984). Their use has the advantage that no disturbance of the plants is involved prior to the period of growth under investigation, but destructive sampling with its attendant statistical complications is usually necessary at the time of measurement. A further drawback is that this method is not applicable to all taxa, although the growth of species with innate markers can sometimes be used to study the growth of others with which they are associated (Longton, 1972b; Vitt & Pakarinen, 1977). Marginal zonation provides an effective marker of growth in some crustose lichens (Fig. 2.1), with the advantage that the same thalli can be scored repeatedly without destructive sampling. Careful preliminary study is necessary, however, as the zones in some Antarctic species are not formed on an annual basis (Hooker, 1980a).

Variations on Clymo's (1970) cranked wire technique are also widely employed with bryophytes. These involve placing vertical markers such as stainless steel pins into moss colonies, or the substratum beneath them, with a known length initially projecting above the surface. Measurements are then made at intervals from the colony surface to the top of the marker. This method is particularly applicable to densely tufted, erect-growing mosses such as predominate in polar regions, and could probably be adapted for use with some fruticose lichens. It has the advantage that repeated measurements can be made on the same plants with little disturbance. A drawback is that movement of the markers relative to the plants through frost action or other influences could pass unnoticed and lead to error. In their studies on *Dicranum elongatum*, Kallio & Heinonen (1975) attempted to overcome this problem by using plastic markers equipped with numerous small teeth intended to reduce movement within the colony.

Another method applicable to erect-growing mosses involves stretching a nylon net over the colony surface and subsequently measuring the distance of shoot penetration, a technique also applied to *Dicranum elongatum* (Kallio & Heinonen, 1975). Initial disturbance is again slight, but destructive sampling is often necessary when recording the results. Mesh size must be chosen carefully in relation to shoot size, and doubt always exists as to whether the net has inhibited growth or been forced upwards to some extent by pressure of the growing shoots. A somewhat analogous method currently under investigation is to spray biologically inert, cellulose specific, fluorescent dyes onto bryophyte colonies to provide a permanent record of the initial position of the colony surface (Russell, 1984).

Several workers have cut moss shoots or lichen podetia to a known length, inserted them among plants in the field, and then harvested and measured them after a period of growth. This method has the drawback that the initial disturbance may influence growth either through the removal of the lower parts of the plants or because the shoots fail to develop the normal capillary relationships with their neighbours. Results obtained by R. I. Lewis Smith (1982b) on South Georgia confirm that subsequent growth of *Tortula robusta* is reduced if the shoots are initially cut to less than 3 cm, but suggest that normal elongation occurs in longer stems. However, the optimal length inevitably varies with the size and growth form of the plants under investigation. Another difficulty is that an undetermined amount may be lost from the base of the cut shoots by decomposition or breakage. This can be overcome by enclosing groups of plants at normal spatial density in net bags (R. I. Lewis Smith, 1982b) or perforated plastic tubes (Kärenlampi, 1971a; Lindsay; Sonesson, Persson, Basilier & Stenström, 1980), which provide reference points for the original position of the shoot apices. If this reference is initially placed level with the surface of the host colony the influence of disturbance on growth of the experimental shoots can be assessed.

Regular measurement of *Cladonia* podetia with reference points marked in indelible ink has given results similar to those of the innate marker method, thus confirming the validity of the latter approach (Andreev, 1954). Marking individual moss shoots, for example with a nylon thread tied a known distance behind the apex, indicates stem elongation in some cases, particularly in large, loosely tufted species, but care must be taken to ensure that the marker does not interfere with capillary water movement along the shoot. The length of the green part of the shoot has been used as an indication of growth increments, but the results are suspect unless the relationship between the green/brown interface and shoot age is confirmed, as this relationship varies seasonally, between populations and between species (Longton, 1980). Hooker (1980d) attempted to determine weight increase in individual thalli of *Usnea* spp on Signy I by air-dry weighing before and after a period in the field, but most thalli lost weight during the experimental period, presumably due to fragmentation or wind abrasion.

Size increase in crustose lichens has been recorded by repeated measurement of the distance between the thallus margin and reference points such as specific rock crystals in the substratum, by tracing thallus outline at intervals onto clear plastic film, or by regular photography. Thallus diameter varies with water content and thus measurements are usually made on air-dry plants. All these methods have the advantage of providing direct measurements, repeated for the same thalli with minimum disturbance. The tracing and photographic methods indicate changes in area as well as linear dimensions, but none of them gives information on changes in weight. The tracing method is accurate to within 0.5 mm and photography to within 0.1 mm (Miller, 1973). The latter is therefore the more popular, and Hooker & Brown (1977) have presented a detailed account of the procedure. This technique can also be applied to subfruticose lichens such as *Pseudephebe miniscula* that form discrete circular colonies.

Growth curves such as those in Fig. 6.10 can be constructed by recording increase in diameter over periods of one to several years for thalli covering a wide range of initial sizes, as described by Miller (1973). Growth rate in species forming circular colonies can also be assessed by measuring the diameters of the largest thalli on substrata that became available for colonisation at known dates, be they recently constructed buildings, or glacial moraines exposed several thousand years ago as determined by dendrochronology or radiocarbon dating. Depending on the range of dated substrata available in a given climatic region, the results can be expressed either as growth curves (Fig. 6.11) or as mean annual radial growth rate. The age of other substrata in the region can then be estimated from the maximum diameter of thalli of the same species growing upon them (Beschel, 1950, 1961). The methods employed in licheometry are fully discussed in Lock, Andrews & Webber (1979).

Annual growth increments

Bryophytes Annual growth increments have been determined for a number of ecologically important mosses in the cool- and cold-Antarctic and from Stordalen mire in Swedish Lapland, but only scattered data are available from elsewhere (Table 6.2). The estimates for increase in shoot length range from 170 mm yr^{-1} in *Sphagnum riparium* in a mire pool at Stordalen to less than 1 mm yr^{-1} in *Bryum inconnexum* from dry, sandy ground in the frigid-Antarctic. Much of the variation may be related to differences in summer temperature and other climatic factors associated with the major vegetational zonation in polar regions (Table 1.3), and in water availability in different habitats within a given climatic zone.

The influence of climate is seen most clearly in Table 6.2 where individual species have been studied in more than one zone, e.g. hydric species of *Brachythecium*, *Calliergidium* and *Drepanocladus* in the cool- and cold-Antarctic. The relationship to water availability is evident from comparisons of xeric (*Andreaea*, *Dicranoweisia* and *Racomitrium* spp), mesic (*Chorisodontium* and *Polytrichum* spp) and hydric species from cool- and cold-Antarctic islands, of *Sphagnum* spp within the mire complex at Stordalen, and of *Pohlia wahlenbergii* from stream edge and drier stream bank habitats on South Georgia. There is striking similarity in the range of values recorded for *P. wahlenbergii* on South Georgia (cool-Antarctic) and Disko I (cool-Arctic).

The annual dry weight increase of a moss shoot is the product of annual elongation and dry weight per unit length, and the latter is likely to decrease with increasing shoot density. Thus in Table 6.2 the lowest dry weight increase per shoot occurs in compact growth forms of dry habitats, such as *Andreaea acutifolia* on Marion I (0.02 mg yr^{-1}), where mean annual elongation is only 2 mm and density reaches 280 shoots cm^{-2}. The highest values occurred in plants forming loose colonies in wet habitats such as *Sphagnum fimbriatum* on South Georgia (11.4 mg yr^{-1}) where mean elongation is 34 mm yr^{-1} and density only 6 shoots cm^{-2}.

Table 6.2 *Representative data on annual shoot increments in polar bryophytes*

Zone and locality	Growth form and habitat	Species	Mean annual increment		Spatial density (shoots cm^{-2})	Method
			Length (mm)	Dry weight (mg)		
Frigid-Antarctic						
Lang Horde Syowa Coast[a]	Cushions on dry sandy ground	*Bryum inconnexum*	<1	—	—	Innate marker
Cold-Antarctic						
Signy I[b]	Cushion on rock	*Dicranoweisia grimmiaceae*	3	—	—	Innate marker
Signy I[c,d]	Tall turfs in mesic moss banks	*Polytrichum alpestre*	2–5	0.4–1.1	30–111	Innate marker
Signy I[e]	Hummocks at edge of melt stream	*Brachythecium austro-salebrosum*	26	—	—	Innate marker
Signy I[e]	Carpets on wet seepage slopes	*Calliergidium austro-stramineum*	10–32	—	—	Innate marker
Signy I[e]	Carpets on wet seepage slopes	*Drepanocladus uncinatus*	11–16	—	—	Innate marker
Cool-Antarctic						
Marion I[f]	Cushions in dry fellfield	*Andreaea acutifolia*	2	0.02	280	Nets
Marion I[f]	Mats in dry fellfield	*Racomitrium lanuginosum*	5	1.4	4.5	Several
South Georgia[g]	Tall turfs in mesic moss banks	*Chorisodontium aciphyllum*	3–9	0.4–1.1	83	Cut shoots

Location	Habitat	Species			Method
South Georgia[d]	Tall turfs in mesic moss banks	*Polytrichum alpestre*	2–8	1.3–2.6	—	Innate marker
South Georgia[g]	Tall turfs in mesic grassland	*Polytrichum alpinum*	20	4.8	20	Cut shoots
South Georgia[g]	Carpet in late snow patch flush	*Brachythecium austro-salebrosum*	38	3.1	18	Cut shoots
South Georgia[g]	Carpet in frequently flooded bog	*Calliergidium austro-stramineum*	67	1.3	50	Cut shoots
Marion I[f]	Carpet in wet mire	*Drepanocladus uncinatus*	23	1.8	16	Several
South Georgia[g]	Tall turf at edge of stream	*Pohlia wahlenbergii*	89	1.9	42	Innate marker
South Georgia[g]	Stream bank	*Pohlia wahlenbergii*	47	1.7	42	Innate marker
South Georgia[g]	Tall turf in lawn in wet mire	*Sphagnum fimbriatum*	34	11.4	6	Cut shoots
Mild-Antarctic					
Falkland Is[d]	Tall turf in wet grassland	*Polytrichum alpestre*	3–9	2.0–3.7	—	Innate marker
Cool-Arctic					
Devon I[h]	Tall turf in hummocky sedge moss meadow	*Meesia triquetra*	4	0.5	—	Innate marker
Devon I[h]	Tall turf in wet sedge moss meadow	*Meesia triquetra*	6–10	0.6–0.9	—	Innate marker
Devon I[h]	Tall turf by stream in wet sedge moss meadow	*Meesia triquetra*	15	0.9	—	Innate marker
Disko I, Greenland[i]	Tall turfs by stream	*Pohlia wahlenbergii*	40–92	1.4–3.6	—	Innate marker
Mild-Arctic					
Rankin Inlet, NWT[j]	Tall turf in dwarf shrub–lichen tundra	*Polytrichum alpestre*	2–5	1–2	—	Innate marker

(contd)

Table 6.2 Representative data on annual shoot increments in polar bryophytes (contd)

Zone and locality	Growth form and habitat	Species	Mean annual increment		Spatial density (shoots cm^{-2})	Method
			Length (mm)	Dry weight (mg)		
Churchill, Manitoba[j,k]	Tall turfs in spruce woodland	*Polytrichum alpestre*	2–14	1.7–3.8	4–11	Innate marker
Boreal forest						
Stordalen, Sweden[l]	Tall turfs on moderately dry hummock in mire	*Dicranum elongatum*	1	—	—	Cranked wire
Stordalen, Sweden[l]	Tall turfs in moderately wet hummock in mire	*Dicranum elongatum*	3	—	—	Cranked wire
Stordalen, Sweden[l]	Tall turf in lawn (wet) in mire	*Drepanocladus schulzii*	8	—	—	Marked shoots
Stordalen, Sweden[l]	Tall turfs on moderately dry hummocks in mire	*Sphagnum fuscum*	1	—	—	Cranked wire
Stordalen, Sweden[l]	Tall turfs on moderately wet hummocks in mire	*Sphagnum fuscum*	3	—	—	Cranked wire
Stordalen, Sweden[l]	Tall turfs in carpets (wet) in mire	*Sphagnum lindbergii*	47	—	—	Cranked wire
Stordalen, Sweden[l]	Tall turfs in carpets (wet) in mire	*Sphagnum riparium*	170	—	—	Cut shoots

[a] Matsuda (1968)
[b] Webb (1973)
[c] Collins (1976a)
[d] Longton (1970)
[e] Collins (1973)
[f] Russell (1984)
[g] R. I. Lewis Smith (1982b)
[h] Vitt & Pakarinen (1977)
[i] Clarke, Greene & Greene (1971)
[j] Longton (1974b)
[k] Longton (1979b)
[l] Sonesson & Johonsson (1973)

The most comprehensive data on environmentally related variation in annual increments apply to *Polytrichum alpestre*, in which mean elongation was found to range from 2–5 mm in mesic moss banks on Signy I in the cool-Antarctic to 2–14 mm in mild-Arctic spruce woodland. Annual increments in *P. alpestre* reach 53 mm in spruce bogs near Pinawa in southern Manitoba (Longton, 1970, 1979b). The tendency for mean annual stem elongation to decrease with increasing climatic severity is correlated with even more striking decreases in the mean dry weight and number of leaves produced annually by each shoot, and in leaf length (Fig. 6.13). Leaf length variation is principally in the green, photosynthetic limb (Fig. 6.3), the mean length of the sheathing base ranging from 1.3 mm on Signy I to 2.0 mm in southern Manitoba (Longton, 1970, 1974b).

The variation in *P. alpestre* indicated in Fig. 6.13 is generally continuous, with overlap between the mean values for colonies in different climatic zones. It thus appears to represent a topocline on a grand scale, and one may predict that the discontinuity between the results for Churchill and southern Manitoba would be filled by data from intervening boreal forest localities. The cause of the variation in annual growth increments between colonies at a given locality has not been investigated in detail, but it is readily apparent that where *P. alpestre* occurs in mires with well developed microtopographic variation the growth increments are longer in moist hollows than on the drier hummocks. Sexual dimorphism is also involved as, at a given locality, leaves are slightly shorter on male than on female plants (Longton, 1974b). However, annual growth increments may also vary substantially between plants of the same sex in different parts of apparently uniform moss banks, as indicated in the data for female plants from Signy I in Fig. 6.13. Such local variation in growth could be related to the cryptic pattern in the distribution of the dominant species in the banks discussed in Chapter 3.

Variation in the size of annual increments in *P. alpestre* is inversely related to spatial shoot density. Collins (1976a) showed that on Signy I increase in density from 30 to 120 shoots cm^{-2} in different turfs was correlated with a reduction in mean segment weight from 1.2 to 0.4 mg, and with comparable decreases in the length of the annual increments and in weight per unit length. In colonies at Churchill, where mean segment weight was more than 2 mg, spatial density was as low as 8–10 shoots cm^{-2} while a density of only 0.3 shoots cm^{-2} was recorded in a bog in southern Manitoba where *P. alpestre* occurred as isolated stems

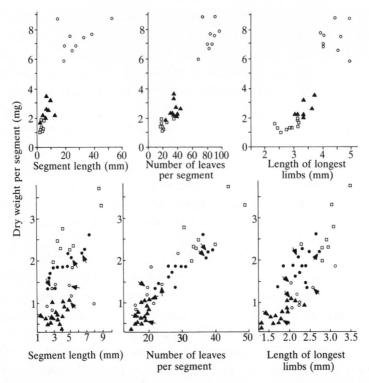

Fig. 6.13. Variation in annual growth increments in northern hemisphere (above) and southern hemisphere (below) colonies of *Polytrichum alpestre*. Points are means for samples of 20 segments. Each sample was taken from a different colony except where indicated by arrows, which refer to three southern-hemisphere colonies. Northern-hemisphere samples were collected at mild-Arctic sites near Rankin Inlet, NWT (□), at more southerly, warmer mild-Arctic sites near Churchill (▲), and at sites near Pinawa in the south of the boreal forest zone (○). Southern-hemisphere samples were collected at mild-Antarctic sites on the Falkland Is (□), at cool-Antarctic sites on South Georgia (●), and at cold-Antarctic sites on Signy I (▲) and the Argentine Is (○). After Longton (1970, 1974b).

projecting through a *Sphagnum* turf and had a mean segment weight of 4.6 mg (Longton, 1979b).

Lichens and lichenometry   Annual growth increments in fruticose lichens have been most intensively studied in the Soviet Union (Andreev, 1954). Average annual linear growth rate (page 230) in *Cladonia* spp was shown to range from about 2.5 to 5.0 mm in low altitude tundra and from 5.0 to 7.5 mm in boreal forest communities. Values as low as 1.2 mm were recorded in montane tundra. The decrease in growth

Table 6.3 *Average annual linear growth rate in* Cladonia sylvatica *in lichen tundra and birch–lichen woodland in the Bol'shezemel'skaya Tundra region, USSR*

Vegetation zone	Growth rate (mm)
Arctic tundra	2.5
Northern tundra	2.6
Southern tundra	2.8
Tundra–sparse forest	3.3
Sub-tundra forest	4.2

Data from Andreev (1954).

rate from south to north (Table 6.3), which parallels that in *Polytrichum alpestre*, was attributed to a corresponding decrease in the length of the snow-free growing season. It results in the far-northern colonies being of low stature and closely appressed to the soil. A decrease in growth rate is also evident as decreasing oceanicity from west to east reduces the length of time that the lichens are sufficiently hydrated to maintain active growth (Table 6.4).

Table 6.4 *Longitudinal variation in the average annual linear growth rate (mm) of* Cladonia *spp in northern USSR*

		Locality*	
		Western Bol'shezemel-	Eastern Bol'shezemel-
	Malozemel'skaya	'skaya	'skaya
Species	(68° N, 52° E)	(68° N, 55° E)	(60° N, 60° E)
Cladonia rangiferina	4.8	4.8	3.6
Cladonia stellaris	4.9	4.1	3.5
Cladonia sylvatica	4.2	3.5	3.3

* The localities cover extensive areas and thus the coordinates are approximations.
The data, from Andreev (1954), are means for several vegetation types in each locality.

Similar trends are evident in data from other areas. Pegau (1968) reported values of 4.3–5.5 mm in oceanic tundra and 5.5–5.8 mm in spruce–lichen woodland for the average annual linear growth rate of three species of *Cladonia* in southern Alaska, whereas growth averaged only

3.4–4.1 mm at forest sites in the more continental interior of Canada (Scotter, 1963). A mean annual linear growth rate of 5.3 mm has been recorded for *Cladonia rangiferina* in oceanic, cool-Antarctic tundra near sea level on South Georgia, compared with only 4.6 mm at a nearby site at 150 m elevation. Lindsay (1973a) attributed this difference to the colder, more exposed conditions at the higher altitude.

Even lower growth rates have been reported in fruticose lichens in the cold-Antarctic. Podetia of *Sphaerophorus globosus* were shown by a photographic technique to increase in length by about 2 mm yr^{-1} in mature colonies, compared with 0.5–0.8 mm yr^{-1} in juvenile stands. In contrast, a reduction in annual growth rate with increasing thallus age was demonstrated in *Usnea fasciata* on dated substrata, mean thallus height being equivalent to an annual growth rate of 3.0 mm, 1.7 mm and 1.2 mm on substrata aged 5, 12 and 19 years respectively (Hooker, 1980d). The base of the thallus in *U. fasciata* remains permanently attached to the substratum and does not gradually decompose as in species of *Cladonia* and *Sphaerophorus*. Hooker suggested that the thalli of *U. fasciata* rapidly reach a maximum height whereafter growth occurs principally as branching.

Radial growth rate in crustose lichens has been shown to vary widely, both between species, and in relation to climatic variation between localities and from year to year in a given taxon. *Rhizocarpon* spp have particularly slow growth rates, as indicated in Fig. 6.11 which shows growth curves for *R. geographicum* in several localities constructed on the basis of radiocarbon dating and other techniques. The curves all show an early 'great period' followed by a longer linear phase at slower growth rates (page 225). Presumed colony diameter after 1500 years ranges from 150 mm in Swedish Lapland to only 50 mm on Baffin I, with intermediate values for alpine localities.

Such low growth rates are commonly expressed in lichenometric studies as a lichen factor, defined as the increase in thallus diameter in millimetres per century. Lichen factors reported for *R. geographicum* at Arctic sites range from 2 to 45 during the 'great period' and from 2 to 4 during the linear phase (Webber & Andrews, 1973). Values of 8–16 have been given for the lichen factor of this species in the cold-Antarctic without reference to age (Birkenmajer, 1981; Hooker, 1980b; Lindsay, 1973a). Decreasing precipitation, temperature and length of growing season are all thought to reduce the annual growth rate of *R. geographicum* (Calkin & Ellis, 1980).

Radial growth rates of other crustose and foliose species in general exceed those of *R. geographicum*, as indicated by estimates of annual increase in colony diameter for lichens near the margin of the Barnes Ice Cap, Baffin I, of 0.06 mm in *R. geographicum*, 0.14 mm in *R. jemptlandicum*, and 0.17 mm in *Umbilicaria proboscidea* (Andrews & Webber, 1964). Similarly, Beschel & Weidick (1973) recorded thalli up to the following diameters on stone surfaces exposed during construction, 65 years previously, of a cairn in west Greenland: *Rhizocarpon tinei*, 5.5 mm; *Physcia caesia*, 12 mm; *Umbilicaria hyperborea*, 28 mm. However, growth of *U. lyngei* on Devon I was not measurable over a period of two summers (Chapter 5).

On Signy I, mean radial growth rate determined by photography repeated over 4–5 years was 0.2 mm in *Caloplaca cirrochroa*, 0.5 mm in *Buellia latemarginata* and *Xanthoria elegans*, and 1.2 mm in *Acarospora macrocyclos* (Hooker, 1980b). Annual growth rates in several of these species varied in relation to the slope and aspect of the substratum, factors that are likely to influence both temperature and water availability. There was also considerable variation between years in some taxa, large annual increments being correlated with warm conditions during February when growth is particularly active (page 245). There is also some evidence (Hooker, 1980a) that radial growth of *Xanthoria elegans* on the nearby Coronation I was temporarily stunted by an unusually severe winter.

Since the pioneer work of Beschel (1950, 1961) the technique of lichenometry has been widely to aid in interpreting Quaternary geomorphological events. Thus on the South Shetland Is measurements of thallus diameter in *Rhizocarpon geographicum* have been used to date neoglacial moraines, geomorphological features on a now-dormant volcano, and raised beaches associated with isostatic uplift (Birkenmajer, 1980a,b, 1981). In the Arctic, Calkin & Ellis (1980, 1981) employed both *R. geographicum* and faster growing species of *Pseudephebe* to date glacial features of differing age in the central Brooks Range, Alaska. They have also reported that recession of an Alaskan glacier is exposing undisturbed, lichen-covered boulders surrounded at the base by dead mosses. Radiocarbon dating of the mosses indicated that an ice advance engulfed the site 1200 ± 180 yr BP, while diameter measurements of the preserved lichens suggested that an ice-free period of at least 1500 years preceded this most recent advance. Near the ice margin the relict lichens were brightly coloured and morphologically undamaged, but with increasing distance from the retreating ice the condition of the thalli deteriorated indicating that the plants had not survived beneath the ice in a viable condition.

On Signy I, a study of predominantly bryophytic vegetation exposed by ice retreat suggested that there have been at least three minor ice advances and retreats since AD 1450, but that no major climatic changes have occurred during the past 5000 years. There was again no evidence of plants surviving an extended period of burial beneath ice on Signy I, or in a comparable study on Anvers I (Collins, 1976a; Fenton, 1982b; R. L. Lewis Smith, 1982a). Many other examples of the application of lichenometry and related techniques are cited in Lock, Andrews & Webber (1979). However, the method has several inherent sources of error stemming from the customary selection of the largest thallus at a site, and from difficulty in accounting for the effects of local and temporal climatic variation and other factors. The problems are outlined by Jochimsen (1973), and evaluated more fully in recent papers by Innes (1982, 1986) and Proctor (1983).

Seasonal patterns of growth

There have been few comprehensive investigations of vegetative phenology in polar bryophytes. Most of the relevant studies have indicated a period of active growth during the early part of the snow-free summer period (e.g. Sonesson *et al.*, 1980; Vitt & Pakarinen, 1977), but only in *Calliergidium austro-stramineum* on Signy Island (Collins, Baker & Tilbrook, 1975) and *Chorisodontium aciphyllum* on South Georgia is there evidence of extension growth beginning before snow clearance. In *C. aciphyllum*, pre-melt growth occurred at only one of two sites investigated, and it coincided with unusually high temperatures of up to 10 °C beneath the snow. Even so, growth in *Polytrichum alpestre* associated with the *Chorisodontium* did not begin until after snow clearance (R. I. Lewis Smith, 1982b).

Seasonal aspects of growth have been studied in detail in *P. alpestre* (Longton, 1970, 1979b; Longton & Greene, 1967). The data were obtained by comparing the uppermost segments in samples of shoots collected at intervals from specific colonies. In Fig. 6.14, which relates to a colony on South Georgia, stem extension refers to the distance between the top of the uppermost segment and the comal leaves of the second, while segment length includes also the basal region of the uppermost segment which is surrounded by leaves of the second. It can be seen that the apex of a new segment first appeared above that of the previous one in mid-November in 1964, approximately one month after the clearance of winter snow cover, growth in length and in dry weight

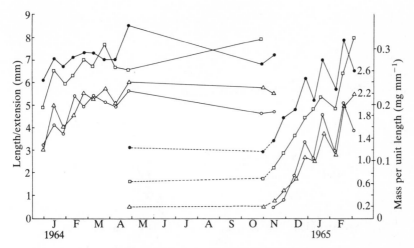

Fig. 6.14. Seasonal variation in the dimensions of the uppermost segments in samples from a colony of *Polytrichum alpestre* on South Georgia during 1963–65, showing segment length (●), dry weight (▲), mass per unit length (□) and stem extension (○). Data are means ($n = 10$). Dotted lines refer to segments with their apices enclosed by the comal leaves of the previous segment. There was continuous snow cover from mid-May to mid-October.

continuing during the summer period. Mass per unit length of the new segment also rose steadily from November to February.

The data suggest that the bud which gave rise to the 1964–65 segment began to enlarge before its apex extended beyond the 1963–64 segment but there is no evidence of growth or dry weight increase during the period of snow cover from May to October. Indeed, there appears to have been little extension growth of the 1963–64 segment in *P. alpestre* from February 1964 onwards, but the growing period is difficult to determine exactly due to irregularities in the data.

In this early study, sampling was undertaken by collecting blocks of turf from the colony at intervals and subsequently scoring ten shoots from each block. The irregularities are thought to represent variation in annual growth rate in different parts of the colony (page 237). Smoother curves are obtained when this error is reduced by comparing the mean extension of the uppermost segment expressed as a percentage of the mean for the second segment in successive samples, as in Fig. 6.15 which compares the seasonal pattern of growth in *P. alpestre* on South Georgia and on Signy I. It is again evident that shoot extension began shortly after snow clearance in spring in both cases. Elongation occurred principally during December, January and the early part of February even

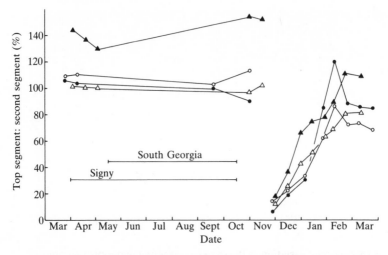

Fig. 6.15. Seasonal variation in the dimensions of the uppermost segments in samples from colonies of *Polytrichum alpestre* on South Georgia in 1961–62 and on Signy I in 1965–66, showing mean data for stem extension (▲●) and number of leaves (△○) on the uppermost segments as a percentage of the second segments. Triangles, South Georgia (*n* = 10); circles, Singy I (*n* = 20). The horizontal bars indicate periods of continuous snow cover.

though snow cover did not become established again until April on Signy I and May on South Georgia. The pattern of leaf production closely paralleled that of shoot elongation.

A similar pattern has also been recorded for *P. alpestre* in southern Manitoba, stem extension there occurring principally in May, June and July following snow clearance during April. Growth in weight continued through August and September, however, suggesting that photosynthesis remains active after elongation ceases. It seems likely that division of the apical cell, with further division and limited differentiation of its derivatives, also continues during late summer, as a bud with numerous leaf primordia is present in the shoot apex during autumn. Studies in boreal forest populations have shown that this is capable of giving rise to substantial shoot growth without further involvement of the apical cell, since growth segments 6–8 mm long and with up to 26 leaves were formed during spring on shoots bearing developing archegonia in an apical position (Longton, 1979b).

The seasonal pattern of growth in *P. alpestre* is reminiscent of that in some perennial flowering plants in which most shoot growth takes place early in the growing season through the development of buds formed during the previous autumn. There is limited experimental evidence that the seasonal growth pattern in *P. alpestre*, as in flowering plants, is partly under inherent control. Thus conditions remain favourable for growth during late summer, at least in oceanic regions such as South Georgia, and boreal forest plants cultured under simulated summer conditions behaved similarly to plants in the field, with extension growth ceasing in midsummer (Longton, 1979b). This point requires further investigation, however, as British plants collected in late summer immediately resumed growth in cultivation.

As with mosses, there is little evidence of lichen growth under winter snow, but lichens and mosses nevertheless appear to differ in their principal periods of growth. Hooker (1980b) showed by repeated photography that radial expansion in several appressed lichens on Signy I occurred principally towards the end of the snow-free period (Fig. 6.16). Podetial

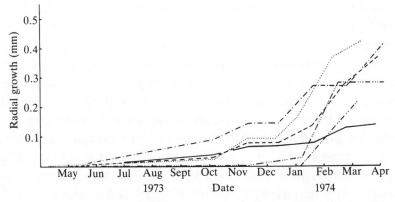

Fig. 6.16. Seasonal pattern of growth in six colonies of *Xanthoria elegans* on Signy I. After Hooker (1980b).

elongation in two fruticose species showed a similar pattern, growth in all the lichens investigated being slow during spring and early summer when stem elongation in mosses appears to be most active. The slow growth rate of lichens in early summer, under favourable conditions of temperature, irradiance and hydration, may be attributable to the condition of the overwintering plants. The low photosynthetic capacity of some species in spring has already been noted (Chapter 5), and Hooker (1980b) has suggested that spring growth is limited by recovery from desiccation associated with freezing, and depletion of respirable substrate during winter.

The latter point is supported by the observation of Gannutz (1970) that the respiratory capacity of cold-Antarctic lichens, but not of the mosses, is considerably lower in spring than later in summer. Again there is conflicting evidence, however, as some lichens are considered to be metabolically active in spring. A full understanding of seasonal aspects of metabolism and growth in both mosses and lichens requires studies of growth and reproductive phenology in the field to be combined, in a range of species, with Kershaw's (1985) response matrix approach to seasonal changes in the capacity and environmental relationships of both CO_2 exchange and growth.

Environmental and inherent control of growth
Climatic factors

The seasonal pattern of growth in most bryophytes and lichens is considered to be controlled primarily by climatic factors, notably temperature and precipitation, acting directly, or through their influence on the physiological condition of the plants. Reference has also been made to striking correlations between annual growth increments and summer temperature, precipitation and habitat water availability. Experimental evidence has been discussed that the field temperature regime is likely to depress growth in *Dicranum elongatum* at Kevo (page 212), and the same applies to frigid-Antarctic plants of *Bryum argenteum* (Fig. 6.1), and *Ceratodon purpureus* in which Kanda (1979) reported much slower growth at 0 °C than at 10 °C or 15 °C. The importance of water availability for the growth of fruticose lichens is emphasised by the observation that local variation in podetial length, internode length and diameter in *Cladonia* spp in northern Ontario is positively correlated with soil moisture (Kershaw & Rouse, 1971), while rainfall was found to be the dominant factor controlling the growth of *C. stellaris* at Kevo (Kärenlampi, 1971a).

Water supply in many polar habitats is most favourable early in the growing season following snow melt, particularly in continental climates with low summer precipitation. The importance of this factor in regulating bryophyte growth is suggested by Vitt & Pakarinen's (1977) data for *Meesia triquetra* in sedge moss meadows on Devon I (Table 6.5). At the two drier sites, approximately half the annual shoot extension took place before mid-July, during the first two weeks of the summer snow-free season, but rapid extension growth was maintained later in the season at a streamside site where mean shoot elongation of 10 mm was recorded during 15 days in late July. As in some populations of *P. alpestre*, substantial dry weight increase continued in *M. triquetra* after the period of maximum shoot extension, and this factor, as well as the increasing length

Table 6.5 *Growth of* Meesia triquetra *in different habitats on Truelove Lowland, Devon I, Canadian cool-Arctic*

Habitat and sampling date	Mean length of current season growth (mm)	Mean dry weight of current season growth (mg shoot^{-1})	Mean shoot dry weight per unit length (mg cm^{-1})
Relatively dry site in sedge moss meadow			
13 July	3.2 (53%)*	0.16 (28%)	0.50
22 August	6.0	0.58	0.96
Central part of sedge moss meadow			
13 July	4.4 (45%)	0.23 (27%)	0.51
22 August	9.7	0.84	0.86
Streamside			
12 July	3.9 (26%)	0.14 (16%)	0.36
27 July	14.8	0.88	0.60

* Figures in parentheses indicate growth of the first sampling data as a percentage of that on the second.
Data from Vitt & Pakarinen (1977), based on samples of at least 20 shoots.

of newly formed leaves, is likely to contribute to the rise in dry weight per unit length of the shoots as the growing season progresses (Fig. 6.14; Table 6.5).

There have been few investigations of the influence of light irradiance on the growth of polar bryophytes, apart from that of Sonesson *et al.* (1980) on *Sphagnum riparium* at Stordalen. In this study the rates of increase in length and dry weight were found to be slightly greater where irradiance was reduced to 50% of ambient levels by nylon gauze than in unshaded controls, but deeper shade resulted in markedly reduced growth. It was difficult to distinguish between the effects of reduced illumination and the associated modification in the temperature regime, but the results support the conclusion reached in Chapter 5 that light levels in open habitats are unlikely to be seriously limiting for polar mosses during summer.

Mineral nutrient availability

It was suggested in Chapter 3 that mineral nutrient availability might be more severely limiting on net assimilation and growth in inland areas than at sites with a strong marine influence, and the limited experimental data so far available support this view. Thus growth of *Tortula robusta* at coastal sites on South Georgia was found to be similar in plants supplied with local stream water or with dilute Hoagland's solution (Clarke, Greene & Greene, 1971). Conversely, growth of *Sphagnum riparium* in an inland, oligotrophic mire at Abisco was more strongly correlated with overall nutrient content of the surrounding water than with the other major environmental variables (Sonesson *et al.*, 1980).

In Alaska, Oechel & Sveinbjörnsson (1978) found that addition of dilute nutrient solutions resulted in no increase in moss growth or photosynthesis on the costal plain at Barrow, although graminoids were stimulated by this treatment. However, in inland tundra at Eagle Creek, Alaska, rates of gross photosynthesis in several mosses were substantially higher in field plots treated two years previously with NPK than in untreated controls. Growth was not measured in the mosses at Eagle Creek: it was found to be increased by the fertiliser treatment in vascular plants, in which, counter to prediction, photosynthesis rates were greatest in the controls (Bigger & Oechel, 1982). Fertilisation also increased moss cover during recolonisation of disturbed sites on organic substrata at Eagle Creek (Chapin & Chapin, 1980).

Available nitrogen and phosphorus are the elements most likely to be limiting in polar environments. Kallio & Heinonen (1973) reported that NAR in *Pleurozium schreberi* from Kevo was increased by nitrogen fertilisation (as nitrate) applied over several weeks in cultivation, whereas *Racomitrium lanuginosum* showed no appreciable response. In lichens, Carstairs & Oechel (1978) found that artificial application of nitrogen (as nitrate), phosphorus and calcium to *Cladonia stellaris* at Schefferville resulted in no significant change in either element content of the plants or in NAR. The form in which nitrogen is supplied in such experiments is likely to be of importance. Among frigid-Antarctic plants, *Bryum algens* grew well on a range of nitrogen compounds. However, the alga *Prasiola crispa* and the isolated mycobiont of *Lecanora tophroeceta* showed optimum growth when supplied with ammonia and other reduced nitrogen compounds, a result which may account for the preference of these organisms for sites influenced by marine birds (Schofield & Ahmadjian, 1972).

Inherent control of growth

Growth in bryophytes and lichens is generally regarded as essentially opportunistic, occurring whenever environmental conditions are favourable unless previous stress has resulted in the plants being physiologically unable to respond. The inherent control of growth phenology suspected in *Polytrichum alpestre* (page 245) would, if confirmed, represent a striking exception. However, there is little doubt that growth rates during favourable conditions are under strong genetic influence. It is clear from cultivation studies that the annual increments in *Andreaea* spp and other lithophytes are unlikely to approach those recorded in some hydric mosses (Table 6.2) even if optimal conditions were to prevail continuously. Such intrinsic differences in growth rate are thought to be partly responsible for inter-specific differences in plant height, and to some extent in growth form, that permit the exploitation of a wide range of habitats (Chapters 3 and 5).

Inherent differences in growth rate occur within as well as between species, as indicated by experimental studies on *P. alpestre* (Longton, 1974b). The shoots considered in Table 6.6 were obtained by placing

Table 6.6 *Mean (n = 10) dimensions of 200-day-old shoots of* Polytrichum alpestre *from two localities raised under two temperature regimes*

Light/dark temperature regime (°C)	Locality	Shoot dry weight (mg)	Shoot length (mm)	Number of leaves per shoot	Limb length of longest leaves (mm)
23/15	Pinawa	3.5	35	54	4.5
23/15	Churchill	5.2	17	88	3.8
15/5	Pinawa	2.6	24	38	3.1
15/5	Churchill	2.5	21	42	2.2

Analysis of variance showed that between locality differences were significant at least at the 5% level for all parameters at 23/15 °C, and for leaf limb length at 15/5 °C. Between treatment differences were significant for shoot length and leaf limb length in Pinawa plants, and for all parameters except shoot length in Churchill plants.
Data from Longton (1974b).

pieces of turf on their side in a light humid environment. Shoots then arose as branches from the tomentose, basal regions of older shoots, and developed entirely under experimental conditions. In can be seen that, in addition to temperature-related differences between shoots from each original colony, the experimental shoots from Pinawa were longer, and bore longer leaves than those from Churchill, particularly under the warmer growth regime, in line with differences between field-grown shoots (Fig. 6.13). In contrast to the field situation, however, the Churchill plants produced more leaves per shoot under warm conditions than those from Pinawa. Comparable results were obtained in a second experiment in which plants from Pinawa, Churchill and Signy I were raised under a light/dark temperature regime of 20/10 °C, the differences in shoot length and leaf length being particularly striking (Fig. 6.17). Similar differ-

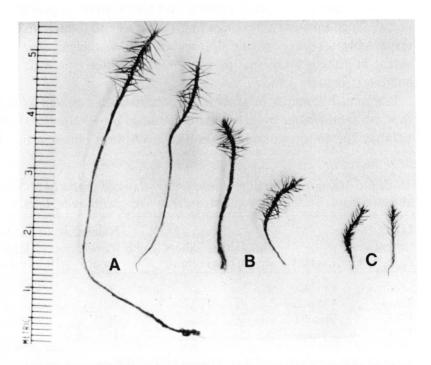

Fig. 6.17. Shoots of *Polytrichum alpestre* raised under collateral cultivation by vegetative propagation from plants collected at Pinawa (A), Churchill (B) and on Signy I (C). Reproduced from Longton (1974b) by courtesy of the Editor, *Journal of the Hattori Botanical Laboratory.*

ences were evident between plants raised under greenhouse conditions from spores collected from several temperate and polar sites, including Churchill and Pinawa (Longton, unpublished data).

The experimental results thus suggest the field variation in *P. alpestre* (Fig. 6.13) arises through a combination of complementary, inherent and plastic responses which result in the growth of progressively shorter shoots, having shorter leaves, and forming denser colonies with increasing climatic severity within the polar regions. This result is likely to be beneficial in terms of both temperature and water relations, and it may be noted that the typical habitats of *P. alpestre* change from wet mire in north temperate and boreal forest localities to mesic slopes or dry heathland in the Arctic and Antarctic.

Other species show different patterns of response. It was noted earlier that growth in *Bryum argenteum* is severely depressed at low temperatures, but the response is strikingly similar in frigid-Antarctic, mild-Arctic and boreal forest provenances (Fig. 6.1). Moreover, plants of both frigid-Antarctic and tropical origin are capable of very slow growth in a light/dark temperature regime of 5 °C/−5 °C (Longton, 1981). There is little evidence of inherent adaptation in frost or heat resistance in *B. argenteum* (Chapter 5), and although bulbils or deciduous shoot apices are characteristic of many non-fruiting polar populations, similar asexual propagules develop in some fruiting colonies in temperate regions (Longton & MacIver, 1977).

Frigid-Antarctic specimens of *B. argenteum* are characterised by shorter shoots bearing obtuse leaves with shorter midribs than temperate material. The differences in shoot length disappeared, but those in leaf morphology were retained in collateral cultivation at a range of temperatures (Longton, 1981, unpublished data). In contrast, *B. argenteum* from Macquarie I produced obtuse, nerveless leaves at 4 °C and apiculate leaves with a prominent midrib at 21 °C (Seppelt & Selkirk, 1984). There is thus evidence of considerable variation, both plastic and inherent, between populations of *B. argenteum*, but the pattern and ecological significance remain to be assessed. To date, the most consistent feature of possible adaptive value is that frigid-Antarctic provenances appear to be particularly vigorous under a range of temperature regimes, particularly with regard to shoot production in agar cultures initiated as single shoot apices (Table 6.7).

This feature could be viewed as favouring establishment during short, cold summers. Similarly, some Arctic hepatics are said to grow more rapidly and luxuriantly in culture than temperate species, and endemic mosses in the circumpolar Arctic element (Chapter 1), far from being dwarfed as in polar populations of *Polytrichum alpestre* and many vascular plants, are as large as, or conspicuously larger than, their temperate cogeners (Brassard, 1974; Steere, 1978b). However, Dodge (1973) claimed that many Antarctic lichens are much smaller than related temperate species, with some crustose species bearing almost spherical aereolae of possible benefit under conditions of continuous solar radiation at low irradiance (page 111). In the absence of cultivation studies, the extent

Table 6.7 *Mean (n = 5) colony diameter and mean numbers of leafy shoots in 0.5 mm wide transects across the centre of 5-week-old cultures of* Bryum argenteum *grown under different temperature regimes*

Day/night temperatures	9/1°C	17/10°C	22/15°C	30/30°C
	Mean colony diameter (mm ± sd)			
Vermillion Falls, Indiana	1 ± 0.4	3 ± 1.0	22 ± 2.7	15 ± 5.7
Churchill, Manitoba	2 ± 0.8	6 ± 1.2	18 ± 1.9	10 ± 2.7
McMurdo, Victoria Land	2 ± 0.6	6 ± 0.9	26 ± 5.0	16 ± 2.9
	Mean number of shoots (± sd)			
Vermillion Falls, Indiana	2.6 ± 0.9	5.6 ± 3.0	11.0 ± 6.1	3.0 ± 3.3
Churchill, Manitoba	1.5 ± 0.6	11.4 ± 7.1	16.0 ± 4.2	1.4 ± 1.1
McMurdo, Victoria Land	4.6 ± 2.1	35.0 ± 9.2	30.4 ± 11.7	7.3 ± 1.5

Cultures were initiated by placing single shoot apices of representing clones from three provenances in the centre of agar plates under axenic conditions. Data from Longton & MacIver (1977).

of a genetic basis for such variation in lichens cannot yet be assessed. The question of adaptation in response to the selection pressure of polar environments is discussed further in relation to reproductive biology in Chapter 8.

7

Cryptogams in polar ecosystems

Energy flow

Production and phytomass
Definition and estimation Bryophytes and lichens are major elements in the tundra communities described in Chapter 2 in terms of cover and also energy flow, mineral nutrient cycling and other dynamic aspects of polar ecosystems. The part played by bryophytes has been briefly reviewed (Longton, 1984). Here we shall explore the role of both groups in greater detail, beginning with the contribution of cryptogams to production and phytomass.

Phytomass is the amount of plant material present, and thus potentially available to consumers and decomposers. Vascular plant phytomass comprises above- and below-ground components, the latter including living and dead, and the former green, other living, standing dead and litter. Only photosynthetic (green) and non-photosynthetic components can realistically be distinguished in many cryptogams because of difficulties in determining how far living tissue extends into the colonies (Chapter 6), and most published estimates of living phytomass probably represent the green component. Mosses and lichens decompose from the base with little surface litter deposition. Phytomass varies seasonally, particularly in flowering plants; the value recorded near the end of the growing season is that most commonly reported.

Above-ground phytomass is determined by harvesting and dry-weighing plants from sample plots of known area, those for cryptogams normally comprising cores through colonies of one or an assemblage of species. The results are extrapolated to phytomass per unit area assuming continuous cover, and this value multiplied by percentage cover indicates phytomass of the species concerned in the community. Below-ground phytomass is assessed from cores through the phanerogamic rooting zone.

Approximations of above-ground phytomass over large areas can be obtained from a combination of extensive aerial photography and more restricted on-ground sampling (Andreev, 1971), an approach used to generate the data in Table 7.1.

Net primary production is the phytomass incorporated into a plant or plant community during a specified time less that lost due to plant respiration; it thus represents the product of net assimilation. Vascular plant production may be estimated by harvest techniques (Milner & Hughes, 1968) as:

$$P_n = B + L + G \tag{7.1}$$

where P_n = net production during time $t_1 - t_2$
B = change in living phytomass during $t_1 - t_2$
L = plant losses by death or shedding during $t_1 - t_2$
G = plant losses by grazing during $t_1 - t_2$

Extensive sampling may be required to give valid results in tundra communities (Sonesson & Bergman, 1980), and different approaches are again adopted for cryptogams due to difficulties in determining living phytomass. In bryophytes, most data discussed below refer to current shoot production, i.e. the dry weight of new growth during $t_1 - t_2$, new growth being determined as outlined in Chapter 6. Assuming no sporophyte production or loss of new growth during $t_1 - t_2$, current shoot production (P_s) is related to net production (Longton, 1972b) by:

$$P_n = P_s + NA_0 + T_d - T_u \tag{7.2}$$

where NA_0 = net assimilation during $t_1 - t_2$ by phytomass originally present at t_1
T_d = translocation downwards from new to originally existing phytomass during $t_1 - t_2$
T_u = translocation upwards from originally existing to new phytomass during $t_1 - t_2$

NA_0, T_d and T_u are seldom determined, but could sometimes be significant, particularly in *Polytrichum* spp (Chapter 6). Despite such uncertainties, this approach is preferable to estimating bryophyte productivity as an arbitrary percentage of phytomass (Longton, 1972b; Vitt & Pakarinen, 1977).

Estimation of lichen production is complicated by the occurrence of intercalary growth and variation in growth rate with age. Andreev (1954) summarised extensive Soviet data for fruticose lichens, but these refer

Table 7.1 *Approximate live, above-ground phytomass ($g\,m^{-2}$) in the eastern Russian Arctic*

	Cool-Arctic		Mild-Arctic		
	Northern tundra	Southern tundra	Northern tundra	Southern tundra	Woodland tundra
Predominant vegetation types	Fellfields and bare ground	Fellfields	Dwarf shrub heath and willow scrub	Dwarf shrub heath, willow scrub and mire	Open woodland and tundra
Bryophytes	22 (18%)	120 (40%)	220 (30%)	288 (37%)	335 (19%)
Fruticose lichens	0	2 (1%)	22 (3%)	66 (9%)	103 (6%)
Graminoids	29 (24%)	103 (34%)	86 (12%)	51 (7%)	49 (3%)
Herbs	29 (24%)	54 (18%)	39 (5%)	7 (1%)	5 (<1%)
Shrubs, dwarf shrubs and woody, creeping perennials	39 (33%)	24 (8%)	357 (49%)	359 (47%)	394 (22%)
Trees	0	0	0	0	892 (50%)
Total	119	303	724	771	1778

Data from Andreev as cited by Alexandrova (1970)

to increase in living phytomass as a resource, which differs from net production as understood here. Mean annual increase in living phytomass
in *Cladonia* spp is zero during the renewal phase, as growth is balanced
by death and decomposition (Chapter 6), even though annual net production made available to heterotrophs may be considerable. Net production
in fruticose lichens has been estimated from successive weighings of air-
dry thalli, or cores of podetia, with the material returned to field sites
between weighings. The mass of lichenometrically dated thalli gives estimates of mean annual production in lichens such as *Usnea* spp in which
the basal part of the thallus remains alive indefinitely (Hooker, 1980c,
d; Lindsay, 1975). Detailed procedures for calculating energy flow
through lichens were described by Kärenlampi (1971b).

Net annual production can also be predicted from microclimatic data
and laboratory-determined responses of CO_2 exchange to environmental
variables. Reasonable agreement between harvest and prediction estimates has been reported (Collins, 1977), even though the latter have
seldom been based on comprehensive micrometeorological data or
seasonal response matrices (Chapter 5).

Frigid-Antarctic   Non-vascular plants are totally responsible for
primary production and phytomass in arid, frigid-Antarctic ecosystems.
Overall, phytomass is low due to the scattered nature of the vegetation
(Chapter 2), but it may be considerable within moss and lichen colonies
(Table 7.2). Total phytomass reaches $1100\,g\,m^{-2}$ in short moss turfs on
Ross I, and must be greater in small moss cushions at Mawson where
similar values were recorded for green phytomass alone. Community phytomass is generally low due to the sparse plant cover, but $938\,g\,m^{-2}$ was
recorded in an exceptional stand of *Bryum antarcticum* on Ross I. Phytomass of $50{-}950\,g\,m^{-2}$ for epilithic fruticose lichens and $177\,g\,m^{-2}$ for scattered green shoots of *B. algens* beneath pebbles (Chapter 2) has been
reported at Birthday Ridge (Table 7.2). Estimates of organic matter in
sandstone colonised by endolithic lichens in southern Victoria Land reach
$177\,g\,m^{-2}$ of rock surface (Friedmann, 1982).

Decomposition is slow in these communities, and thus relatively high
phytomass does not imply comparable levels of production, of which
no estimates based on harvest techniques are yet available. Physiological
and microclimate data indicated net annual production as $250\,g\,m^{-2}$ in
turfs of *Bryum argenteum* on Ross I, or $5\,g\,m^{-2}$ for the open community,
but $100\,g\,m^{-2}$ was considered a more realistic figure for the turfs for reasons discussed in Longton (1974a). In contrast, Ino (1983) predicted mean

Table 7.2 *Phytomass (g m⁻²) of frigid-Antarctic cryptogams*

Locality	Species	Colony phytomass	Community phytomass	Notes	Source
Mawson Station	*Grimmia lawiana* and *Bryum algens*	1097	197	Data refer to green phytomass	Seppelt & Ashton (1978)
Ross I, southern Victoria Land	*Bryum antarcticum*	1012–1108	14 (–938)	Data refer to green + brown phytomass. Figures in brackets refer to communities with unusually high percentage cover of moss	Longton (1974a)
	Bryum argenteum	241–602	5–17 (–397)		
Birthday Ridge, northern Victoria Land	*Bryum algens*	—	177	Scattered green shoots beneath pebbles	Kappen (1985a)
	Neuropogon sulphureus	—	51–372	Epilithic species	
	Usnea pictata	—	950		
Dry valleys of southern Victoria Land	Endolithic lichens	—	46–177	Data refer to phytomass per unit area of rock surface	Friedmann (1982)

annual net production as only $4\,\mathrm{g\,m^{-2}}$ in turfs of *Bryum pseudotriquetrum* near Syowa Station, with negative values as low as $-16\,\mathrm{g\,m^{-2}}$ during years when meteorological data suggested that respiration exceeded gross photosynthesis: the moss was colonised by cyanobacteria and appeared moribund.

Cold-polar regions   Bryophytes and lichens are dominant in above-ground primary production in cold-polar ecosystems, but angio-sperm roots may form a considerable proportion of total phytomass in the cold-Arctic. Above-ground living phytomass averages only $10\,\mathrm{g\,m^{-2}}$ throughout the east European cold-Arctic (Alexandrova, 1970), but with considerably higher values in specific communities. In the example con-sidered in Table 7.3, bryophyte and lichen phytomass was about 100 and $200\,\mathrm{g\,m^{-2}}$ respectively, with most regarded as living. Angiosperms contributed only 4% of living above-ground plant material.

Vascular plants are absent from most cold-Antarctic communities (Chapter 2), but bryophyte productivity in wet and mesic habitats is remarkably high. Harvest and prediction techniques have yielded esti-mates of $315–660\,\mathrm{g\,m^{-2}\,yr^{-1}}$ in tall turfs of *Chorisodontium* and *Polytri-chum* spp, and $223–893\,\mathrm{g\,m^{-2}\,yr^{-1}}$ in carpets of *Calliergidium*, *Calliergon* and *Drepanocladus* spp (Baker, 1972; Collins, 1973, 1977; Davis, 1983; Longton, 1970). These figures also approximate to community producti-vity as the species concerned form large, almost pure stands. Although net annual production is broadly similar in the tall turf and carpet commu-nities, annual respiration in the turfs is approximately ten times that in the carpets, suggesting that a greater proportion of gross primary produc-tion is converted into new moss growth in the latter (Davis, 1981). Net annual production of bryophytes in these Antarctic communities is com-parable with total net productivity in some types of temperate vegetation, and is thought to reflect free availability of water and mineral nutrients, combined with lack of competition from flowering plants (Longton, 1980).

Phytomass is also large in the moss turfs. That of green shoots may be $300–1000\,\mathrm{g\,m^{-2}}$, with total phytomass $20\,000–30\,000\,\mathrm{g\,m^{-2}}$ above permafrost, and up to $46\,000\,\mathrm{g\,m^{-2}}$ for undecomposed shoots to the base of a bank 1 m deep. *Polytrichum alpestre* remains alive to a depth of at least 25 cm in boreal mires (Longton, 1972b); a figure of 20 cm for Signy I would indicate living phytomass of about $10\,000\,\mathrm{g\,m^{-2}}$ in the moss banks (Davis, 1981; Smith, 1984a).

Carpet- and tall turf-forming mosses cover less than 20% of the non-glaciated surface of Signy I (Chapter 2). Macrolichen phytomass reaches

Table 7.3 *Phytomass (g m^{-2}) in tundra communities in the eastern Russian Arctic*

Zone	Cold-Arctic	Cool-Arctic	Mild-Arctic	
Locality	Franz Josef Land	New Serbian Is	Vorkuta	Komi ASSR
Vegetation type	Cryptogamic tundra	Dry meadow	*Salix–Betula* scrub	*Betula* scrub
Living phytomass				
Angiosperms: above-ground	15 (4%)	174 (38%)	460 (37%)	1992 (38%)
Angiosperms: below-ground	71	1249	3349	9083
Bryophytes	98 (31%)	200 (44%)	582 (47%)	3178 (60%)
Fruticose and foliose lichens	127 (41%)	64 (15%)	198 (16%)	115 (2%)
Crustose lichens	71 (23%)	15 (3%)	—	—
Gelatinous lichens and algae	5 (1%)	—	—	—
Dead phytomass				
Above-ground	22	169	895	59
Below-ground	372	926	12 763	27 680

Data from Alexandrova (1970).
Percentages refer to live, above ground phytomass
*Means for three slightly different communities

800–1750 g m^{-2} in *Umbilicaria, Himantormia* and *Usnea* spp, with productivity up to 250 g m^{-2} in *Usnea fasciata* (Smith, 1984a), in the more widespread, open communities of lichens and small acrocarpous mosses on drier terrain. Lichen production in terms of relative growth rate ranges from 34 mg g^{-1} yr^{-1} in *Sphaerophorus globosus* to 84 mg g^{-1} yr^{-1} in *Cladonia rangiferina*. It is strongly age-related in *Usnea fasciata*, falling from 50–200 mg g^{-1} yr^{-1} in young thalli to 35 mg g^{-1} yr^{-1} in mature, and 1–5 mg g^{-1} yr^{-1} in old individuals (Hooker, 1980c, d).

Cool-Antarctic   Mosses and lichens contribute substantially to production and phytomass, and total chlorophyll content (Tieszen & Johnson, 1968), in many tundra communities containing abundant vascular plants. Primary productivity in three cool-Antarctic grass heaths and herbfields ranges from 840 to 1635 g m^{-2} yr^{-1} and total phytomass may exceed 9500 g m^{-2}: below-ground roots and rhizomes often form the major component as is typical of tundra vegetation (Table 7.4). Bryophytes contribute up to 30% of above-ground production and 50% of above-ground phytomass, but the lichen component is apparently small. The ratios of annual production : phytomass in bryophytes exceed those in some Arctic communities (page 263), indicating more rapid turnover, and phytomass may vary seasonally. In an extreme case, phytomass of *Tortula robusta*, the dominant moss in an *Acaena magellanica* herbfield on South Georgia, fell from 500 g m^{-2} in early spring to 125 g m^{-2} in midsummer as increasing *Acaena* leaf area restricted moss photosynthesis and the lower parts of the *Tortula* shoots decomposed. Decreasing leaf area of the *Acaena* permitted renewed moss growth later in the summer, *Tortula* phytomass being restored to 425 g m^{-2} in autumn (Fig. 7.1). Dense stands of tussock grassland may support little growth of cryptogams (Table 7.4), but deep moss banks comparable with those in the cold-Antarctic have developed locally in open tussock grassland on South Georgia and may be highly productive (page 266).

Mild- and cool-Arctic   Net primary productivity, and in many cases phytomass, are strikingly lower under the relatively dry conditions in the cool-Arctic than on cool-Antarctic islands. Andreev's integrated data for east European northlands (Table 7.1) indicate that total living phytomass, and that of bryophytes and lichens, increase progressively southwards, with bryophytes contributing 30–40% of the total in the mild-Arctic and in the south of the cool-Arctic. As on cool-Antarctic islands,

Table 7.4 *Representative data for annual net production and phytomass (g m^{-2}) in cool-Antarctic communities*

Vegetation type	Locality	Net annual production					Phytomass				
		Vascular plants					Vascular plants				
		Above ground	Below ground	Bryophytes	Lichens	Total	Above ground	Below ground	Bryophytes	Lichens	Total
Grass heath *Festuca contracta*	South Georgia	340	350	150	2	842	425 (+1598)	1642	500	12	4177
Herbfield *Acaena magellanica*	South Georgia	885	500	250	0	1635	1300 (+517)	7536	221	0	9574
Pleurophyllum hookeri	Macquarie I	314	550	146	4	1014	139 (+266)	1920	393	9	2727
Tussock grassland *Poa foliosa*	Macquarie I	1890	3670	21	0	5581	912 (+2592)	4800	6	0	8310

Data from Smith & Walton (1975) for South Georgia and Jenkin (1975) for Macquarie I.
Production data for bryophytes are probably current shoot production (page 254) but the methods of analysis are not described in every case. Data for above-ground vascular plant phytomass show living (+ standing dead). Other phytomass data are totals of living plus attached dead.

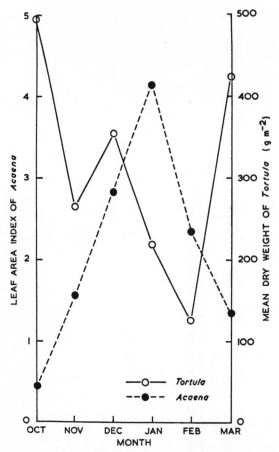

Fig. 7.1. Seasonal variation in leaf area of *Acaena magellanica* and phytomass
of *Tortula robusta* in an *Acaena magellanica* herbfield on South Georgia.
Reproduced from Walton (1973) by courtesy of the IBP Tundra Biome Steering
Committee.

the larger fruticose lichens are a minor component of phytomass in the
region as a whole. Both total above-ground phytomass, and the proportion contributed by bryophytes and lichens, are higher in some specific
vegetation types, as indicated in Table 7.3, and by the data of Vassiljevskaja *et al.* (1975) for the Agapa region of the western Taimyr. Here,
live above-ground phytomass in different communities ranged from 140
to 900 g m^{-2}, with bryophytes contributing 30–70% and lichens up to 20%.

Production and phytomass estimates for Fennoscandian and North
American tundra (Table 7.5) indicate that total annual productivity in
the cool-Arctic tends to increase along a xeric → hydric gradient. In
wetlands, total net productivity appears normally to be in the range

$100-300 \, \mathrm{g \, m^{-2} \, yr^{-1}}$ from Devon I south to Stordalen mire. Productivity in sedge meadows on Devon I rises with increasing water availability in accordance with the general trend. Bryophytes represent 10–45% of total production, and exceed the above-ground vascular plant component at the wetter sites. Turnover is slower in bryophytes than in the above-ground vascular phytomass, as reflected in the high bryophyte phytomass. This is seen particularly at Demster, Alaska (Table 7.5), where, however, most of the bryophyte phytomass $(4520 \, \mathrm{g \, m^{-1}})$ was brown and non-photosynthetic. Annual bryophyte production in wet meadows at Barrow was estimated as $22-39 \, \mathrm{g \, m^{-2}}$, and green moss phytomass as $19-37 \, \mathrm{g \, m^{-2}}$ (Rastorfer, 1978; Webber, 1978). Rates of CO_2 incorporation of up to $26.5 \, \mathrm{mg \, g^{-1} \, day^{-1}}$ have been recorded in *Polytrichum alpinum* at this site (Oechel & Sveinbjörnsson, 1978). Although insignificant in many wetlands (Table 7.5), lichens comprise up to 14% of the above-ground phytomass in wet meadows at Barrow (Webber, 1978; Williams *et al.*, 1978).

Among mesic communities, total productivity in the cool-Arctic dwarf shrub heaths and grass heaths considered in Table 7.5 is consistently less than $150 \, \mathrm{g \, m^{-2} \, yr^{-1}}$. Lichens again contribute insignificantly to production and phytomass, although they may be visually conspicuous (Bliss & Svoboda, 1984). The bryophyte component varies, even between stands of a graminoid steppe on Elef Ringness I, but mosses are often dominant in terms of both production and phytomass. A striking feature of these data is that bryophyte phytomass exceeds estimated annual production by a factor of about 70:1, indicating very slow turnover. Unfortunately, the methods of determining bryophyte production were generally not described. Above-ground phytomass in three dry to mesic graminoid communities at Barrow ranges from 50 to $360 \, \mathrm{g \, m^{-2}}$, with green moss contributing up to 67%. Lichens are more abundant at Barrow than in the grass heaths considered in Table 7.5, and form up to 44% of the above-ground phytomass (Webber, 1978).

High lichen phytomass is also characteristic of lichen heath and lichen woodland in mild-Arctic and northern boreal regions. Live lichen phytomass reaches $300-450 \, \mathrm{g \, m^{-2}}$ in alpine lichen heath at Hardangervidda, Norway (Wielgolaski & Kjelvik, 1973). Living phytomass of $28-85 \, \mathrm{g \, m^{-2}}$ for lichens and $7-66 \, \mathrm{g \, m^{-2}}$ for mosses has been reported from birch and pine woodland at Kevo, where lichens predominate in the drier sites but are subject to overgrazing by reindeer (Kärenlampi, 1973; Kallio, 1975). According to Ahti (1977), lichen phytomass may exceed $300 \, \mathrm{g \, m^{-2}}$ in mature *Cladonia stellaris* woodland, while Scotter (1962) gives the

Table 7.5 *Representative data for annual net production and phytomass (g m⁻²) in representative Arctic communities*

Vegetation type	Locality	Annual Net Production					Phytomass					Source
		Vascular plants		Bryophytes	Lichens	Total	Vascular plants		Bryophytes	Lichens	Total	
		Above ground	Below ground				Above ground	Below ground				
Wet meadow												
Cotton grass–dwarf shrub tundra	Demster, Alaska	87	—	69	<5	—	66(+102)	2372	4753ˣ	69	7362	Wein & Bliss (1974)
Cotton grass–dwarf shrub tundra	Eagle Creek, Alaska	39	—	25	<5	—	169(+250)	6684	362ˣ	25	7490	Wein & Bliss (1974)
Ombrotrophic mire	Stordalen, Sweden	59	24	70	<3	155	180(+−)	400	300*	12	—	Rosswall et al. (1975)
Wet sedge moss meadow	Devon I	46	130	103	0	279	78(+120)	1295	1097	0	2592	Bliss (1977)
Hummocky sedge moss meadow	Devon I	45	104	33	0	182	86(+187)	2023	908	0	3208	Bliss (1977)
Frost-boil sedge moss meadow	Devon I	58	119	15	0	193	112(+202)	1332	1100	0	2748	Bliss (1977)
Grass heath												
Graminoid steppe	Elef Ringness I, Canada	13	13	32	<1	58	13(+74)	88	2128	20	2323	Bliss & Svoboda (1984)
Graminoid steppe	Elef Ringness I, Canada	4	4	1	<1	9	4(+120)	519	76	9	728	Bliss & Svoboda (1984)

Moss–graminoid meadow	King Christian I, Canada	5	5	32	<1	42	41+	23	2136	10	2210	Bliss & Svoboda (1984)
Dwarf shrub heath												
Cassiope tetragona heath	Devon I	18	90	20	4	132	159(+228)	1041	423	48	1899	Bliss (1977)
Fellfield												
Cushion plant–lichen fellfield	Devon I	15	3	2	3	23	89(+298)	57	15	49	508	Bliss (1977)
Cushion plant–moss fellfield	Devon I	27	5	20	2	54	126(+192)	50°	600	23	991	Bliss (1977)
Barren												
Papaver radicata barren	Devon I	0.5	1.0	0.1	0	1.5	310(+8.2)	0.9	2.4	0	15	Bliss, Svoboda & Bliss (1984)

Production data for bryophytes are probably current shoot production (page 254) but methods of analysis are not described in every case. Data for above-ground vascular plant phytomass are living (+ standing dead). Other phytomass data are living plus attached dead. Total production and phytomass include small algal components in some cases.

*Green phytomass only
×Phytomass to 10 cm depth
°Live roots only
+Sum of living plus standing dead

phytomass of *Alectoria jubata* and other arboreal lichens in northern Saskatchewan as 145 kg ha^{-1} on *Picea mariana* and 88–472 kg ha^{-1} on *Pinus banksiana*. Over 50% of the lichens on the spruce occurred within 3 m of the ground and were considered accessible to caribou (page 283).

The importance of cryptogams also varies in xeric, cool-Arctic fellfields, as indicated for Devon I in Table 7.5. Total annual production appears to be under 60 g m^{-2}, with mosses making a major contribution to both production and phytomass in some cases, and lichens forming a significant component of above-ground phytomass. Lichen phytomass reaches 100–170 g m^{-2} on rock outcrops on Devon I (Richardson & Finegan, 1977). However, Table 7.5 emphasises the sparse total production, with insignificant moss and lichen components, in the extensive barrens that characterise the more arid parts of the North American cool-Arctic.

Colony production in mild- and cool-polar cryptogams Shoot production in individual bryophyte colonies on cool-Antarctic islands is comparable with that in the cold-Antarctic. Particularly high annual shoot production, locally up to 1000 g m^{-2}, characterises bryophytes in both wet and mesic habitats on South Georgia, while on Marion I a clear gradient has been demonstrated from permanently wet drainage lines through mire and mesic slope communities to the more xeric fellfields (Table 7.6). A comparable gradient has been reported on Devon I, but under the drier climatic conditions at this cool-Arctic site maximum annual production was only 350 g m^{-2} for mosses by a stream (Vitt & Pakarinen, 1977).

Similarly, annual shoot production in *Polytrichum alpestre* has been recorded as 450–500 g m^{-2} in open tussock grassland on South Georgia compared with only 100–150 g m^{-2} in mild-Arctic spruce woodland at Churchill (Longton, 1970, 1979b). There is thus little doubt that bryophyte productivity is strongly controlled by water availability as determined by both climatic and topographic factors, in line with the correlation between water availability and annual shoot elongation (Chapter 6). Both mean annual shoot elongation and mean dry weight increase per shoot are higher in *P. alpestre* at Churchill than on South Georgia or in highly productive turfs in the cold-Antarctic, the reduced annual production at Churchill resulting from low spatial density of shoots within the colonies.

There are few data for production in individual lichen colonies. Andreev (1954) showed that phytomass increase in stands of *Cladonia* spp during the accumulation phase (Chapter 6) may average 17–27 g m^{-2}

Table 7.6 *Representative data for current shoot production (g m^{-2}) in colonies of cool-Antarctic bryophytes*

Locality	Species	Annual shoot production	Source
South Georgia			
Late snowpatch flush	*Brachythecium subplicatum*	1017	R. I. Lewis Smith (1982b)
Late snowpatch flush	*Brachythecium austro-salebrosum*	554	Smith (1982b)
Stream bank	*Pohlia wahlenbergii*	782	Clarke, Greene & Greene (1971)
Stream	*Pohlia wahlenbergii* and *Tortula robusta*	1028	Clarke *et al.* (1971)
Graminoid mire	*Calliergidium austro-stramineum*	642	Smith (1982b)
Graminoid mire	*Sphagnum fimbriatum*	716	Smith (1982b)
Graminoid mire	*Tortula robusta*	325	Smith (1982b)
Mossbank in open tussock grassland	*Polytrichum alpestre*	462	Longton (1970)
Sheltered part of mossbank	*Chorisodontium aciphyllum*	913	Smith (1982b)
Exposed part of mossbank	*Chorisodontium aciphyllum*	292	Smith (1982b)
Acaena herbfield	*Tortula robusta*	111	Smith (1982b)
Moist rock face	*Andreaea* cf. *alpina*	104	Smith (1982b)
Marion Island			
Drainage line	*Brachythecium subplicatum*	754	Russell (1985)
Mire	*Blepharidophyllum densifolium*	326	Russell (1985)
Slope	*Brachythecium rutabulum*	113	Russell (1985)
Fellfield	*Ditrichum strictum*	27	Russell (1985)

yr^{-1} in Soviet tundra and taiga. The productivity on *Cladonia rangiferina* on South Georgia has been estimated as 98–127 g m^{-2} yr^{-1} (Lindsay, 1975), or up to 1130 g m^{-2} yr^{-1} (Smith, 1984a), both estimates again considerably exceeding the production figures for Arctic lichens in Tables 7.4 and 7.5. These plants tend to be abundant in drier habitats than bryophytes, and their low productivity in much of the Arctic supports the view (Chapter 5) that many lichens spend considerable periods during the growing season too dry to be physiologically active.

Nutritional value of cryptogamic phytomass

Energy contents of 17–20 kJ g^{-1} dry weight have been reported for green phytomass in polar bryophytes (Table 7.7). Rastorfer (1976) suggested that values for Arctic mosses are slightly above those for comparable, lower latitude material, in agreement with an established trend in vascular plants. Energy content is generally lower in brown phytomass, and within a given community that in vascular plants is on average slightly greater than in bryophytes, as on Signy I (Table 7.7) and in alpine tundra at Hardangervidda (Table 7.8). In mosses on Marion I, energy content on a dry weight basis was found to increase along a hydric → xeric gradient, this relationship being the reverse of that for production, but the position was less clear in terms of ash free dry weight because ash content was greatest in the wetter habitats (Russell, 1985). A broadly similar situation exists on Devon I (Fig. 7.2), but some hydric mosses have energy contents on a dry weight basis as high as those from dry habitats, e.g. *Bryum cryophilum* on Devon I and *Tortula robusta* on South Georgia (Table 7.7).

The highest energy contents in lichens have been recorded in nitrogen-fixing species, with values up to 20.7 kJ g^{-1} dry weight (Kallio, 1975). Results for other lichens are generally similar to, or rather lower than those of associated bryophytes, as at Hardangervidda (Table 7.8). Similarly on Devon I, values of 14.6 kJ g^{-1} in *Cetraria cucullata* and 16.7–18.2 kJ g^{-1} and exceptionally 19.2 kJ g^{-1} in *Thamnolia vermicularis* (Richardson & Finegan, 1977) were below typical results for the mosses (Table 7.7). Both energy and organic nutrient contents in the photosynthetically active portions of mosses and lichens show less seasonal variation than those in leaves and other vascular plant organs (Skre, Berg & Wielgolaski, 1975), due perhaps to restricted translocation.

The low energy content of cryptogams is reflected in the concentrations of major organic nutrients. Table 7.9 shows that protein content is lower in cryptogams than in most vascular plants at Hardangervidda, ranging from 6–12% dry weight in green moss, and from 3–4% in lichens, compared with values up to 28% for monocots. Comparable differences have been recorded on Svalbard (Staaland, Brattbakk, Ekern & Kildemo, 1983). On Devon I, protein content in 35 mosses ranged from 2% to 16%, with the highest levels in green parts of wetland species (Pakarinen & Vitt, 1974). Crude protein content in many lichens at boreal sites in North America was under 2%, but reached 4.5% in arboreal species of *Usnea* and 8–22% in terricolous, nitrogen-fixing species (Scotter, 1965, 1972).

Soluble sugar contents at Hardangervidda were also consistently lower in mosses (1–10%), and particularly in lichens (<1%), than in vascular plants (Table 7.9). Comparable values (3–8%) have been reported in cold-Antarctic mosses (Rastorfer, 1972), and in *Dicranum fuscescens* at Schefferville where a late season maximum suggested involvement in frost resistance (Chapter 5). The lichen *Thamnolia vermicularis* on Devon I contained 2% soluble sugars, principally sugar alcohols, with little variation during the snow-free period (Richardson & Finegan, 1977). Similar compounds predominate among the soluble sugars in lichens on Signy I, where concentrations do vary seasonally and are again high enough to suggest a role in frost resistance (Chapter 5). Starch and other storage polysaccharides occur in comparable concentrations in mosses, lichens and vascular plants at Hardangervidda, although the polysaccharide content is consistently greatest in monocots (Table 7.9).

Ether-soluble compounds, principally lipids, represented 1–2% dry weight in mosses, and 1% and 5% in two species of lichen on Svalbard, compared with 7% in woody plants and 2–3% in other phanerogams (Staaland *et al.*, 1983). Crude fat concentrations in many North American boreal forest lichens (1–3%) were below those in woody plants (3–11%), but fat contents of 3–6% were recorded in some arboreal lichens (Scotter, 1965, 1972). On Devon I, ether-soluble compounds formed 2–4% dry weight of the mosses tested, with no significant difference between green and brown material (Pakarinen & Vitt, 1974). A similar result was reported for lipid concentration in green and brown portions of *Dicranum fuscescens* at Schefferville by Hicklenton & Oechel (1977b) using methanol-chloroform extraction, and this method indicated 7–18% lipids in cold-Antarctic mosses (Rastorfer, 1972). The occurrence in mosses of triglycerides and unusual fatty acids was noted in Chapter 6.

In contrast to the compounds so far discussed, the lignin content recorded at Hardangervidda was consistently higher in bryophytes than

Table 7.7 Representative data for energy content (kJ g⁻¹) and ash content (%) in polar bryophytes

Zone and locality	Species	Habitat	Energy content		Ash content	Notes
			Dry weight	Ash free dry weight		
Cold-Antarctic *Signy I*	*Brachythecium austro-salebrosum*	Associated with *Deschampsia antarctica*	16.90	—	—	Collins, Baker & Tilbrook (1975). Energy contents in different parts of *D. antarctica* = 18.64–20.40 kJ g⁻¹
Cool-Antarctic *South Georgia*	*Chorisodontium aciphyllum*	Moss bank	19.18	19.51	1.7	Smith & Walton (1975)
	Polytrichum alpestre	Moss bank	19.26	19.55	1.5	Data refer to current year growth in mid-growing season
	Sphagnum fimbriatum	*Rostkovia* mire	17.71	18.25	3.0	
	Tortula robusta	*Rostkovia* mire	19.85	20.47	3.0	
Marion I	*Brachythecium subplicatum*	Drainage line	14.80	15.60	5.2	Russell (1985) Data are approximations read from graphs and refer to a mixture of green and brown material
	Blepharidophyllum densifolium	Mire	15.70	16.40	4.0	
	Brachythecium rutabulum	Mesic slope	16.20	16.70	2.8	
	Ditrichum strictum	Fellfield	16.40	16.50	0.5	

Cool-Arctic
Devon I

Bryum cryophilum	Hydric	19.64	20.87	5.9	Pakarinen & Vitt (1974). Data refer to green parts sampled 27 July–6 August
Calliergon giganteum	Hydric	17.09	19.87	14.0	
Meesia triquetra	Hydric	17.35	19.19	9.6	
Aulacomnium acuminatum	Mesic	18.46	18.90	2.3	
Tomenthypnum nitens	Mesic	18.66	19.32	3.4	
Hylocomium splendens	Xeric	18.68	19.36	3.5	
Racomirirum lanuginosum	Xeric	19.05	19.28	1.2	

Barrow

Campylium stellatum	Low centre polygon	18.31	19.23	4.9	Rastorfer (1976). Data refer to green parts sampled on 15 August
Polytrichum commune	High centre polygon	18.73	19.24	2.8	

Table 7.8 Composite data for energy content of phytomass from Norwegian tundra sites ($kJ\ g^{-1}$ dry weight)

Plant parts	Shrubs	Forbs	Monocots	Pteridophytes	Bryophytes	Lichens
Green	20.37	19.03	18.56	17.43	17.14	16.25
Brown above ground	19.91	18.56	18.44	17.39	16.46	—
Roots and rhizomes	19.36	18.02	18.23	14.22	—	—

Data from Wielgolaski & Kjelvik (1975)

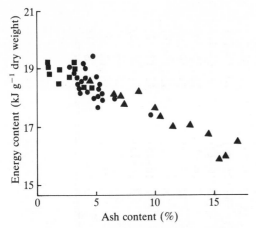

Fig. 7.2. Relationship between ash content and energy content in mosses from wet sedge moss meadows (▲), hummocky sedge moss meadows (●) and rock outcrops (■) on Truelove Lowland, Devon I, Canada. Data from Vitt & Pakarinen (1977).

in green parts of other plants (Table 7.9), with maxima in late summer. The substance concerned was almost certainly not true lignin, but a lignin-like phenolic compound known to occur in high concentrations in bryophyte cell walls (Hébant, 1977). Similar compounds are probably responsible for the higher concentrations of crude fibre reported in mosses (23–30%) than in vascular plants on Svalbard (Staaland *et al.*, 1983), as crude fibre includes lignin-like compounds as well as much of the cellulose and part of the hemicellulose. Crude fibre content of South Georgian mosses varied from only 15–23% in *Tortula robusta* to 36% in *Polytrichum alpestre* (Fig. 7.5). The occurrence of lignin in lichens (Table 7.9) is also doubtful.

Crude fibre content was low in lichens on Svalbard (10–17%), with comparable levels in several lichens at Inuvik and in northern Saskatchewan (Scotter, 1965, 1972; Staaland *et al.*, 1983). In contrast, Scotter's studies suggested that crude fibre ranges from 20–35% in *Peltigera* spp and 30–70% in *Cladonia* spp, the latter being considerably above the levels common in vascular plants. Crude fibre is generally regarded as roughage of little food value, but in lichens it apparently includes significant amounts of the readily digestible carbohydrates lichenin and isolichenin (Scotter, 1972). These compounds are included in the holocellulose component and may contribute towards the high concentration of holocellulose recorded in lichens from lichen heath at Hardangervidda (Table 7.9), in which species of *Cladonia* are abundant (Wielgolaski &

Table 7.9 *Mean concentrations (and seasonal amplitude) of some organic compounds (% dry weight green material) in plants from different habitats at Hardangervidda (Norway)*

Habitat	Plant group	Lignin	Holocellulose	Ethanol-soluble sugar	Storage polysaccharide	Protein
Birch woodland	Trees and shrubs	14 (7–18)	32 (26–35)	13 (8–15)	6 (5–7)	11 (9–14)
	Dwarf shrubs	20 (19–20)	28 (25–33)	11 (4–15)	8 (7–10)	8 (7–10)
	Monocotyledons	11 (5–18)	45 (41–55)	12 (6–17)	13 (8–17)	8 (6–16)
	Bryophytes	35 (23–46)	33 (24–43)	4 (1–8)	8 (3–15)	7 (6–8)
Willow scrub	Shrubs	17 (8–23)	28 (20–39)	12 (4–22)	8 (6–13)	18 (13–24)
	Monocotyledons	13 (6–20)	38 (30–48)	13 (6–18)	14 (10–20)	11 (7–14)
	Forbs	20 (14–26)	22 (17–29)	14 (11–17)	10 (5–16)	13 (8–20)
	Bryophytes	41 (36–51)	28 (16–41)	5 (3–7)	7 (5–12)	8 (7–8)
Lichen heath	Dwarf shrubs	21 (16–26)	26 (23–31)	13 (10–17)	10 (5–19)	7 (6–7)
	Monocotyledons	19 (19–19)	44 (37–51)	10 (7–16)	17 (11–20)	10 (9–12)
	Lichens	15 (10–21)	68 (60–73)	0.6 (0.4–0.8)	13 (10–15)	4 (3–4)
Dry meadow	Dwarf shrubs	24 (18–29)	27 (23–32)	14 (11–18)	7 (4–9)	13 (11–16)
	Monocotyledons	13 (7–27)	39 (32–46)	11 (6–15)	12 (7–18)	14 (8–28)
	Forbs	19 (17–25)	27 (18–38)	8 (5–9)	8 (4–12)	15 (11–18)
	Bryophytes	35 (30–37)	37 (32–42)	2 (1–4)	11 (6–18)	12 (12–12)
Wet meadow	Dwarf shrubs	26 (18–32)	21 (15–25)	13 (10–17)	8 (6–15)	17 (12–25)
	Monocotyledons	15 (7–23)	41 (38–48)	9 (5–15)	14 (8–18)	13 (8–16)
	Forbs	19 (17–22)	23 (17–32)	13 (9–18)	8 (4–12)	18 (12–22)
	Bryophytes	38 (36–40)	32 (31–33)	5 (3–10)	5 (3–7)	10 (8–11)

Data from Skre, Berg & Wielgolaski (1975)

Kjelvik, 1973). The range of holocellulose concentrations in mosses is comparable with that in vascular plants (Table 7.9; Pratt & Smith, 1982).

Mineral contents also vary between the major life forms. Protein levels are calculated as nitrogen content × 6.25, and thus it is already clear that nitrogen contents are greater in vascular plants than in cryptogams, except in nitrogen fixing lichens. Lichens are commonly low in several other elements, including phosphorus, potassium, calcium and magnesium (Nieboer, Richardson & Tomassini, 1978). The low calcium content in *Cladonia amaurocraea* at Barrow is particularly noticeable (Table 7.10), but higher values were recorded in *Masonhalea richardsonii* and other lichens. Similarly low calcium contents of 0.05–0.15% were recorded in *Cladonia* spp from northern Canada, with levels in *Cetraria nivalis* higher (0.3–0.4%) throughout the year (Scotter, 1965, 1972). A comprehensive survey of Canadian Arctic lichens has confirmed that variation between species and between sites may be considerable, but mean calcium content was again particularly low in species of *Cladonia*, a fact attributed in part to their preference for acidic substrata (Puckett & Finegan, 1980). Element contents in bryophytes also vary widely, but for some elements they may be comparable with, or rather greater than those in some types of vascular plant. Thus calcium is typically higher in mosses than in foliage of associated monocots, but maximum values occur in herbaceous dicots (Fig. 7.5; Staaland *et al.*, 1983).

Herbivory

Energy flow is primarily via the detritus pathway in many terrestrial ecosystems, the situation in polar regions conforming to this pattern (Table 7.11; Fig. 7.3). Moreover, it is generally believed that lichens, and particularly bryophytes, are less freely consumed by herbivores than vascular plant foliage. Suggested reasons include the slightly lower nutrient value of the cryptogams, the presence of secondary compounds acting as feeding inhibitors, and the low digestibility of mosses resulting from lignin-like phenolics and other cell wall compounds. Some lichens also possess toxic chelating agents, or heavily agglutinated cortical hyphae which could confer physical protection against invertebrates (Frankland, 1974; Gerson, 1982; Gerson & Seaward, 1977; Longton, 1984; Rundel, 1978). The view that mosses and lichens are not readily eaten is also generally applicable to polar ecosystems (Chernov, 1985), but consumption of these plants by some boreal and Arctic vertebrates occurs on a scale apparently without parallel in temperate and tropical regions.

Table 7.10 *Element content (% dry weight) of plant material from Barrow, Alaska*

Species	Date	N	P	K	Ca	Mg	Fe	Zn
Cladonia amaurocraea	July	0.47	0.07	0.15	0.06	0.07	0.02	0.001
Masonhalea richardsonii	July	0.52	0.04	0.13	0.59	0.06	0.01	0.001
Peltigera aphthosa	July	2.38	0.11	0.57	0.19	0.13	0.03	0.002
Thamnolia vermicularis	July	0.83	0.08	0.34	0.50	0.05	0.02	0.003
Campylium stellatum	Aug	1.45	0.23	0.55	1.00	0.42	0.11	0.006
Polytrichum hyperboreum	Aug	0.90	0.18	0.66	0.30	0.18	0.04	0.005
Polytrichum commune	Aug	1.50	0.21	0.88	0.15	0.13	0.05	0.005
Dupontia fisheri (leaf blade)	July–Sept	3.00	0.40	1.63	0.22	0.15	0.01	0.005
Dupontia fisheri (leaf sheath)	July–Sept	2.37	0.42	2.17	0.08	0.12	0.01	0.004
Carex aquatilis (leaf blade)	July–Sept	3.32	0.28	1.46	0.25	0.18	0.01	0.006
Petasites frigidus (leaf blade)	July–Sept	1.51	0.26	3.16	0.84	0.47	0.01	0.009

Data, from Rastorfer (1974), Ulrich & Gersper (1978) and Williams *et al.* (1978), are based on one or two samples per species for mosses and lichens. Data for flowering plants are means of up to 64 samples. Samples comprised whole lichens, the green portions of mosses, and angiosperm leaves as indicated.

Table 7.11 *Estimated annual uptake (g dry weight m⁻² yr⁻¹) of food
sources by primary consumers in two moss communities on Signy I*

Food source	Tall moss turf community	Moss carpet community
Algae	92.5⎫ 10.5⎭	146.1⎫ 14.9⎭
Bryophytes and lichens (living)	0.05	0.15
Bacteria, fungi and other microflora	95.3⎫ 192.6⎭	147.3⎫ 278.6⎭
Dead organic matter	20.8⎫ 5.5⎭	3.9
Living animals	0.11	0.16
Total	208.8	297.6

Pairs of figures indicate estimates derived from different assumptions about the
diet of particular groups of consumers.
Data from Davies (1981)

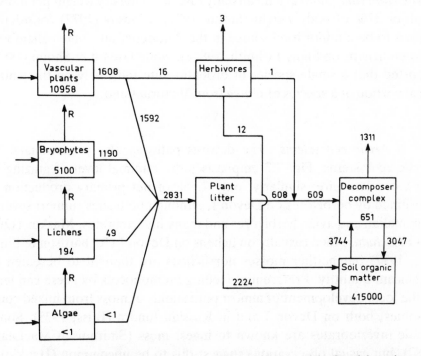

Fig. 7.3. Energy flow through an ombrotrophic part of the Stordalen mire,
Abisco, Sweden. Boxes represent biomass, and lines annual energy flow, in
kJ m⁻². R is respiration. Reproduced from Svensson & Rosswall (1980) by
courtesy of Springer-Verlag.

Antarctic ecosystems   The dominance of the detritus pathway is particularly pronounced in southern polar regions where, except in the mild-Antarctic, native land-feeding vertebrates are represented by only a few species of birds on cool-Antarctic islands (Burger, 1985). Arthropods, particularly collembola and mites, are the most conspicuous invertebrates, but other groups such as rotifers, nematodes and tardigrades are also represented, and protozoa often form the bulk of the animal biomass (Davis, 1981). Table 7.11 shows that direct feeding by such animals on bryophytes and lichens in cold-Antarctic moss turf and carpet communities was negligible, despite the high moss productivity. The principal food of the resident fauna comprised bacteria, fungi and algae (Davis, 1981).

Limited grazing of Antarctic cryptogams does occur. Some collembola on Signy I consume both algae and moss peat (Block, 1985). Moss gametophytes were eaten by several frigid-Antarctic collembola in laboratory feeding trials (Wise, Fearon & Wilkes, 1964; Pryor, 1962), and coleoptera consume both gametophytes and sporophytes of mosses on South Georgia (Smith & Walton, 1975). In feeding trials on Marion I, a weevil ate *Brachythecium rutabulum* at a mean daily rate of 1.7 mg dry weight per individual, or 37% of body weight (Smith, 1977). Lindsay (1977) considered lichens to be a minor food source in the Antarctic, but some collembola feed on lichens on Signy I (Smith, 1984a), while Gressit & Shoup (1967) reported that a single species of orobatid mite fed freely on the white, basal portion of a species of *Usnea* from Victoria Land.

Arctic ecosystems   The detritus pathway also predominates in Arctic ecosystems. Fig. 7.3 emphasises the minimal level of grazing in the Stordalen mire; similarly, only 2.5% of net primary production is consumed by herbivores on Devon I, although both sites support several mammalian and avian herbivores and many invertebrates. Muskox (*Ovibos moschatus*) feed casually on lichens on Devon I (Richardson & Finegan, 1977) but neither mosses nor lichens are thought to be eaten in significant quantity. Preferential feeding on monocots by geese can lead to the local development of almost pure stands of moss from mixed communities, both on Devon I and in Russian tundra (Bliss, 1975). Some Arctic invertebrates are known to ingest moss (Smirnov in MacLean, 1980), but casual observations suggest this to be uncommon (Haukioja, 1981).

There are, however, reports that mosses are more freely consumed in Arctic regions than at lower latitudes by vertebrates including several species of geese and lemmings. Prins (1982) has speculated that the animals benefit from arachidonic acid, a highly unsaturated fatty acid with four double bonds so far unreported from angiosperms. It occurs in some algae and pteridophytes, but the highest concentrations are found in mosses, particularly those of wet habitats, in which it may comprise 35% of the fatty acid content. Prins suggested that unsaturated fatty acids could confer several benefits on geese and other animals in cold environments, for example by increasing limb mobility at low temperature, and protecting cell membranes against cold. Arctic geese are migratory, returning to the tundra in spring when lipid concentrations in some mosses are at their highest (Chapter 6). Moss consumption could then benefit adult birds and also young to which unsaturated fatty acids are transferred via the eggs.

Lemmings Mosses form a small part of the diet of several Arctic rodents, including a ground squirrel, *Spermophilus parrayi* (Batzli & Sobaski, 1980), and a vole, *Clethrionomys rutilis*. The vole appears to eat moss freely in early summer when its normal diet of berries is unavailable (West, 1982). Collared lemmings (*Dicrostonyx groenlandicus*) have been observed to feed on *Polytrichum* gametophytes during summer on both Ellesmere I (Longton, 1980) and Devon I. In feeding trials animals from Devon I ate only capsules when offered fruiting *Funaria arctica*, possibly attracted by the high fat content of the spores (Pakarinen & Vitt, 1974).

Moss gametophytes are more freely consumed by species of *Lemmus*. Both *L. sibiricus* in North America and *L. lemmus* in Fennoscandia eat graminoids, generously supplemented by mosses particularly in winter (Bunnell, MacLean & Brown, 1975; Wielgolaski, 1975c). In Soviet tundra, Tiskov (1985) considers both *L. lemmus* and *L. amurensis* to be obligate bryophages, consuming 70–90% moss, but Chernov (1985) contends that any consumption of moss by these species is accidental.

Populations of *L. sibiricus* fluctuate widely through cycles of 4–6 years, those at Barrow varying from normal levels of around 0.5 individuals ha^{-1} to peaks of 100–200 individuals ha^{-1} (Batzli *et al.*, 1980). Plant consumption reaches 700 kJ m^{-2} yr^{-1} during peak years, and in contrast to the situation on Devon I, this may amount to some 25% of above-ground or 10% of total net primary production. Mosses, particularly species of *Calliergon*, *Dicranum* and *Polytrichum*, form 5–20% of the diet during summer and 30–40% in winter (Bunnell, MacLean & Brown, 1975). However, lemmings from Barrow lost weight when fed a diet of moss. Their daily food ingestion was only 20% of that of comparable animals offered tundra graminoids, the mosses being significantly less digestible than the monocots, and giving low energy yields (Table 7.12). Digestibility of mosses by lemmings is nevertheless higher than by other herbivores, and this may represent a nutritional adaptation (Batzli, White & Blunnell, 1981).

Given these results, it is not immediately clear why *L. sibiricus* should eat moss. Possible explanations include a shortage of more nutritious food, particularly in winter, and the potentially beneficial effects of unsaturated fatty acids proposed by Prins (1982). Mosses could also provide an important source of mineral nutrients (Bunnell, MacLean & Brown, 1975). Overall digestibility of tundra monocots by *L. sibiricus* is relatively low (Table 7.12), and Batzli & Cole (1979) suggested that the increased food intake necessary to satisfy energy requirements may improve the animals' nutrient status. Some elements are almost completely digested when in short supply, and several, including calcium, magnesium and iron, occur in higher concentrations at Barrow in mosses than in the leaf sheaths of *Dupontia fisheri*, another major winter food source (Table 7.10).

Grazing pressure can cause massive damage to bryophyte and lichen vegetation during peak years, through both consumption, and disturbance as expanding lemming populations seek graminoid shoot bases beneath winter snow (Bunnell, MacLean & Brown, 1975; Chernov, 1985; Oksanen & Oksanen, 1981). Recovery is slow, and exclosure experiments at Barrow suggest that lemming grazing may, in the long term, favour the creation of grass-sedge meadows rather than stands of moss on well-drained sites (Batzli *et al.*, 1980; Batzli, White & Bunnell, 1981).

Caribou and reindeer   Lichens are vitally important to caribou and reindeer, *Rangifer tarandus*. The taxonomy of this species is controversial, but four ecological groups of populations may be recognised, i.e. barren-ground caribou, reindeer, woodland caribou and finally caribou and reindeer on cool-Arctic islands. Barren-ground caribou currently number about a million individuals in North America (Bliss, 1975). They undergo extensive migration between summer feeding grounds in the tundra and winter ranges close to or south of treeline. Most Eurasian reindeer, with 700000 individuals in Fennoscandia alone, are at least partially domesticated. They are grazed in a range of habitats, often with seasonal movement between tundra and woodland in parallel with the barren-ground caribou. Essentially wild herds also occur, notably in the

Table 7.12 *Mean (±1 s.e.) food consumption and digestibility by brown lemmings (Lemmus sibericus) on five different diets*

Diet	Number of animals	Ingested dry matter (mg g^{-1} day^{-1})	Ingested energy (J g^{-1} day^{-1})	Digestibility % Dry matter	Energy	Digestible energy (J g^{-1} day^{-1})	Metabolisable energy (J g^{-1} day^{-1})
Commercial rabbit pellets	6	198 ± 17	3675 ± 315	54.5 ± 1.8	56.9 ± 1.8	2117 ± 223	2062 ± 227
Leaves and stems of *Dupontia fisheri*	9	356 ± 21	6875 ± 391	33.3 ± 2.2	35.1 ± 1.9	2436 ± 214	2318 ± 210
Leaves of *Carex aquatilis*	5	368 ± 41	7434 ± 924	33.2 ± 0.5	34.0 ± 0.5	2516 ± 298	2390 ± 286
Leaves of *Eriophorum angustifolium*	9	344 ± 12	6871 ± 248	36.8 ± 0.6	39.3 ± 0.5	2696 ± 97	2596 ± 105
Bryophytes, mostly *Calliergon* and *Drepanocladus* spp	7	72 ± 2	1260 ± 34	22.9 ± 2.2	25.4 ± 2.0	319 ± 25	294 ± 29

Data, from Batzli & Cole (1979), refer to subadult animals of 30–40 g weight.

Soviet Union, and on Hardangervidda where the animals occupy non-wooded habitats throughout the year. Reindeer have been extensively introduced elsewhere. The other forms occur in smaller numbers. Woodland caribou spend the winter in woodland and the summer in nearby tundra with only limited migration, while island reindeer and caribou in parts of Greenland, Spitzbergen and the Canadian Arctic Archipelago are confined to cool-Arctic tundra throughout the year (Bliss, 1975; White *et al.*, 1981).

R. tarandus is a generalist feeder. Its diet varies between herds, presumably depending on availability of palatable species, but lichens are consistently important in winter (Fig. 7.4) when they commonly represent

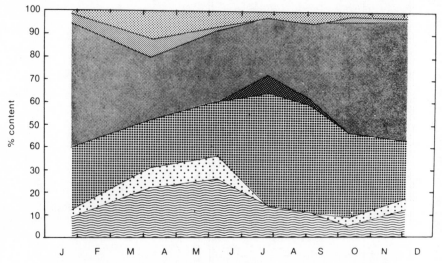

Fig. 7.4. Seasonal variation in the representation of different types of plant in reindeer rumen-content at Hardangervidda, Norway, in 1970. Plant types from top to bottom: mosses, epilithic lichens, terricolous lichens, forbs, graminoids, litter, woody plants. Reproduced from Gaare & Skogland (1975) by courtesy of Springer-Verlag

60–70% of food intake in barren-ground caribou and domesticated reindeer (Boertje, 1984; Chernov, 1985; Thompson & McCourt, 1981). The animals reach the lichens by digging craters in relatively deep but soft woodland snow, or in the shallow but harder snow cover of exposed lichen heaths. *Cladonia* and *Cetraria* spp appear to be preferred to *Stereocaulon* spp and other nitrogen fixing lichens (Kallio, 1975), but there are conflicting reports concerning the preference of *R. tarandus* for particular species. Gaare & Skogland (1975) consider *Cladonia stellaris* less acceptable than related lichens, but Rundel (1978) cites earlier reports that *C. stellaris* is preferred to *C. arbuscula* and *C. rangiferina*, both of which contain the bitter-tasting and possibly inhibitory depside fumar-protocetraric acid. Arboreal species of *Alectoria*, *Evernia* and *Usnea* are also eaten, and reindeer herders in Fennoscandia fell lichen-covered trees to provide an emergency food supply during periods of adverse feeding conditions (Hustich, 1951).

Reindeer on Hardangervidda eat minor amounts of epilithic lichens in spring, but lichen consumption then declines as graminoids, willow leaves and forbs become available. Lichens continue to represent more than 25% of the diet during summer at Hardangervidda (Fig. 7.4), but their intake is negligible at midsummer in many herds. Less is known about woodland caribou diet, but arboreal lichens are important in winter (Richardson & Young, 1977; White *et al.*, 1981).

Bryophyte consumption by *R. tarandus* is variable, but normally low. Mosses are actively avoided by most North American populations, but they form about 30% of the winter diet of the Porcupine caribou herd in Alaska and the Yukon (Thompson & McCourt, 1981; White *et al.*, 1981). Rumen content of reindeer on Hardangervidda contains up to 12% moss in winter (Fig. 7.4), but the moss is thought to be unavoidably ingested during grazing on lichens there, and in the Soviet Union (Chernov, 1985; Gaare & Skogland, 1977).

Island caribou are distinctive because their winter diet contains a low proportion of lichens and high proportions of monocots and mosses. This may be largely attributable to low availability of preferred lichens, and an extensive study in Canada indicated some correlation between the abundance of lichens in rumen contents and in the vegetation. Mean lichen content of rumen samples ranged from 2% to 15% for a given island but varied between years and reached 38% in some individuals, with *Alectoria*, *Cetraria*, *Pertusaria* and *Thamnolia* the most important genera. The moss component of these samples averaged 13–58%, with individual values up to 86% (Thomas & Edmonds, 1983). Similarly, mosses comprised 32–39% of the rumen content of reindeer on Svalbard, with no lichens recorded (Reimers, 1977).

The preferred lichen species in some respects represent a low quality food, being deficient in protein, lipids and several essential mineral elements. The experiments of White, Jacobsen & Staaland (1984) suggest that lichens in the alimentary canal may bind some minerals and thereby restrict uptake by the animals. There are reports that captive animals lose weight on a diet of lichens. A supplement of nitrogen and other minerals reverses this trend, in part by increasing lichen consumption, and leads to enhanced milk production and calf survival in pregnant females (Jacobsen, Hove, Bjarghor & Skjenneberg, 1981; Staaland, Jacobsen & White, 1984). However, *Cladonia* spp contain abundant carbohydrate (page 273). They are readily digestible (Table 7.13), and

Table 7.13 *Digestibility of forage plants on a summer caribou range at Atkasook and Prudhoe Bay, Alaska*

Plant group	Number of species	Digestion (%)		In vitro
		In vivo		*In vitro*
		10 hr	48 hr	48 hr
Graminoids				
Eriophorum spp	3	37 ± 8	86 ± 1	49 ± 2
Carex spp	2	30 ± 1	56 ± 9	44 ± 8
Grasses	5	37 ± 3	68 ± 7	56 ± 4
Herbaceous dicots	8	72 ± 3	83 ± 2	51 ± 2
Deciduous shrubs	6	55 ± 7	71 ± 7	45 ± 4
Evergreen shrubs	5	31 ± 8	44 ± 7	26 ± 4
Lichens				
Highly preferred	4	50 ± 1	89 ± 2	45 ± 9
Preferred	5	24 ± 3	55 ± 7	28 ± 4*
Non-preferred	7	34 ± 13	73 ± 8	19 ± 5†
Mosses	6	—	11 ± 3	11 ± 2

Data, from White & Trudell (1980), are means ±1 s.e. for the numbers of species indicated.
*Data for four species
†Data for six species
— No data

form an effective, available energy source in winter when the animals must maintain a high basal metabolism to generate body heat, while utilising fat reserves and breaking down muscle to compensate for a low nitrogen diet (Bliss, 1975; Klein, 1982; Person, Pegau, White & Luick, 1980; Richardson & Young, 1977). The body reserves are replenished during spring feeding on vascular plants, particularly on northward migration following the thaw when young, highly nutritious leaves are consumed (Skogland, 1984).

The low protein content of forage lichens can be interpreted in several ways. Klein (1982) stresses that caribou supplement winter protein intake by active selection of green vascular plant material whenever available, and he reports that the animals consume nitrogen fixing species of *Pelti-gera* and *Stereocaulon* in limited amounts, as well as the generally preferred cladonias. On the other hand, Batzli, White & Bunnell (1981) note that *R. tarandus*, like other ruminants, is able to recycle nitrogen from urea to the rumen. Indeed, reindeer have specialised kidneys capable of concentrating urea when dietary protein is low. These authors also argue that low protein intake in winter may be advantageous in decreasing urinary and faecal water loss, and in turn the energy required to heat ingested frozen water.

Some temperate invertebrates graze selectively on lichens with relatively low contents of nitrogen and other essential elements, a situation parallel with the preference of *R. tarandus* for *Cladonia* spp rather than nitrogen fixing species. Lawrey (1983) speculated that lichens with a favourable nutrient status may be best able to produce chemical inhibitors to feeding, and Rundel (1978) cited several cases of lichens unpalatable to caribou apparently possessing such compounds (page 283). Many lichens also contain bacteriostatic agents, and the microbial population in the reindeer rumen changes reversibly from bacteria, when feeding is primarily on vascular plants, to protozoa when lichens are being eaten (Skogland, 1984). Whether this contributes to the greater digestibility of lichens by *R. tarandus* than by most herbivores remains to be determined, but on present evidence the premise that *Rangifer* evolved in response to a lichen-based Arctic food niche unoccupied by other herbivores (Klein, 1982) has many attractions.

In contrast to lichens, digestibility of mosses by *R. tarandus* is consistently low (Table 7.13; Person *et al.*, 1980; Thomas & Kroeger, 1981). High moss consumption by island caribou in Canada appears to be correlated with deep snow cover and other difficult feeding conditions, and to result in malnutrition (Thomas & Edmonds, 1983). There is some evidence that island reindeer on Svalbard have evolved morphological adaptations to a low quality diet including abundant moss, notably an unusually large caecum–colon complex (Staaland, Jacobsen & White, 1979).

Rangifer are extensive rather than intensive feeders. They move continuously, feeding in a given area for only a few weeks at a time, and while some areas are visited annually, others are grazed only spasmodically. They are thus well adapted to terrain where plant growth is slow and overgrazing inhibits forage production for many years. Work reviewed by Bliss (1975) and Moser, Nash & Thomson (1979) indicates winter lichen consumption by one individual as about 5 kg dry weight day^{-1}. Virgin lichen range in Alaska can yield 3360 kg ha^{-1} of forage, but grazing at high intensity would almost completely remove the lichen cover with the result that restoration would take 30–50 years. Recovery is more rapid where the lichens are not grazed below the actively growing portions, giving a three year grazing cycle for Soviet reindeer where 30–35% of the lichen cover is removed, and a three to five year cycle in Alaska with a 45% harvest (Bliss, 1975).

Consumption is probably the main cause of winter damage to lichens, but Oksanen (1978) considered trampling to be more destructive during summer, in the absence of protective snow. In his study area in Arctic Norway, reindeer had removed the *Cladonia* layer almost completely from summer range, while the winter range appeared to be in a subclimax stage controlled by grazing, with a thin (2–4 cm) and rather uneven *Cladonia* cover.

Interaction between *Rangifer* and lichens is also evident on islands where reindeer have been introduced. On St Matthew I in the Bering Sea, an introduced population virtually eliminated forage lichens at the time of maximum animal numbers, and then crashed to near extinction during one particularly severe winter (Klein, 1968). In contrast, introduced reindeer on South Georgia have survived for 75 years despite loss of most of the *Cladonia* cover from the feeding grounds. Their success results from the presence of nutritious vascular plants such as *Acaena magellanica*, and particularly *Poa flabellata* which is available in winter. The slow recovery of *Cladonia* spp from overgrazing is demonstrated by exclosures where lichen cover had barely increased after eight years, whereas *A. magellanica* and *P. flabellata* had recovered extensively (Kightly & Smith, 1976; Leader-Williams, Scott & Pratt, 1981; Vogel, Remmert & Smith, 1984).

Decomposition

The relatively low proportion of net primary production consumed by herbivores in many tundra ecosystems results in most phytomass passing into the detritus pathway as dead organic matter (DOM). The relationship between rates of net production and decomposition thus largely determines whether a system is balanced, with inputs of carbon and energy matched by losses through community respiration, or is subject to an accumulation of DOM. Although decomposition rates vary considerably, the more productive tundra communities tend to have larger pools of DOM in relation to net primary production than lower latitude systems, implying that decomposition is more inhibited than production by polar environments.

Low temperature, short growing season and lack of large, litter-consuming invertebrates at many sites are among the major factors restricting decomposition, particularly where permafrost results in its almost complete cessation close to the ground surface. The supply and seasonal distribution of moisture also exert a major control, breakdown being most rapid at sites that are wet but not sufficiently waterlogged to create anaerobic conditions. Thus within the polar regions, the highest rates of decomposition occur in moist substrata under relatively warm, oceanic conditions, as on cool-Antarctic islands, and the lowest in regions with cold, dry climates and in permanently waterlogged mires.

Vegetation quality also profoundly influences decomposition, breakdown being favoured by low lignin content and high concentration of soluble constituents and mineral nutrients (Heal, Flanagan, French & MacLean, 1981). Foliage is reduced more rapidly than wood, and mosses tend to decompose slowly, presumably for various combinations of the factors already discussed in relation to herbivory. In addition, *Sphagnum* spp produce bactericides (Rosswall, Veum & Kärenlampi, 1975), and their high cation exchange capacity may further reduce decomposition rates in many mires by increasing acidity (page 292).

The result is that bryophytes tend to accumulate in peat and other forms of DOM disproportionately to their contribution to net production. Attention has already been drawn to the low turnover rates of bryophyte phytomass in several communities: *Tortula robusta* on South Georgia is clearly an exception (page 260), a fact that may be related to its relatively low crude fibre content (Fig. 7.5). The low decomposition rate of bryophytes compared with vascular plant foliage at Stordalen mire is evident in Table 7.14. Data concerning lichens are scarce: low decomposition rates comparable with those of wood were recorded at Fennoscandian sites (Table 7.14), and Williams *et al.* (1978) considered lichen decomposition at Barrow to be slow. While these results may be typical, Wetmore (1982) reported high decomposition rates for corticolous species placed on the ground in a boreal spruce bog.

Detailed studies have been undertaken of decomposition in Antarctic moss turf and carpet communities (Baker, 1972; Wynn-Williams, 1980, 1984; Yarrington & Wynn-Williams, 1985). Relatively high annual rates of decomposition, from 4% to 25% of net primary production, were recorded near the surface of waterlogged, soligenous moss carpets, with peat respiration rates negatively correlated with water content. Significant

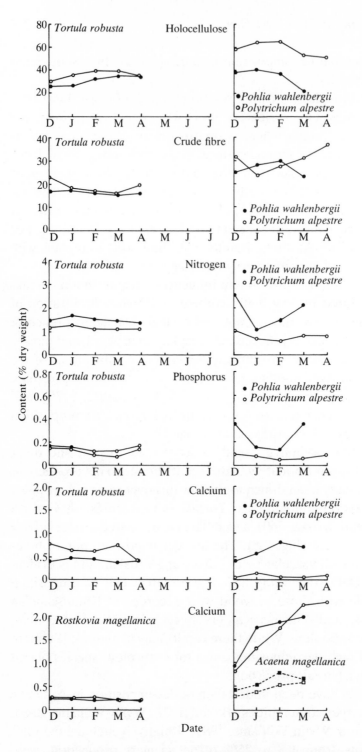

Fig. 7.5. Seasonal variation in the contents of some elements and organic constituents in three mosses on South Georgia, with data for calcium content in *Rostkovia magellanica* (a monocot) and *Acaena magellanica* (a dicot). Multiple lines for a given species represent data from several sites. Data from Pratt & Smith (1982).

Table 7.14 *Decomposition rates of plant material at Fennoscandian tundra sites*

Site and community	Plant material	Percentage decomposition after		
		One year	Two years	Three years
Kevo, pine woodland	*Pinus sylvestris* wood	5	7	8
	Pinus sylvestris needles	18	33	45
	Betula tortuosa wood	4	6	7
	Cladonia stellaris	6	17	28
Hardangervidda, lichen heath	*Juncus trifidus*	7	13	19
	Carex bigelowii	20	36	49
	Lichens	5	13	18
Stordalen, ombrotrophic mire	*Rubus chamaemorus*	24	37	48
	Empetrum nigrum	6	20	26
	Betula nana leaves	21	28	—
	Betula nana twigs	8	16	—
	Eriophorum vaginatum	37	—	—
	Sphagnum fuscum	5	—	—
	Sphagnum balticum	4	8	—
	Sphagnum lindbergii	7	20	—
	Dicranum elongatum	7	—	—
	Drepanocladus schulzei	5	—	—

Data from Rosswall, Veum & Kärenlampi (1975).

methanogenesis, indicative of anaerobic respiration, was recorded and turnover in the carpets was increased by erosion caused by wind and particularly by birds. Decomposition rates of the moss decreased, but those of artificially introduced cotton strips increased, with depth. The difference could be related to greater early loss of soluble compounds by leaching, and of readily degradable compounds by microbial action, from the moss than from the cotton with its more uniform chemical composition.

Decomposition was less rapid in the drier, semi-ombrogenous moss turfs, ranging from 0.3–2.0% of net annual production near the surface to 0.1% in permafrost. The low rates in the active layer appear anomalous as the peat remains moist but seldom waterlogged, aerobic conditions prevail, and the temperature regime is slightly warmer in the turfs than the carpets (Chapter 4). The turf peat is rather more acid than that beneath the carpets, but quality of the mosses as a substrate for decomposers is regarded as the most important factor (Davis, 1981). Walton (1985b) attributed the slow decomposition of *Polytrichum alpestre* peat on South Georgia to high levels of holocellulose and crude fibre (Fig. 7.5), and high carbon:nitrogen ratios. The turfs are also less prone to erosion than the carpets, the net result being greater accumulation of peat (Fig. 2.5).

Nutrient cycling
General features of polar nutrient cycles

Nutrient cycling between soil and living organisms occurs in all terrestrial ecosystems. There are also inputs to the system from particulates in dust, ions in solution in precipitation, weathering of rock and soil particles and biological nitrogen fixation. Losses occur through denitrification, leaching and surface runoff, the latter also being responsible for local redistribution. Available nitrogen and phosphorus are released only in small quantities by weathering. They are the elements most commonly limiting for plant growth, particularly in pioneer communities during primary succession when input of nitrogen and phosphorus commonly exceeds output as the soils develop.

Aerial deposition is particularly important in oceanic regions due both to heavy precipitation and high concentrations of several elements in rain and marine spray. Dowding, Chapin, Wielgolaski & Kilfeather (1981) report phosphorus deposition (mg m^{-2} yr^{-1}) as 172 on Macquarie I (annual precipitation 926 mm), compared with only 9 at Kevo (390 mm), 4.1 on Devon I (142 mm) and 1.2 at Barrow (124 mm). Other elements such as calcium occur in low concentrations in precipitation, even in oceanic regions (Smith, 1978), and rock weathering provides the principal input.

Release of nutrients by chemical weathering is retarded by cold, particularly when combined with aridity, and the same factors restrict nitrogen fixation. Rates of nutrient cycling in polar systems are further depressed by slow decomposition and the limited extent of herbivory, which result in nutrients accumulating in an unavailable form in DOM. This applies especially to wet habitats where the organic layer resulting from relatively high productivity combined with particularly low rates of decomposition results in permafrost extending close to the soil surface (Chapter 4). While tending to conserve nutrients from leaching, permafrost also restricts their release from organic matter. Where there is neither peat nor permafrost, the predominantly coarse-textured mineral soils permit free leaching in areas of high precipitation, as on cool-Antarctic uplands, whereas in arid regions some elements tend to accumulate at the soil surface (Chapter 3). Nutrient cycling is thus profoundly influenced by water supply (Van Cleve & Alexander, 1981).

These effects, and the youthfulness of polar ecosystems, result in available nitrogen and phosphorus being limiting in many tundra soils. Indeed, low rates of nutrient cycling could impose restrictions on productivity as severe as those attributable to direct effects of temperature on metabolism in the producer organisms (Chapin, 1983; Heal *et al.*, 1981). The relative importance of these factors is not clear, as nutrient cycling has so far been less intensively studied than energy flow (Dowding *et al.*, 1981).

Nutrient uptake in cryptogams

The range of elements required by mosses, lichens and higher plants appear to be similar, but the mechanisms of uptake and retention differ substantially. These processes in cryptogams have been extensively reviewed (Brown, 1982, 1984; Brown & Beckett, 1985a; Longton, 1980; Nieboer, Richardson & Tomassini, 1978), and only the essential features need be summarised here.

Nutrient elements in solution are taken up by plants via the same route as water. They are therefore likely to enter directly into the photosynthetic parts of most bryophytes and lichens, with internal movement in a transpiration stream important only in endohydric bryophytes (Chapter 5). Bryophytes and lichens appear to be efficient in taking up nutrients from dilute solutions (Babb & Whitfield, 1977; Williams *et al.*, 1978). They are believed to absorb ions from precipitation, enriched by leachate from a vascular canopy in some habitats. This is thought to be the principal source of nutrients for many species, including epiphytes and loose, weft-forming mosses of boreal woodland. Apart from endohydric mosses, nutrients in soil moisture and surface runoff appear to be of greatest significance to small plants growing appressed to the substratum, or those forming dense colonies in which upward external water movement occurs by capillarity or by diffusion in response to the humidity gradients discussed in Chapter 4. The relative importance of atmospheric and soil nutrient sources is not clear, but as many bryophytes and lichens, representing a range of growth forms, show some degree of preference in respect of soil or rock chemistry (Chapter 3), it is probable that few species are nutritionally fully independent of their substratum.

Element content of cryptogams may thus be expected to reflect the composition of both the substratum and of precipitation and other forms of aerial deposition, and such relationships have been demonstrated among Antarctic bryophytes. Thus on South Georgia, *Pohlia wahlenbergii* and *Marchantia berteroana* in flushed sites have relatively high element contents, *Tortula robusta* and *Brachythecium rutabulum* of mires have intermediate values, with low levels in *Polytrichum alpestre* and *Chorisodontium aciphyllum* from acid peat banks. On Signy I the concentration of sodium in a given species tends to decline with increasing distance from the sea (Smith, 1984a).

Absorption of some elements, including nitrogen, phosphorus and potassium, is mainly intracellular, with the uptake process at least partially active. Thus more than 75% of total potassium has been shown to occur as K^+ ions in solution within the cytoplasm in both bryophytes and lichens. Certain toxic ions such as Ag^{2+} are also readily absorbed. Some intracellular uptake obviously occurs in the case of all metabolically essential elements. However, much of the element content of mosses and lichens resides outside the cell membrane as ions bound to extracellular exchange sites or as particulates on plant surfaces, and has no immediate effect on metabolism.

Extracellular cation uptake is a rapid, essentially physicochemical process occurring in dead as well as living material, and involving reversible binding to anionic exchange sites in the cell walls. In mosses, these sites are thought to lie in uronic acid molecules which in *Sphagnum* spp may comprise up to 30% of plant dry weight, but it is less clear which compounds are involved in lichens. The exchange sites have a greater affinity for divalent cations (e.g. Ca^{2+}, Mg^{2+}), and particularly for borderline metals (*sensu* Nieboer & Richardson, 1980) such as Cu^{2+}, Pb^{2+} and Ni^{2+}, than for monovalent ions. Bryophytes tend to have a greater cation exchange capacity than lichens, and this may explain their generally higher contents of elements such as calcium and magnesium (page 275). Exchange capacity is particularly high in species of *Sphagnum* growing under extremely nutrient-poor conditions on hummocks in ombrotrophic mires. Uptake of metallic cations involves displacement of lower affinity cations, notably H^+, a factor contributing to the acidity of *Sphagnum* bogs. It remains to be determined whether binding of essential ions to cell wall exchange sites is important as a prelude to internal uptake.

The location and mobility of an element within a plant is reflected in the relative concentrations in green and brown tissue. The greatest concentrations of nitrogen, phosphorus and potassium, which occur primarily in the cytoplasm, are commonly found in green, metabolically active tissues, while calcium and magnesium tend to accumulate in the brown, partially senescent region. These relationships generally apply in Fig. 7.6, and have also been demonstrated for mosses on both Devon

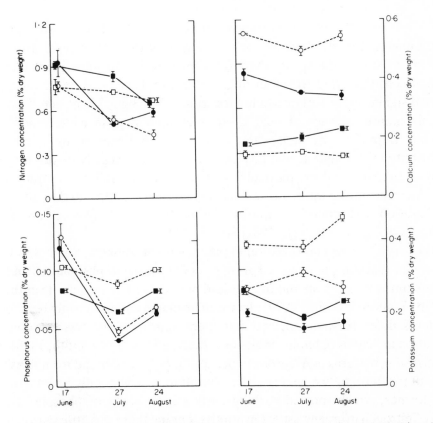

Fig. 7.6. Seasonal variation in element content in green (open symbols) and brown (closed symbols) portions of *Aulacomnium* spp (left) and *Polytrichum* spp (right) in Alaskan tundra. Circles refer to nitrogen and phosphorus, and squares to calcium and potassium. Data from Chapin *et al.* (1980).

I and Signy I (Smith, 1978, 1984a; Usher & Booth, 1985; Vitt & Pakarinen, 1977).

Particulates become trapped in considerable quantity on external surfaces of both mosses and lichens. In lichens they also accumulate in internal spaces between hyphae in the cortex, and particularly in the loose-textured medulla. They are derived both from the substratum and from atmospheric dust. Weathering of such particles may continue to some extent within the lichen thallus (Chapter 3), resulting in release of ions. It should be noted that particulate matter, although physiologically inactive, can contribute significantly to the element contents recorded for mosses and lichens.

Seasonal aspects of nutrient cycling

Availability of soil nutrients varies seasonally. Phosphorus and potassium commonly show an early season peak thought to result from leakage following the spring thaw from microbial, and some plant and animal cells that were frozen during winter. Hooker (1977) has demonstrated significant loss of intracellular cations from Signy I lichens following six months storage at −20 °C. Ions accumulating in winter snow, leached from litter by melt water or released by decomposition, may contribute to the high availability of some ions in spring. Nitrogen may show a similar trend or, particularly at wet sites, its soil concentration may increase during the growing season. This effect is attributed to nitrogen fixation and input from marine animals in coastal areas (Booth & Usher, 1985; Dowding *et al.*, 1981; Smith, 1985a).

There is also seasonal variation in plant element contents. Some South Georgian vascular plants have spring peaks of nitrogen, phosphorus and potassium, with calcium and magnesium maxima in late summer (Smith, 1985b; Walton & Smith, 1979). In vascular plants and in bryophytes the patterns differ between species, but there appears to be less seasonal variation in lichens (Chapin, Johnson & McKendrick, 1980; Smith, 1978). Contrasting patterns in *Aulacomnium* spp and *Polytrichum* spp from Alaskan tundra are shown in Fig. 7.6. Variation was generally greater in the former, which showed striking early season peaks of phosphorus, and of nitrogen in green tissue. On South Georgia, there was little seasonal variation in *Polytrichum alpestre* or *Tortula robusta*, but in *Pohlia wahlenbergii* nitrogen, phosphorus and potassium fell to minimum levels in midsummer (Fig. 7.5), a time when this species shows particularly rapid growth (Chapter 6). This suggests that fluctuations in element concentrations may result from variation in growth rates as well as in rates of uptake and loss.

Role of cryptogams in nutrient cycling

It is clear that bryophytes and lichens can be expected to exert a major influence on nutrient cycling. They are likely to enhance retention within the soil–plant system of nutrients dissolved in rainwater by incorporating part of the input directly into shoot or thallus tissue. By retaining precipitation in extracellular capillary spaces, so that it evaporates, or drains slowly into the underlying soil, bryophytes and lichens may further increase retention of dissolved nutrients and dust deposited in precipitation, and also reduce leaching of nutrients present in soil moisture. Dead organic matter accumulating beneath moss and lichen colonies may also reduce loss of nutrients through leaching, first by chemical association, second by retaining water, and third by favouring the development of permafrost. Cryptogams, particularly lichens, appear to stimulate chemical weathering (Chapter 3), while some lichens and microorganisms associated with bryophytes contribute to nitrogen fixation.

Conversely, nutrient immobilisation in living tissue and in dead bryophyte phytomass may reduce availability to other organisms. Soluble products are released from the cytoplasm following loss of membrane integrity in dead mosses, and some ions may be released from cell wall exchange sites should the chemical environment change. However, much of the nutrient content of DOM is retained, either bound to the cell walls or in insoluble structural material. Thus 50% of the calcium in dead plant material in a mesic meadow on Devon I was in the bryophyte component, with a turnover time estimated at 22 years (Dowding *et al.*, 1981). Bryophytes and lichens can thus be expected to increase total nutrient pools, but to reduce rates of nutrient cycling. Overall, their abundance should thus tend to increase the stability but reduce the productivity of polar ecosystems.

The potential significance of cryptogams in polar nutrient cycling is widely accepted (Babb & Whitfield, 1977; Stoner, Miller & Miller, 1982; Van Cleve & Alexander, 1981; Williams *et al.*, 1978), but experimental confirmation is still required to substantiate much of the preceding speculation. However, the importance of mosses and lichens in nitrogen fixation is well established, and there is scattered evidence of their efficiency in nutrient retention.

The ability of mosses from Signy I to remove elements from percolating nutrient solutions has been demonstrated experimentally (Allen, Grimshaw & Holdgate, 1967), and lichens at Kevo have been shown to absorb NH_4—N from rainfall (Crittenden, 1983). In Alaska, the studies of Weber & Van Cleve (1984), in which ^{15}N compounds were applied in solution to ground vegetation in spruce woodland, suggested that the weft-forming feather mosses *Hylocomium splendens* and *Pleurozium schreberi* play a vital role in the nitrogen economy of the community by retaining and immobilising both NH_4—N and NO_3—N, and releasing it slowly to the soil horizons where vascular plant roots are located. Overall, however,

Oechel & Van Cleve (1986) believe that nutrient immobilisation by an increasing abundance of moss lowers vascular plant productivity as succession passes from deciduous forest to stages dominated by *Picea glauca* and eventually by *P. mariana*. Pakarinen (1978, 1981) has shown that net annual uptake of nitrogen, phosphorus and potassium by *Sphagnum* spp, and of phosphorus by lichens, in Finnish ombrotrophic mires may exceed annual receipt in precipitation, his data also indicating substantial accumulation of calcium and magnesium by the mosses. Rainfall events at Abisco have been shown to stimulate nitrogenase activity in *Sphagnum fuscum*, and in this case utilisation by *Sphagnum* spp of NO_3—N in precipitation is thought to reduce availability to associated vascular plants (Woodwin, Press & Lee, 1985). Finally, removal of lichen cover, formed mainly of *Cetraria delisei*, from a lichen heath on Svalbard resulted in a decrease in soil organic matter, in nitrogen and other macronutrients, and in soil respiration (Sendstad, 1981).

Nitrogen fixation

Regular input of combined nitrogen is essential to the functioning of tundra ecosystems, particularly in mires and other communities where an increasing pool of nitrogen compounds is immobilised in DOM. Precipitation and biological fixation are the principal sources, with the latter believed to be of the same order as, or rather greater than, the former except in areas with a strong marine influence (Alexander, Billington & Schell, 1978; Kallio, 1975). Although nitrogen input by biological fixation is small in tundra compared with temperate systems, it nevertheless assumes disproportionate importance (Dowding *et al.*, 1981). During fixation, atmospheric nitrogen is first converted to ammonia, a process mediated by the enzyme nitrogenase. Fixation rates can be measured directly by supplying ^{15}N, but most polar studies have used the indirect, and possibly less accurate assay in which acetylene is supplied in excess and the rate at which it is reduced by nitrogenase to ethylene is recorded by gas chromotography.

Nitrogen fixing bacteria, either free living or associated with vascular plants, are only locally abundant in polar soils (Van Cleve & Alexander, 1981). Under these circumstances, the major contributors to biological fixation are, in dry habitats, lichens with cyanobacterial photobionts and, under wetter conditions, cyanobacteria and other microorganisms growing epiphytically on mosses. The latter provide a favourable habitat marked by relatively high temperature and low oxygen concentration (Alexander, 1975; Alexander, Billington & Schell, 1978; Kallio, 1975).

Lichens are of particular importance in relatively dry woodland near Kevo, and Kallio (1975) has estimated that fixation may reach 384 mg $N m^{-2} yr^{-1}$ in *Stereocaulon paschale* ($3 g N m^{-2} yr^{-1}$ in closed stands), and 104 mg $N m^{-2} yr^{-1}$ in *Nephroma arcticum* ($4 g N m^{-2} yr^{-1}$ in closed stands). Biological fixation was estimated by Alexander, Billington & Schell (1978) to yield an average of 69 mg $N m^{-2} yr^{-1}$ in tundra communities near Barrow: the highest rates were recorded in lichens, and in *Nostoc* and other cyanobacteria associated with mosses.

Cyanobacteria growing epiphytically on *Sphagnum* spp and other mosses are considered major contributors of fixed nitrogen in minerotrophic mires in the northern boreal zone of Fennoscandia (Granhall & Lid-Torsvik, 1975; Basilier, Granhall & Senström, 1978). Fixation by heterotrophic bacteria predominates in an ombrotophic part of Stordalen mire, being largely confined to the uppermost 5 cm of the *Sphagnum* cover (Rosswall & Granhall, 1980). *Nostoc* spp within the hyaline cells of *Sphagnum* provide one of the most important nitrogen inputs to some Fennoscandian boreal forests, while significant fixation by bacteria growing on feather mosses also occurs (Basilier, Granhall & Senström, 1978; Granhall & Lindberg, 1976). In Alaskan woodland, Billington & Alexander (1978) reported the highest rates of fixation in lichens, but they considered that cyanobacteria and other organisms epiphytic on feather mosses were of greater significance due to the extensive moss cover.

Acetylene reduction by cyanobacteria associated with mosses occurs in some communities on cool-Antarctic and cold-Antarctic islands at rates similar to those reported in the Arctic (Christie, 1987; Smith & Russell, 1982). There are also important nitrogen inputs from precipitation and marine animals at these oceanic Antarctic sites (Smith, 1984a). In the frigid-Antarctic there is evidence of significant fixation by cyanobacteria in soil and particularly on damp moss (Davey & Marchant, 1983; Nakatsubo & Ino, 1986), but not in endolithic communities (Friedmann & Kibler, 1980).

Low temperature appears to be a major factor limiting nitrogen fixation by lichens and free living cyanobacteria in polar regions, particularly in the cold- and frigid-Antarctic, and the rates vary diurnally with maxima during the day (Alexander, Billington & Schell, 1978; Davey, 1983; Kallio & Kallio, 1978; Lindsay, 1978; Smith & Russell, 1982). However, rates of nitrogenase activity in lichens, like those of photosynthesis (Chapter 5), are controlled by complex interactions between temperature, irradiance, water availability and other factors (Kershaw, 1985). Such relationships are fully understood for few species, and the reliability of estimated annual fixation rates such as those cited above must be assessed against this background.

Despite these uncertainties, there is little doubt of the importance of fixation by lichens and microorganisms associated with mosses in the nitrogen economy of tundra and boreal woodland. Crittenden (1983), working at Kevo, demonstrated that while *Stereocaulon paschale* and *Cladonia stellaris* both effectively absorb NH_4—N from rainfall, leaching of organic nitrogen compounds is six times as great on an area basis, and twice as great in terms of dry weight, from *S. paschale* as from *C. stellaris*, a species not involved in nitrogen fixation. Crittenden also cited evidence that fixed nitrogen may be transferred from lichens to mosses growing nearby, although some authors consider lichens to conserve much of their nitrogen within the thallus until eventual death and decomposition (Eckardt, Heerfordt, Jorgensen & Vaag, 1982; Lindsay, 1978).

In Alaska, Alexander, Billington & Schell (1978) suggested that newly fixed nitrogen from epiphytic cyanobacteria is particularly important in short-term nitrogen turnover during the growing season: their experiments with ^{15}N indicated that fixed nitrogen, released as NH_4—N, is rapidly transferred to the associated mosses and to vascular plants. Similarly, Ino & Nakatsubo (1986) concluded that actively growing moss at Syowa Station is supplied with nitrogen fixed by epiphytic cyanobacteria with little mobilisation of nitrogen from older parts of the plant. Conversely, bryophytes compete strongly with tundra vascular plants for available nitrogen (Van Cleve & Alexander, 1981).

Biological fixation is likely to be particularly important during early stages of primary succession when immature soils are almost devoid of available nitrogen, an effect demonstrated on the recently formed volcanic island of Surtsey (Chapter 8). Here, an association between the moss *Funaria hygrometrica* and cyanobacteria, particularly *Anabaena variabilis*, occurred widely during early colonisation. Experimental studies suggested that development of the *Funaria* was enhanced by nitrogen compounds and growth regulators released by the *Anabaena*, while the moss had a growth promoting effect on the latter (Rodgers & Henriksson, 1976).

Other effects of bryophytes and lichens

In addition to having a major impact on energy flow and nutrient cycling in polar ecosystems, bryophytes and lichens also interact with other organisms in a variety of different ways. For example, they provide a favourable habitat for invertebrates (Tilbrook, 1970), and lichens yield nesting material for birds and small mammals (Chernov, 1985). Both groups also exert a significant influence on soil moisture and temperature regimes (Edward & Miller, 1977; Rouse, 1976; page 127), and on the distribution of permafrost (Chapter 4). The latter is particularly important, for it affects not only rates of nutrient cycling, but also depth of rooting zone, soil drainage and other edaphic factors, and through these the distribution and successional relationships of vascular plant communities (Benninghoff, 1952; Chernov, 1985; Viereck, 1970).

The complexity of these relationships is illustrated by Cowles's (1982) study of the influence of lichen cover, principally of *Cladonia stellaris*, on black spruce near Schefferville. Removal of the lichens resulted in lower soil moisture content and increased soil temperatures in summer, while aqueous leachate from the lichen colonies was shown to reduce rates of tree branch elongation. However, fertilisation with nitrogen and phosphorus elicited a greater response on spruce growth in plots with intact lichen cover than in similar plots with the lichens removed. Germination and tree seedling establishment seldom occurred within areas of lichen cover, but establishment was particularly successful under moister conditions among mosses around the edge of lichen stands and in polygonal cracks within them (Chapter 2). Cowles concluded that, in this instance, the net effect of the lichens on tree growth was beneficial.

The effects of bryophytes and lichens on vascular plant establishment are not clear cut. Experiments of Bell & Bliss (1980) suggested that moss colonies may be particularly important sites for germination and seedling establishment in arid, cool-Arctic environments, but Sohlberg & Bliss (1984) demonstrated interspecific differences in response under similar conditions, the distribution of some species being positively correlated with moss cover and others with bare soil. Similar differences were evident in temperate chalk grassland where moss cover can inhibit germination but favour seedling establishment in a given species (Keizer, Van Tooren & During, 1985). In lichens, there is evidence that species such as *Cladonia stellaris* restrict tree seedling establishment through allelopathic effects on mycorrhizal development, and Rundel (1978) suggested that such taxa may be important in determining canopy characteristics in boreal lichen woodland.

Human influence on polar cryptogams

Many tundra communities such as the cold-Antarctic moss banks appear to be stable in the sense that they change relatively little with time, in marked contrast to much of the boreal forest with its regular cycle of burning and recovery (Chapter 3). However, stability in this sense may render a system more sensitive to fluctuating environmental conditions, whether due to climatic change or human interference, than systems normally subject to moderate levels of oscillation. For this and other reasons, tundra may be regarded as fragile, or unstable in the sense of being 'unable to absorb serious perturbation and return to a stable state, usually the *status quo ante*' (Dunbar, 1973). Polar terrestrial ecosystems have to date been little modified by human impact: they may nevertheless be vulnerable to damage by both local activity and by airborne pollutants from population centres in temperate regions. This is particularly true in the Arctic, where widespread temperature inversions lead to pollutants accumulating as Arctic haze, and to their retention in snow-cover (Benson, 1986).

Introduced biota

Being species poor, polar communities are susceptible to infiltration by introduced plants and animals able to exploit unfilled niches. This is particularly true of southern oceanic islands where the existing species poverty results in part from dispersal barriers. Many vascular plants and several lichens have been introduced (Lindsay, 1973b; Walton, 1975). Some of the former have become so widely naturalised as to give cause for concern. However, their impact has so far been less dramatic than that of mammalian herbivores that were introduced to the Falkland Is and cool-Antarctic islands, principally by nineteenth-century whalers and sealers. Particularly troublesome are rabbits, whose selective grazing has been responsible for widespread changes in the specific composition of vegetation, and in places for its almost complete destruction. Rabbits have eliminated *Poa foliosa* and *Stilbocarpa polaris* as dominants from parts of Macquarie I, in places initiating severe soil erosion due to the loss of binding root systems (Costin & Moore, 1960). Elsewhere on the island, *Marchantia berteroana* is an abundant colonist of soil laid bare by grazing, and it appears not to be consumed (Selkirk, Costin, Seppelt & Scott, 1983).

Significant changes in angiosperm vegetation must have serious repercussions for associated bryophytes and lichens. They have seldom been documented, but massive reduction in fruticose lichen cover by reindeer grazing has been reported on parts of South Georgia and the Kerguelen archipelago (Imshaug, 1972). Some Arctic islands, e.g. St Matthew I (page 286), are also susceptible to vegetational modification by introduced herbivores.

Physical disturbance

Polar vegetation is particularly prone to the various forms of physical disturbance that are inevitably associated with human habitation, industrial activity, tourism and scientific research. Being weakly anchored to the substratum, moss and lichen cover is readily removed by trampling and vehicle movement, especially when it consists of dry, brittle lichens. Kappen (1984) describes Antarctic lichen vegetation being destroyed through trampling by a few hundred visitors or blown away by the turbulence created by a landing helicoptor. Recovery is slow because of low growth rates. Indiscriminate scattering of coal ash and other rubbish near a scientific station on Goudier I in the cold-Antarctic destroyed lichen communities and may have resulted in the loss of species from the island (Smith & Corner, 1973). Mosses and lichens proved especially susceptible to destruction by simulated oil spills in mild-Arctic tundra, and recovered more slowly than the vascular plants (Freedman & Hutchinson, 1976; Wein & Bliss, 1973), while a series of interrelated factors arising from dust accumulation and improved drainage reduced growth rates in a species of *Sphagnum* near a newly constructed road through Alaskan tundra (Spatt & Miller, 1981).

Lightning-induced fire is a natural feature of boreal woodland, but fire frequency appears to have increased with the advent of European settlers leading to destruction of barren-ground caribou winter range. Forage lichens become established relatively late in post-fire succession. However, they do not always persist in climax forest (Chapter 3), and thus in the long term regular burning could enhance the value of boreal woodlands as winter pasture. The extent to which a dramatic fall in caribou numbers in recent decades is attributable to increased fire frequency remains controversial (Klein, 1982; Scotter, 1970).

Removal of moss and peat cover, for example by passage of tracked vehicles in summer, can have a major impact on energy budgets, particularly in the mild-Arctic with its thick vegetation cover, often initiating melting of permafrost (Haag & Bliss, 1974). This creates ditches which continue to deepen and widen for several years and may, in fact, never heal, leading to the so-called thermocarst effect (Dunbar, 1973). Thermocarst was widely associated with early oil exploration in Arctic North America but with care it can largely be avoided, for example by confining major vehicular movement to winter. The intensity of oil-related activity in the Arctic has declined since the mid-1970s. Even if economic conditions were to stimulate renewed interest it is probable that the Arctic tundra could absorb physical disturbance of the type so far discussed without major change because it occurs on such a vast scale that local perturbation would have little effect on the system as a whole (Dunbar, 1973).

This is not the case in the Antarctic, where well-developed biotic communities are only locally distributed, occurring particularly on level, coastal terrain free from permanent ice-cover. Such sites are likely to be favoured for operational bases in any future commercial enterprises involved, for example, with mineral extraction. The possible impact of such development has been discussed by several authors including Cameron (1972) and E. Schofield (1972) who both considered particularly the effect on cryptogams, and more recently by Sage (1985).

Atmospheric pollution

Sulphur dioxide   Bryophytes and lichens are, in general, more readily damaged than vascular plants by sulphur dioxide and other water-soluble pollutants. Their extreme susceptibility arises because, in the absence of an effective cuticle (Chapter 5), ions in solution pass freely into photosynthetic cells from the external environment. Thus sensitivity is greatest when the plants are in a hydrated state. Sulphur dioxide is one of the most widespread atmospheric pollutants and, together with its sulphite and sulphurous acid derivatives, it is also one of the most damaging, particularly at low pH (Grace, Gillespie & Puckett, 1985; Winner & Koch, 1982).

Increased membrane permeability, ultrastructural damage to chloroplasts and other organelles, and reductions in chlorophyll content and rates of photosynthesis are among the deleterious effects on bryophytes and lichens of experimental sulphur dioxide fumigation or treatment with sulphite solutions (Holpainen & Kärenlampi, 1984; Moser, Nash & Clark, 1980; Rao, 1982). The results could be especially damaging to cryptogams subject to periodic desiccation (Lechowicz, 1982b), as the photosynthetic gain during periods of moisture availability may be insufficient to offset respiratory losses on rehydration (Chapter 5). Respiration rates are also reduced at high concentrations of sulphur dioxide. Severe atmospheric pollution has resulted in the elimination of many lichen species over wide areas in temperate regions, although susceptibility varies between species (Hawksworth & Rose, 1976). Atmospheric pollution has also been shown to reduce nitrogen fixation by lichens, and by cyanobacteria epiphytic on mosses, in transplants to cities from relatively unpolluted boreal sites (Huttunen, Karhu & Kallio, 1981; Kallio & Varheenma, 1974).

Interactions between plant water content, windspeed, boundary layer resistance and internal resistance to sulphur dioxide uptake in Arctic *Cladonia rangiferina* have been modelled by Grace, Gillespie & Puckett (1985). Their results showed threshold uptake values for stress, indicated by potassium efflux, to be 240 μg SO_2 g^{-1} (lichen dry weight), or 20–30 μg m^{-2} (thallus area) in 24 hours. The sensitivity of Arctic species of *Cladonia* and *Cetraria* to sulphur dioxide was confirmed by experimental fumigation *in situ* at Anaktuvuk Pass, Alaska. Exposure to sulphur dioxide concentrations of about 2600 μg m^{-3} for 37 days in summer resulted in cessation of photosynthesis, with no subsequent recovery. The effect declined with decreasing sulphur dioxide concentration, but 650 μg m^{-3}, the lowest concentration tested, caused a reduction in photosynthetic rate which was still evident after three years (Moser, Nash & Clark, 1980; Moser, Nash & Olafsen, 1983). Mean winter (growing season) concentrations of atmospheric sulphur dioxide above 100 μg m^{-3} are associated with a severe decline in many fruticose and foliose lichens in Britain (Hawksworth & Rose, 1976). As Sigal (1984) has pointed out, however, the effects of fluctuating sulphur dioxide concentrations, and of peak doses as opposed to prolonged, low-level exposure, require evaluation before likely effects of pollution can be assessed.

Schofield & Hamilton (1970) calculated that under conditions of strong temperature inversion such as are common on the Alaskan North Slope, sulphur dioxide concentrations could be expected to reach 130 μg m^{-3} over areas up to 100 km^2 as a result of fossil fuel consumption associated with oil well sites, and they predicted serious effects on the lichen vegetation. While this remains to be confirmed, severe damage to cryptogamic vegetation has been documented near isolated cities and industrial operations in boreal regions (Ahti, 1977). For example, cover of the feather mosses *Hylocomium splendens* and *Pleurozium schreberi* has been severely reduced for at least 4 km downwind of oil refineries in Alberta where sulphur dioxide is the principal pollutant (Winner & Bewley, 1978a).

Metals   There are many reports of mosses and lichens accumulating borderline metals (Nieboer & Richardson, 1980) to substantially above background levels without apparent damage, as in the case of lead in *Pleurozium schreberi* in northwestern Ontario (Rinne & Barclay-Estrup, 1980). This is related to morphological features which favour the entrapment of airborne particles, and to high cation exchange capacity which results in the binding of metal ions to cell wall exchange sites.

Detoxification mechanisms and a degree of physiological tolerance may operate in some species, but intracellular absorption of metal ions supplied at high dosage can have adverse metabolic and physiological consequences for both mosses and lichens, including enzyme inhibition and reduction in the net assimilation rate (Brown, 1984; Brown & Beckett, 1984, 1985b; Rao, Robitaille & LeBlanc, 1977). These effects also appear to have led to depletion of terricolous cryptogams in the vicinity of industrial plants in boreal regions, as at brass mills in Sweden which liberate copper and zinc but little sulphur dioxide (Folkeson, 1984).

Particularly extensive damage is associated with nickel smelters near Sudbury, Ontario (LeBlanc, Rao & Comeau, 1972), and in central Manitoba. The feather moss stratum was found to be dead or moribund at distances of 5–15 km from a fifteen-year-old smelter in Manitoba, and this vegetation layer was absent from sites examined at similar distances from an older smelter, with only limited replacement by tolerant species such as *Ceratodon purpureus*. Depletion of epiphytic lichens and terricolous *Cladonia* spp extended for comparable distances, but vascular plants were much less affected (Longton, 1985b). The smelters emit both sulphur dioxide, and lead, zinc and other borderline metals, and it is not clear which pollutant was the principal cause of the damage, or whether a synergistic effect was involved. However, this situation illustrates the potential consequences of widespread industrial activity of this type in northern regions, and also provides an opportunity to examine the ecological consequences of eliminating the cryptogam stratum from large areas of boreal forest.

Acid deposition   A significant rise in levels of atmospheric pollution originating in north-temperate regions could result in serious disruption of Arctic ecosystems over wide areas. Oxidation products of sulphur and nitrogen combustion, formed during long-range atmospheric transport, result in wet and dry deposition of strong acids. The effects of such deposition at sites remote from urban and industrial centres have received considerable publicity.

Lechowicz (1982b) has shown that maximum photosynthetic rates in Arctic *Cladonia stellaris* are reduced, and the lag period after desiccation extended, in thalli wetted with simulated acid precipitation at pH 4 and a sulphate concentration of $10 \, \text{mg} \, \text{l}^{-1}$, although sulphate is less harmful than other sulphur dioxide derivatives. Acid deposition increases susceptibility to sulphur dioxide. It also increases the solubility of metal compounds, thus enhancing absorption (Rao, Robitaille & LeBlanc, 1977), and it may contain unnaturally high concentrations of nitrogen compounds. Woodwin, Press & Lee (1985) point out that the relationship between nitrate deposition in relatively uncontaminated rain and physiological processes in *Sphagnum fuscum* (page 296) suggests that some mosses may be vulnerable to increased nitrate supply. It is also clear that a significant rise in available nitrogen inputs could profoundly alter the functioning of polar ecosystems, unless offset by reduced rates of biological fixation.

Radionuclides Radioactive fallout as particulates from atmospheric nuclear weapon tests occurred on a global scale during the 1950s and early 1960s, and still continues to a limited extent. Deposition was greatest in temperate and tropical regions, but current levels of radionuclides such as caesium-137 are particularly high in Arctic lichens (Fig. 7.7).

At low latitudes, radioactive fallout quickly reached the soil and was then dispersed by vertical and lateral drainage to deep soil strata and other sinks where it has little influence on biological systems. The effective half-life of the various nuclides, a measure of the speed with which they are lost from ecological systems (Svoboda & Taylor, 1979), was thus relatively short. In the Arctic, lichens, mosses and dense cushion-forming vascular plants were effective in trapping the fallout, with some incorporation of radionuclides into their tissues. The effective half-life of caesium-137 in a given species increases with latitude due to decreasing rates of growth and decomposition. This was demonstrated by Taylor, Hutchison-Benson & Svoboda (1985) in a latitudinal survey, conducted in 1980 from latitude 50° to 80° N in central Canada. This study showed that, at a given latitude, effective half-lives in lichens were commonly double those in cushion-forming vascular plants.

The net result of these patterns of deposition and retention is that, in the Canadian study, the greatest caesium-137 activity on a plant dry weight basis was found between latitudes 60° and 70° N. The highest concentrations were in lichens, especially in dry habitats where growth and decomposition are particularly slow. There were interspecific differences in caesium-137 concentrations, and differences in cover led to further variation in data expressed as activity per unit area. Levels in the moss *Polytrichum juniperinum* were lower than in the lichens (Fig. 7.7). In Sweden, Mattsson & Lidén (1975) demonstrated that the effective half-life of caesium-137 in *Pleurozium schreberi* is considerably less than in

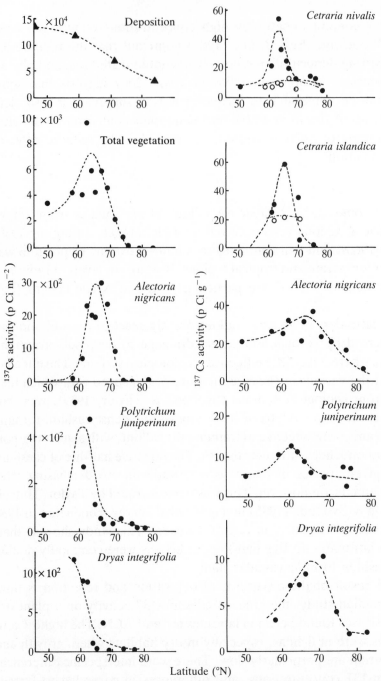

Fig. 7.7. Caesium-137 deposition during the 1950s and 1960s, and radioactivity in 1980, in total vegetation and in different species in dry (●) and mesic (○) habitats along a latitudinal transect in central Canada on a ground area basis (left) and a dry weight basis (right). Data from Hutchison-Benson, Svoboda & Taylor (1985) and Taylor *et al.* (1985).

Cladonia stellaris, although there was evidence of recycling of the radio-nuclide from senescent to actively growing material in both species.

The plants show no evidence of being damaged, but the ability of Arctic lichens to intercept and retain radionuclides, particularly those such as caesium-137 that appear to be absorbed intracellularly, has clear impli-cations for other members of the food-chain lichen → *Rangifer* → man. Spring-harvested caribou at Anaktuvuk Pass, Alaska, during the 1960s had caesium-137 levels about four times those in lichens, with levels in adult Inuit 85% of those in the caribou (Hanson, 1982). At times, Alaskan reindeer contained more than 20 times the levels of radionuclides found in cattle grazing near nuclear test sites in Nevada (Richardson & Nieboer, 1981). Hanson's detailed analysis of the Alaskan data provided no evi-dence of adverse effects on human health.

More recently, fallout from the Chernobyl accident has resulted in substantially enhanced levels of radioactivity in reindeer meat throughout much of northern Scandinavia, a situation that is likely to persist for as number of years given the vital role of lichens in the reindeer diet. There are therefore fears that this single nuclear accident could have the effect of destroying a whole culture based on reindeer husbandry (MacKenzie, 1986). Various possible remedies are being considered such as feeding less heavily contaminated animals on commercial feed for a period before slaughter to reduce radionuclide levels in the meat.

Pollution monitoring and conservation

Low species diversity, simple trophic structure, low rates of growth, decomposition and nutrient cycling, and an abundance of mosses and lichens with their extreme sensitivity to physical disturbance and gaseous pollutants thus appear to render polar ecosystems seriously vul-nerable to disruption by urban or industrial development. The principal threat to Antarctic terrestrial ecosystems stems from the localised distribu-tion and limited extent of well-developed biotic communities, and their occurrence on sites favourable for activity associated with mineral resource extraction. In the Arctic, localised industrial activity might be less generally disruptive, but any widespread development of industry or centres of human population could have serious ecological repercus-sions, given the extent to which moss and lichen vegetation can be de-stroyed by emissions from a single industrial plant.

A further threat to Arctic tundra is posed by airborne pollution originat-ing in temperate regions. Thus nitrate concentrations in precipitation have been steadily rising over the past three decades throughout northwest

Europe, including the most remote sites (Derwent & Nodop, 1986). It should also be noted that the pattern of increased half-life shown by radionuclides in tundra lichens could, in the long term, be repeated for other persistent materials, including certain pesticide residues, although concentrations of chlorinated hydrocarbons currently present in Norwegian mosses and lichens fall from south to north in line with deposition levels (Carlberg, Ofstad, Drangsholt & Steinnes, 1983).

As the latter study indicates, mosses and lichens can provide effective monitors of pollution. Analysis of concentrations in boreal and Arctic cryptogams has been used to assess deposition of persistent materials such as arsenic (Arafat & Gloschenko, 1982) and borderline metals (Beckett *et al.*, 1982; Rühling & Tyler, 1984). Both regional and temporal variation in deposition can be estimated provided that the productivity of the sample species, and the extent of vertical movement of material within the plant colony and its substratum are known (Pakarinen & Tolonen, 1976). Brown (1984) provided a review of the extensive literature and an evaluation of the appropriate techniques. Another approach is dependent on the sensitivity of many cryptogams to sulphur dioxide, fluoride and other gaseous pollutants. The diversity and vigour of bryophyte and lichen floras thus provide an index of pollution levels before the latter become dangerous to other biota, a technique already adopted in a number of studies in boreal regions (Richardson & Nieboer, 1981; Winner & Bewley, 1978b).

There are compelling reasons for seeking to conserve polar ecosystems. Significant numbers of Arctic residents remain partially dependent on the land for subsistence, and it is a lifestyle that they are reluctant to see disappear completely. The relative simplicity of polar ecosystems renders them of considerable value as models to assist in understanding the functioning of the more complex and increasingly threatened systems operating at lower latitudes. The genetic resource represented by polar biota is of potential value, for example in any future attempts to extend agriculture further into cold or arid regions. Nor should the amenity value of polar regions as wilderness areas offering a variety of immensely satisfying recreational pursuits be undervalued.

Science, through its increasing understanding of the biology of cryptogams and other organisms and of their interaction with abiotic components of local ecosystems, can monitor pollution levels and predict the probable consequences of various forms of human impact on the polar regions. It cannot do so with precision, nor always with certainty, but this is no reason to ignore the warnings. Biologists can suggest areas of particular sensitivity and importance for conservation (Usher & Edwards, 1986), while research can reveal ways of ameliorating the undesirable effects of impact, as where knowledge of the relationship between vegetation and permafrost was used to alleviate the thermocarst problem. Scientists have also been largely responsible for operating the Antarctic Treaty which, through its provisions for international cooperation in research, for banning all military activity and for preserving the Antarctic flora and fauna (Sage, 1985) stands as a model for civilised international behaviour in the Arctic and throughout the rest of the world.

The cynical view is that the Treaty was enacted at a time when the possibility of exploiting Antarctic mineral resources appeared remote, and that its provisions for conservation will come under increasing pressure when technical advances ultimately make the extraction of the proven coal and probable oil and other reserves commercially attractive. For the future, therefore, ecologists cannot be content merely to improve the precision and reliability of predictions concerning the ecological consequences of human impact, and to develop techniques for reducing its worst effects. They and others concerned for the environment must also address the more challenging problem of persuading society to assign to environmental protection a level of priority comparable with that given to economic development. There is little virtue in enhancing economic prosperity to the extent of robbing the global ecosystem of the natural diversity that can make life so rewarding an experience.

8

Reproductive biology and evolution

Reproductive processes and propagules
Bryophytes

In this final chapter we shall consider first the reproductive biology of polar cryptogams, as an introduction to a discussion of life strategies and their significance in the origin and evolution of the polar floras. The emphasis is of necessity on bryophytes, as there have been few relevant studies on lichens (Smith, 1984a).

The bryophytes life history involves heteromorphic alternation of a haploid gametophyte and a diploid sporophyte, the latter permanently attached to and, nutritionally, partially dependent upon the gametophyte (Fig. 8.1). Gametophytes are monoecious, dioecious or occasionally heteroecious, depending on the species. In mosses, antheridia and archegonia develop in groups at the apices of leafy shoots in acrocarpous species, or of reduced lateral branches in pleurocarps. Groups of gametangia with their surrounding bracts are termed inflorescences, and are bisexual, or more commonly exclusively male (perigonia) or female (perichaetia). Female bracts in leafy liverworts fuse to form tubular perianths that eventually surround the developing sporophytes.

Each antheridium liberates many biflagellate sperm, and a single, non-motile ovum develops in each archegonium. Where fertilisation occurs, repeated mitotic division of the zygote and its derivatives gives rise to the sporophyte, normally one per perichaetium. The mature sporophyte comprises a haustorial foot embedded in the gametophyte, and a slender stalk, the seta, on which the sporangium, or capsule, is borne several centimetres above the gametophyte. Meiosis occurs during sporogenesis, and several hundred thousand spores commonly develop synchronously within each capsule. Bryophytes with sporophytes are traditionally, if loosely, described as fruiting. Following spore dispersal, establishment of new gemetophyte shoots is delayed in most mosses by the develop-

Fig. 8.1. The mosses *Bartramia patens* (centre) and *Polytrichum alpinum* on Signy I (cold-Antarctic). *B. patens* has sporophytes, with capsules 2–3 mm in diameter.

ment, after spore germination, of a filamentous protonema upon which leafy shoots later arise.

Gametangial development and fertilisation take place in the relatively favourable microclimate near the shoot apices (Chapter 4). Capsule development and sporogenesis in liverworts occur under similar conditions, for the seta does not elongate until the spores are mature. Setal elongation in mosses precedes expansion of the capsule, meiosis and spore maturation taking place in developing capsules already raised above the gametophytes, but the surrounding air is warmed by convection currents from the ground during sunny weather (Longton & Greene, 1967). At Barrow, this effect is enhanced by the unusually short setae of many species, while meiosis occurs mainly during warm, daytime conditions, and the sporocytes contain copious amounts of lipid which are thought to prevent freezing. The Arctic endemic moss, *Aplodon wormskjoldii*, is unusual because, as in hepatics, elongation of the seta is delayed until after the capsule develops (Steere, 1954, 1973). Despite these features, few polar bryophytes are dependent on sexual reproduction for the maintenance of populations. As in other areas, most gametophytes are capable of asexual propagation by separation of branches to form independent individuals, or by gemmae and other specialised asexual propagules. They can also regenerate from rhizoids and from leaf or shoot fragments.

Lichens

Reproduction in lichens is a complex process, as two organisms are involved. A wide range of propagules may be produced, some containing only one component, and others both. The reproductive processes are poorly understood. In ascolichens, sexual reproduction of the mycobiont results in the formation of ascospores, which are normally dispersed independently of the photobiont. They are produced in asci on the inner surface of cup-shaped apothecia (Fig. 2.2), or less commonly in more or less closed, flask-shaped perithecia. Cytological events leading to ascus formation are assumed to resemble those in non-lichenised ascomycetes.

In these haplomitotic organisms, the female gametangium or ascogonium contains several haploid nuclei. It is equipped with a hair-like trichogyne which makes contact with the male gametangium. This comprises either an adjacent hyphal cell or a spermatium, a minute, uninucleate microconidium. The intervening cell walls dissolve and the nucleus or nuclei from the male gametangium migrate into the ascogonium, thus initiating a dikaryotic phase during which mitotic nuclear divisions occur but male and female nuclei remain in pairs. Eventually, the ascogonium gives rise to several ascus mother cells, each containing one pair of nuclei which fuse to form a diploid zygote. In ascomycetes, unlike bryophytes, karyogamy is followed rapidly by meiosis. A subsequent mitotic division typically results in eight haploid ascospores developing in each ascus, but the number is variable, 16–64 being common in crustose lichens.

Many lichens, particularly Antarctic species (Dodge, 1973), liberate abundant spermatia from embedded, flask-shaped pycnidia. The role of spermatia in sexual reproduction has not been confirmed in lichens, and it is possible that they act, additionally or alternatively, as asexual propagules (Hale, 1983). However, the potential exists for genetic variants to arise in lichenised fungi in the absence of meiosis, through heterokaryosis following fusion of somatic hyphae. This could occur among groups of germinating ascospores, in composite thalli formed by fusion of mature thalli, or in germinating asexual propagules (Bailey, 1976; Jahns, 1973).

Algal cells within the thallus reproduce by simple mitotic division and in other ways. Zoosporogenesis is common in liquid cultures of *Trebouxia* and *Pseudotrebouxia* spp but is usually suppressed in lichens, with aplanospores forming from arrested zoospores and becoming incorporated into the thallus. Motile zoospores are occasionally produced and it is possible that some are released and give rise to independent colonies of these algae, such as are occasionally observed in nature (Slocum, Ahmadjian & Hildreth, 1980). Sexual reproduction is an uncommon event in cultured *Trebouxia* spp and is unknown in nature, but Slocum, Ahmadjian & Hildreth (1980) suggested that it may occur in free living colonies of normally lichenised species. If so, then a process involving photobiont release and sexual reproduction, followed by lichen resynthesis involving germinating ascospores, would provide for genetic recombination through meiosis in both partners.

Lichens produce several types of asexual propagule containing cells of both mycobiont and phycobiont, the most common being soredia and isidia. Soredia are groups of algal cells, 25–100 μm in diameter and closely enveloped by hyphae. They originate in the medulla and algal layer of the thallus and erupt through cracks in the cortex. Isidia are cylindrical, finger-like projections up to 3 mm long, produced more or less irregularly over the upper cortex. They are fragile and easily broken off. Like mosses, fruticose and foliose lichens also reproduce freely by regeneration from detatched lobes or smaller thallus fragments, and by branching in fruticose species.

Reproduction in polar species
Bryophytes
Most bryophyte floras include species not known to produce sporophytes, thus relying exclusively on asexual reproduction, and this behaviour is particularly common in the polar regions. The proportion of moss species recorded in fruit declines from 80–90% in New Zealand and the British Isles to 30–40% on cool-Antarctic islands and in parts of the cool-

Arctic, and below 25% in the cold-Antarctic (Brassard, 1971a; Longton & Schuster, 1983; Smith, 1984a). Sporophytes of only three of some 30 mosses have been observed in the frigid-Antarctic (Selkirk, 1984). Mature spores have been reported in *Pottia heimii* from Syowa Station (Kanda, 1981b), and viable spore production has been confirmed in nine species on Signy I (Webb, 1973). These figures exaggerate the incidence of sporophyte production, as some species fruit only in occasional populations or in particularly favourable years. Brassard (1971a) considered that 28% of the mosses on northern Ellesmere I (cool-Arctic) fruit regularly, while sporophytes are common in only six to eight of approximately 75 species in the cold-Antarctic (Smith, 1984a), and in none of those in the frigid-Antarctic.

Most regularly fruiting polar mosses are both monoecious and acrocarpous (Holmen, 1960; W. B. Schofield, 1972; Webb, 1973). The same tendency is evident, though to a lesser degree, in temperate regions where, however, many dioecious species fruit freely (Longton, 1976). Species that regularly produce sporophytes in temperate regions often fail to do so in polar populations, as in the case of *Pleurozium schreberi* (page 321) and *Polytrichum alpestre* (page 322). Conversely, fruiting is conspicuously abundant in *Voitia hyberborea, Funaria polaris* and several other members of the circumpolar Arctic element.

Several liverworts in the Marchantiales regularly produce sporophytes on Greenland and northern Ellesmere I, but in these areas fruiting is even less widespread, and more strongly associated with monoecism, in leafy hepatics than in mosses. On northern Ellesmere I, sporophytes have been observed in all six monoecious species but in only one of 37 dioecious species. Three leafy liverworts, all monoecious, have been recorded in fruit in the cold-Antarctic, but two of them only at sites influenced by volcanic heat and moisture (Longton & Schuster, 1983).

Production of asexual propagules is, in a few cases, more characteristic of polar than of temperate populations of a given taxon, and this could be viewed as an adaptation compensating for failure in sexual reproduction. An example is the development of protonemal gemmae in cultures of *Amblystegium serpens* from barren populations on Macquarie I (Selkirk, 1981). Similarly, *Bryum argenteum* forms deciduous shoot and branch apices more consistently in the Antarctic than in fruiting temperate populations. Many leafy liverworts produce gemmae but among mosses specialised asexual propagules are formed by a minority of species, and so far they appear to be no more common in polar regions than elsewhere.

Even where they occur, the functional significance of specialised asexual propagules has seldom been confirmed. Young shoots have been observed on detached leaf apices in *Sarconeurum glaciale* from Patagonia (Matteri, 1982), but they were present only on protonema arising from the rhizoids of mature plants in Antarctic specimens (Savicz-Ljubitskaja & Smirnova, 1961). Fossil evidence suggests that dispersed asexual propagules and gametophyte fragments were important in the early establishment of mosses on newly deglaciated terrain in northeastern United States (Miller, 1985).

What is clear is that the majority of bryophytes in the more severe polar regions produce neither spores nor specialised asexual propagules, and are dependent on gametophyte branching (Figs. 6.2, 6.3 and 6.4), fragmentation and regeneration for colony establishment, development and maintenance, and for local dispersal.

Lichens

There are conflicting reports concerning the frequency of asco-spore production in polar lichens. Thomson (1972, 1982) states that the majority of Arctic species produce apothecia, although some only in favourable microhabitats near persistent snow patches. According to Williams *et al.* (1978) most lichens at Barrow lack sexual fruiting structures. Dodge (1973) regards sexual reproduction as 'almost universal' among fungi in cold- and frigid-Antarctic lichens, although he notes that apothecia are rare in most Antarctic species of *Usnea* and *Cladonia*. However, Lindsay (1977) reports that in many Antarctic lichens, including *Usnea fasciata* (Fig. 2.2), apothecia are abundant but mature ascospores rarely develop. Some cold-Antarctic endemics, including species of *Buellia* which Lindsay (1977) considers to have evolved locally, are known to produce spores (Lamb, 1968), a situation comparable with that among endemic Arctic mosses.

Soredia and isidia are rarely produced by crustose lichens, in polar regions or elsewhere. Soredia are common in *Cladonia* and *Usnea* spp, and Smith (1984a) regards asexual propagules as the principal mode of reproduction in Antarctic macrolichens. In contrast, Dodge (1973) and Thomson (1972) consider isidia and soredia to be less widespread among lichens in polar than in temperate regions. The general conclusion appears to be that ascospores play a significant role in the reproduction of polar crustose lichens, while asexual propagules and fragmentation of the thalli, so brittle when dry, by animals and wind-blown material are of major importance in the larger species.

Reproductive phenology

Apothecia of many lichens are perennial, each capable of liberating spores over a period of several years. Seasonal and diurnal periodicity in spore release has been reported in some temperate species (Pyatt, 1973), but little is known about reproductive phenology in polar lichens.

Most bryophytes have distinct annual cycles of gametangial and sporophytic development. The cycles are consistent in general features within a given species, although subject to minor variation in relation to climatic differences between localities, and from year to year at a given site. They differ substantially between species in relation to the reproductive strategies of the taxa concerned. It is thus of interest to determine how such orderly patterns of development are affected by polar climatic regimes.

Smith (1985a) considers that species of *Pottia* and *Pterygoneurum* on Signy I exhibit an ephemeral habit. These taxa are pioneers on recently exposed moraines and substrata subject to cryoturbatic disturbance. They are said to appear rapidly in spring, develop short shoots, gametangia and large numbers of sporophytes, and then become buried by fine clay during freeze–thaw cycles in winter. However, Webb (1973), while commenting on the rapidity of sporophyte development in these taxa, suggested that fertilisation in spring involved gametes from overwintering gametangia.

Typical cycles are more protracted, with overwintering of both gametangia and sporophytes, as in *Polytrichum alpestre* and *P. alpinum* on South Georgia (Figs. 8.2 and 8.3; Longton, 1972a). Here, development of male gametangia is initiated in late summer, and most antheridia overwinter in the juvenile stage, which is defined as being less than half the mature length. Juvenile archegonia are not visible until spring. Rapid gametangial development then leads to fertilisation in midsummer, principally during December. Archegonial swelling, indicative of early stages in sporophyte development, soon follows (Fig. 8.3), but setal elongation is delayed until the following spring. Sporophyte development continues steadily during the second summer with most spores shed in March or April, some 15 months after fertilisation, and two years after the initiation of antheridia (Longton, 1966b, 1972a; Longton & Greene, 1967). It is normal for successive cycles to overlap within a population.

Overwintering of developing gametangia and sporophytes cannot be regarded as a feature necessitated by short summers as it is also characteristic of many temperate species. Indeed, the reproductive cycle of *P. alpestre* on South Georgia is essentially similar to that followed by this species in boreal forest populations, and in Britain where winter tempera-

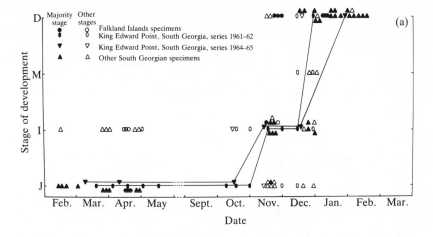

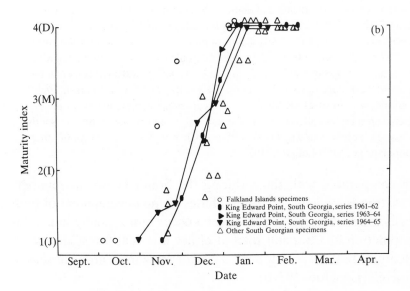

Fig. 8.2. Maturation cycle for antheridia (a) and archegonia (b) of *Polytrichum alpestre* in Antarctic regions. The states of development (Greene, 1960) are: J (Juvenile) – gametangia less than half their mature size; I (Immature) – gametangia larger but undehisced; M (Mature) – antheridia liberating antherozoids and archegonia receptive for fertilisation; D (Dehisced) – older gametangia no longer with viable gametes. The data refer to samples collected at intervals from one population on South Georgia (points joined by lines), and to single samples from other populations. Antheridial data show the developmental stage most frequent in each sample and the other stages present. Archegonial data show mean numerical index values based on all the gametangia present. After Longton (1972a).

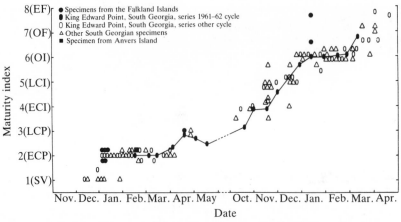

Fig. 8.3. Maturation cycle for sporophytes of *Polytrichum alpestre* in Antarctic regions. The stages of development (Greene, 1960) are: SV (Swollen Venter) – venter expanded indicative of embryo development; ECP, LCP (Early and Late Calyptra in Perichaetium) – calyptra fully or partially enclosed within perichaetial bracts; ECI (Early Calyptra Intact) – the stage of setal elongation; LCI (Late Calyptra Intact) – the stage of capsule expansion terminating with meiosis; OI (Operculum Intact) – the stage of sporogenesis; OF (Operculum Fallen) – the stage of spore liberation; EF (Empty and Fresh) – current cycle capsules with most spores liberated. Data are mean numerical index values based on samples collected at regular intervals from one population on South Georgia (points joined by lines), and on single samples from a range of other populations. After Longton (1972a).

tures are comparable with those during the South Georgian summer. Minor variation is evident, as certain phases in the development of both gametangia and sporophytes in *P. alpestre* occur more rapidly in a mild-Arctic population at Churchill than at either a boreal forest site or on South Georgia, both of which have a longer snow-free growing season than Churchill (Longton, 1979b).

Similarly, gametangia of *Pohlia nutans* mature more rapidly under polar than temperate conditions; sporophyte development is accelerated on Disko I, west Greenland, compared with either Britain or South Georgia, but the cycle is again basically similar in the three areas. In *P. nutans* there is experimental evidence of a lower temperature optimum for gametangial maturation in South Georgian than in British plants, suggestive of inherent adaptation to the prevailing conditions (Clarke & Greene, 1970, 1971).

Some lithophytic acrocarps on Signy I show a complex pattern in that two crops of gametangia develop annually (Webb, 1973). One crop is initiated in December and January, the summer cycle, and a second in

April, the autumn cycle, in *Schistidium antarctici* (Figs. 8.4 and 8.5). Gametangia of both cycles overwinter, with fertilisation occurring over an extended period from September to late November (summer cycle), and from December to January (autumn cycle).

Development of sporophytes in summer cycle archegonia of *S. antarctici* begins in September, and the majority overwinter at a stage where the capsules have still to expand. Capsule development, meiosis and sporogenesis occur during the following growing season and most capsules dehisce in late summer, from February to April. Young sporophytes developing in autumn cycle archegonia are present from January onwards, but most abort during the following winter. Again, this distinctive pattern can not be regarded as a response to the local climate as a somewhat comparable cycle is shown by *Grimmia pulvinata* and other lithophytes in Britain (Miles & Longton, unpublished data).

Failure in sexual reproduction
Causes of failure

In the cases just considered, reproductive phenology of polar mosses follows a normal pattern culminating in release of spores but, as noted earlier, most species fail to produce sporophytes in the more severe polar environments. Spatial separation of male and female plants in dioecious species and imbalance in the production of archegonia and antheridia, with the latter often rare, are the prime causes of a similar failure in some temperate mosses. Other reasons include failure in initiation or maturation of gametangia, inadequate moisture to permit fertilisation, and failure in sporophyte development (Longton, 1976). All these restrictions are thought to operate among polar species, often in combination, and the factors limiting the reproductive success of a given species may vary within its range.

Gametangia are rarely, if ever, produced by many species of *Lophozia*, *Scapania* and other leafy liverworts in the cool-Arctic (Longton & Schuster, 1983). Failure in gametangial maturation appears to be primarily responsible for a local rarity of fruiting in the moss *Pohlia cruda*. Its sporophytes are common in many areas, including Disko I, west Greenland, but on South Georgia gametangial development is irregular, often with two annual crops initiated within an inflorescence; most gametangia fail to mature and sporophytes are inevitably rare (Clarke & Greene, 1970).

Reproductive failure through sporophyte abortion has been reported in several mosses including *Bryum algens* in southern Victoria Land

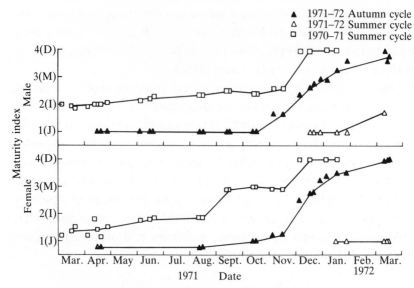

Fig. 8.4. Gametangial maturation cycle in a population of *Schistidium antarctici* on Signy I showing maturity indices, based on the stages defined in Fig. 8.2, for samples collected at regular intervals over several years. The lines join mean index value for all samples collected during each month. After Webb (1973).

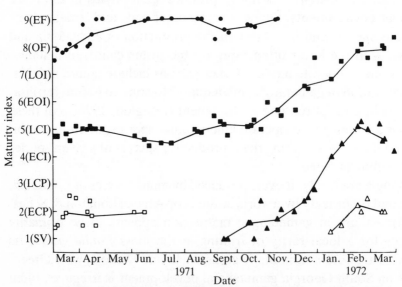

Fig. 8.5. Sporophyte maturation cycle in a population of *Schistidium antarctici* on Signy I showing maturity indices, based on the stages defined in Fig. 8.3, for samples collected at regular intervals over several years. Closed symbols indicate sporophytes from summer cycle gametangia; open symbols indicate sporophytes from autumn cycle gametangia. The lines join mean index values for samples collected during each month. After Webb (1973).

(Greene, 1967) and *Pleurozium schreberi* at Churchill. Control in the first case is likely to be environmental, as gametangia and sporophytes of a moss thought to be *B. algens* (Kanda, 1981b) developed when material from Victoria Land was cultured under favourable laboratory conditions (Rastorfer, 1971b).

Pleurozium schreberi, a dioecious moss, is abundant and common in fruit in the boreal forest zone where, at least at some sites, shoots with antheridia and those with archegonia occur in similar numbers. In contrast, sporophytes are rare towards the edge of the species' range in temperate, and in mild-Arctic regions. In the south this is due to an extreme rarity of shoots bearing antheridia. Shoots with antheridia were also found to outnumber those with archegonia by about 8:1 at sites near Churchill. The imbalance at Churchill arose at least in part from an excess of female plants over males, although the sex ratio among the shoots could not be determined precisely because many individuals remained sterile. Nevertheless, sufficient gametangia of both sexes were produced at Churchill to permit widespread fertilisation, but capsules were present at low density as the majority of young sporophytes aborted prior to setal elongation or capsule expansion (Longton, 1985b). Conditions at Churchill thus appear to be unfavourable for sporophyte development.

The sexual imbalance in some dioecious species is so pronounced that only one type of gametangium develops within a given geographical region. Thus no antheridia of *Bryum argenteum* have been recorded in the frigid-Antarctic (Greene, 1967; Longton & MacIver, 1977), but whether this is caused by absence or sterility of male plants is not clear. In other cases sporophytes fail to develop because male and female plants usually occur in separate colonies, as in the hepatic *Arnellia fennica* on northern Ellesmere I (Schuster, 1966–80). The fertilisation range in most terrestrial mosses is less than 15 cm (Longton, 1976), and both W. B. Schofield (1972) and Steere (1965b) have stressed that lack of water is likely to be severely restrictive of fertilisation in the more arid polar climates. Several of the above factors can be expected to operate least intensively in monoecious, acrocarpous species in which fruiting most commonly occurs (page 314), as they usually produce antheridia and archegonia close together.

Fertility of Polytrichum

Detailed studies of the dioecious mosses *Polytrichum alpestre* and *P. alpinum* have shown that several biological and environmental factors interact in controlling spore production (Longton, 1966b, 1972a; Longton

& Greene, 1967). *P. alpestre* forms continuous banks (Chapter 2) and also occurs in smaller, discrete turfs, both on South Georgia and in the cold-Antarctic. On South Georgia, antheridia and archegonia are freely produced and sporohyte production is prolific, particularly in the continuous banks. The latter comprise a mosaic of predominantly male and predominently female areas, suggesting that the banks developed through coalescence of originally discrete turfs. The male inflorescence in *Polytrichum* spp functions as a splash platform giving rise to an unusually long fertilisation range. Capsules were recorded up to 75 cm from the nearest male plants in the continuous banks on South Georgia, but there and at Churchill most of the smaller colonies were either bisexual and fruiting, or unisexual and barren.

In the cold-Antarctic, many *P. alpestre* turfs are sterile, others regularly produce archegonia, but antheridia are rare. Male inflorescences developed in study colonies on Signy I during some years but not in others, and it is thus clear that their rarity in this long-lived perennial is due, at least in part, to failure in initiation rather than rarity of male plants. Antheridia were confined to sunny, north-facing sites, and there is some evidence that their initiation was correlated with particularly favourable microclimatic temperatures.

Despite the occasional bisexual turfs, sporophytes have been recorded in only one cold-Antarctic colony of *P. alpestre*, on Anvers I. When first examined in 1964–65 (Longton, 1972a), it contained sporophytes of two annual cycles, and fruiting has since been observed on several occasions up to 1977 (R. I. Lewis Smith, personal communication). However, most sporophytes in the original collection had aborted prior to setal elongation, or had subsequently developed abnormally, and there is as yet no evidence that viable spores are produced.

P. alpinum usually occurs in discrete turfs and also produces antheridia, archegonia and sporophytes freely at many lowland sites on South Georgia. In the cold-Antarctic, archegonia are more widespread than antheridia which, as in *P. alpestre*, show a preference for well-insolated sites. Bisexual colonies are not uncommon in such places, but gametangial maturation is delayed by several months compared with South Georgia. Sporophytes are rare, and there is evidence that they develop only following fertilisation in particularly warm years. The majority of capsules are small or abnormally developed, although it is known that viable spores are sometimes produced. Similar, diminutive capsules are characteristic of high altitude sites on South Georgia, and of northern Ellesmere I (Longton, 1972a).

Thus conditions on sheltered coastal lowlands of South Georgia are clearly highly favourable for reproduction in these two species. In contrast, the cold-Antarctic environment is sufficiently severe to limit spore production, particularly in *P. alpestre*, by restricting antheridial initiation and/or the frequency of male plants. There also appear to be limitations on both fertilisation and sporophyte development, perhaps related to retarded gametangial maturation.

Dispersal and establishment

Dispersal

Dispersal mechanisms   Wind is undoubtedly a major factor in the dispersal of bryophytes and lichens. Many species produce spores in the range 5–20 μm in diameter, which are capable of being carried thousands of kilometres by wind. The bryophyte seta ensures that spores of most species are liberated above the laminar boundary layer over the colony, and the wiry nature of the moss seta tends to promote spore release when wind sets the capsules vibrating. Peristome teeth around the capsule mouth, or other features of sporophyte morphology, are also thought to regulate spore liberation in mosses, generally favouring release under warm, dry conditions when convection currents favour ascent.

Studies on temperate mosses with spores less than 20 μm in diameter suggest that although spores are deposited at high density within and immediately outside the parent colony the majority are nevertheless dispersed to distances greater than 2 m, and possibly much further (Miles & Longton, 1987). Liverworts tend to release spores when dry conditions cause shrinkage of the outer layer of the capsule wall, and active discharge by elaters then brings about rapid spore release in many species (Ingold, 1965).

Lichen apothecia are either sessile, or raised on podetia as in *Cladonia* spp. The Antarctic flora includes several species with stipitate apothecia in genera where the ascocarps are normally adnate or sessile (Dodge, 1973). The ascospores of most lichens are actively discharged from the apothecia to distances beyond the laminar boundary layer, and soredia may be removed from the thalli of some species by wind speeds of only 2–3 m s^{-1} (Bailey, 1976). Soredia are larger than many spores, commonly 25–100 μm in diameter, but, like isidia, gemmae and other specialised asexual propagules, as well as small vegetative fragments of both bryophytes and lichens, they are undoubtedly carried moderate distances by wind and may function in gene exchange as well as dispersal (Schuster, 1983).

Soredia are released by a splash-cup mechanism in some species of *Cladonia* (Bailey, 1976). Many types of propagule are probably carried substantial distances when melt water flows across the Arctic tundra in spring (Thomson, 1972), and lichen fragments have been detected in Antarctic melt streams (Chapter 1). Dispersal of propagules across the Arctic Ocean on drift ice has been postulated, with some observational evidence, for both mosses (Polunin, 1955) and lichens; Westman (1973) suggested that lichen soredia might eventually be carried ashore by birds. Long distance transport of propagules by migrating birds has been much discussed but never proved.

Local dispersal   Dispersal of propagules by wind blowing over smooth snow in winter is believed to be particularly significant in local dispersal. Lichen fragments are abundant in Arctic snow banks (Thomson, 1972). Miller & Ambrose (1976) estimated that over 4000 viable bryophyte fragments per cubic metre were present in late summer in a cool-Arctic snow drift that had accumulated during the previous winter. Culture studies indicated that leaf-bearing moss shoot apices, which were present in the greatest numbers, showed higher viability than detached leaves or liverwort fragments.

Rudolph (1970) trapped a variety of wind-blown plant propagules near ground level at Cape Hallett in the frigid-Antarctic. Algal fragments predominated, but fungal spores, lichen soredia and spores, and moss shoot apices, probably of *Bryum argenteum*, were also caught. Most propagules were thought to have originated from nearby vegetation, and the absence of lichen thallus fragments was attributed to the predominance locally of crustose species. Propagules were trapped at the low rate of less than $1 \, hr^{-1} \, m^{-2}$ trapping surface. Only a few algal cells and fungal spores were caught at a site 7 km from the nearest known vegetation.

Problems of long distance dispersal   Despite this observation, there is little doubt about the importance of wind in local transport, but the question of long distance wind dispersal over ocean barriers remains a matter of controversy. From an aerodynamic standpoint, there is no reason why small moss and lichen spores should not be transported by wind over the considerable distances involved. However, Crum (1972) and Schuster (1983) have pointed out that successful establishment following long range dispersal in bryophytes is probably infrequent due to the influence on spore viability of desiccation, freezing and enhanced ultra-violet irradiance during high altitude transport. Problems of sporeling establishment in alien environments and in competition with existing vegetation may also militate against frequent long distance dispersal.

The problems for lichen ascospores, or for spermatia should they function as asexual propagules, are compounded by the fact that, after germination, the young fungal hyphae must establish contact with an appropriate alga before the lichen can be reconstituted. Algal cells from the ascocarp are regularly dispersed with ascospores in a few genera and there are unconfirmed reports that this may occasionally happen in other lichens (Bailey, 1976). It has recently been established that species of *Trebouxia* and the mycobiont of *Xanthoria parietina* are capable of growing independently and the lichen has been resynthesised from the two constituents in culture (Bubrick & Galun, 1986). Free living lichen photobionts seem to be uncommon in polar regions, perhaps because of insufficient study. Exceptions include macroscopic species of *Prasiola*, occurring in a lichenised form as *Mastodia tesselata*, and of *Gleocapsa* on Marion I which may be invaded by fungal hyphae to form the holocarpous lichen *Edwardiella mirabilis* (Henssenn, 1986).

Dispersal in the southern hemisphere The issue of long distance dispersal is particularly germane in relation to southern hemisphere floras and the recolonisation of Antarctic regions following Pleistocene glaciations (Chapter 1). How the present biota arrived, and to what extent present floras are restricted by dispersal barriers, are topics that have been widely debated (Aleksandrova, 1980; Greene, 1964; Skottsberg, 1960; Smith, 1984a). The southern circumpolar region is characterised by powerful, westerly winds, and ice retreat left vacant niches. Lindsay (1977) suggested that ecologically aggressive lichen species reached the Antarctic through long range dispersal of ascospores, and transoceanic dispersal seems the only explanation of the floras currently established on geologically recent oceanic islands such as the South Sandwich Is (Longton & Holdgate, 1979).

Airborne spores appear to be the most probable agents of the long distance transoceanic dispersal thus implied, and it is interesting to note that spores of endemic New Zealand bryophytes were found to be less resistant to desiccation and freezing than those of species with wider distributions in southern temperate and cool-Antarctic regions (van Zanten, 1978). Van Zanten & Pócs (1981) consider that long range dispersal of spores carried in the west wind drift could best explain the wide distribution of many southern species. Schuster (1979), however, has postulated that most widely distributed southern hemisphere hepatics evolved in Gondwanaland and became established on several of the current land masses by stepwise migration prior to the creation of major oceanic barriers by continental drift.

Bipolar distributions   The origin of bipolar distribution patterns (Chapter 1) has also been the subject of considerable speculation (Du Reitz, 1940). Traditionally there have been three principal theories: first and least likely that the taxa concerned had polytopic origins, second that bipolar distribution was achieved by long distance dispersal, and third that stepwise migration occurred via tropical mountain systems. Acceptance of continental drift has raised further possibilities, and Lindsay (1977) speculated that the evolution of bipolar lichens antedated the separation of Gondwanaland from Laurasia. In contrast, Schuster (1983) argues that most bipolar mosses evolved in the northern hemisphere and extended their ranges southwards by long range spore dispersal. He notes that spores of several bipolar mosses were among those proving resistant to desiccation and freezing in van Zanten's experiments, and he suggests that the low incidence of bipolar distribution among hepatics (Chapter 1) reflects a generally low dispersal effectiveness in this group.

Evidence from fumaroles   The effectiveness of dispersal in some bryophytes is supported by the occurrence of temperate or cool-Antarctic taxa, including several hepatics, on isolated patches of volcanically heated ground on the South Sandwich Is (Chapter 3), and by the rapid colonisation by exotic species of fumaroles resulting from eruptions on Deception I, South Shetland Is, during 1967–70 (R. I. Lewis Smith 1984b, personal communication). *Funaria hygrometrica* became established within nine months of the 1967 eruption, and by 1981 the new fumaroles also supported *Leptobryum pyriforme*, *Campylopus canescens* and *C. introflexus*, among other mosses. These four taxa are otherwise unknown from the cold-Antarctic or, in the case of *Campylopus* spp, have been recorded there only from fumaroles on the South Sandwich Is, some 2000 km to the northeast (Fig. 1.2).

The observations prompted Smith to postulate that this part of Antarctica receives a regular influx of viable spores blown from more northerly latitudes but normally unable to undergo germination and establishment under the prevailing conditions, but he could not entirely exclude the possibility that the exotic taxa had been inadvertently introduced to Deception I by man. In East Antarctica, a high altitude fumarole on Mount Erebus supports an unidentified moss protonema, and *Campylopus pyriformis* has colonised fumarolic ground near the summmit of Mount Melbourne (altitude about 3000 m), almost certainly without human influence. The site is a major disjunction for this bipolar species which has a circumpolar distribution in south temperate regions (Broady, Given, Greenfield & Thompson, 1987).

The four exotic species tentatively identified from Deception I all fruit freely in southern temperate regions, producing spores less than 20 μm in diameter. *Campylopus introflexus* has shown a rapid stepwise range expansion following its introduction to Europe some 40 years ago (Richards & Smith, 1975). *C. introflexus* and *Funaria hygrometrica* were also among the species with spores showing the greatest resistance to desiccation and freezing in van Zanten's (1978) experiments, but there are no comparable data for the South Sandwich Is hepatics.

The conclusion thus seems inescapable that dispersal barriers can be held only partly responsible for the poverty of the Antarctic cryptogamic flora. It will be noted, however, that only a restricted assemblage of species has become established under the relatively favourable microclimatic conditions provided by the fumaroles. It is uncertain how far this is due to severe edaphic conditions as opposed to lack of incoming propagules. Smith's (1984c, 1986) experiments on the propagule bank in Antarctic soils should help to clarify this issue.

In contrast to the spores of the Deception I exotics, those of *Archidium alternifolium*, a temperate species recorded on Iceland and Greenland only near fumaroles (Chapter 3), are among the largest in bryophytes (about 200 μm in diameter). The potential of such large spores, as well as soredia, gemmae and other asexual propagules, to be carried long distances by wind is controversial. Both Steere (1965a) and Schuster (1983) consider the present Arctic populations of *A. alternifolium* to represent relicts of a more widespread Arctic distribution during the Tertiary, whereas Seppelt (1978) suggested that *Ulota phyllantha*, a species which rarely produces sporophytes anywhere, had reached Macquarie I as gemmae.

Colonisation of Surtsey Revealing observations on local dispersal and establishment were made on Surtsey, a new island formed off southern Iceland during volcanic eruptions beginning in November 1963 and continuing for three years (Fridriksson, 1975). Surtsey is 35 km from the Icelandic mainland, but small islands in the Westman group are only 5–20 km distant. The substrata are porous and therefore periodically dry, many are unstable, and all were initially low in available nitrogen. The rate of immigration and colonisation varied between the major plant groups. Over 100 species of cyanobacterium were isolated from soil on the island in 1968, but even in 1973 they and algae were seldom evident in the terrestrial vegetation. Twelve species of vascular plant, mostly coastal species, were recorded between 1965 and 1973, growing sparsely on sandy beaches.

No bryophytes were recorded until 1967, when *Bryum argenteum* and *Funaria hygrometrica* appeared, to be followed in 1968 by *Leptobryum pyriforme*, *Ceratodon purpureus* and *Pohlia bulbifera*, all species which typically act as pioneers elsewhere. The number of bryophyte species rose dramatically to 16 by 1970 and 72 by 1972, the flora then including species such as *Racomitrium canescens* and *R. lanuginosum* which form extensive, persistent colonies on lavas on the mainland.

Initial bryophyte colonisation was more rapid on lava than on the less stable tephra and sand favoured by the angiosperms. Cover remained sparse and by 1973 bryophyte biomass was seldom more than $4 \, \mathrm{g \, m^{-2}}$ dry weight, in contrast to the high values common in established polar vegetation (Chapter 7). It is notable, however, that all except six of the bryophyte species recorded between 1967 and 1971 were still present in 1972, indicating a low extinction rate among species reaching the island and undergoing the initial stages of establishment.

Lichens were first recorded on Surtsey in 1970, and only seven species were present in 1972, again principally on lava. These included *Stereocaulon vesuvianum*, a common pioneer species on Icelandic lava flows, and also *Trapelia coarctata*, previously unrecorded from Iceland. Fridriksson (1975) suggested that *Xanthoria candelaria* had been brought to Surtsey by birds, but that most of the cryptogams had arrived as spores blown from the mainland. The slow influx of lichens could be related to the problem of germinating ascospores establishing contact with an appropriate alga during primary colonisation of the new island.

Colony establishment and maintenance

Eight species of moss, including *Funaria hygrometrica* and *Racomitrium lanuginosum*, were fruiting on Surtsey by 1972 and this may accelerate their spread on the island. However, there is evidence from temperate mosses that asexual reproduction may predominate even in species producing spores in abundance (Miles & Longton, 1987). Smith

(1972, 1985a) reports that *Pottia austro-georgica* and *Pterygoneurum* cf *ovatum* regularly reproduce by spores on Signy I, but there have been few other reports of bryophytes or lichens becoming established from spores in polar regions where, as already noted, the majority of species fail to produce sporophytes. In contrast, asexual reproduction can frequently be observed. *Bryum argenteum*, which on Signy I behaves as an annual, reproduces there by deciduous shoot apices (Smith, 1985a). In perennials, branching and regeneration (Chapter 6) are undoubtedly of major importance in colony development, maintenance and recovery from disturbance.

Asexual reproduction may be particularly important in lichens as the propagules contain both mycobiont and photobiont. Many rock surfaces on Signy I are densely covered by unbranched individuals of *Usnea antarctica* only a few millimetres in length, and Hooker (1980d) demonstrated a predominance of small, presumably young thalli on imported timber. As apothecia are rarely produced, he suggested that large numbers of young plants become established from soredia, with relatively few surviving to maturity. There is thus little evidence that the abundance of mature thalli is controlled by problems of establishment.

In boreal regions, *Cladonia stellaris* commonly becomes established from thallus fragments some years after fire (Chapter 3). In contrast to the pattern in *Usnea antarctica*, subsequent clonal development by a variety of different branching patterns results in a logistic increase in the number of individuals, the maximum density of about 900 podetia m^{-2} being reached after a further 30 years (Yarranton, 1975).

The presence of morphologically distinct annual growth increments combined with low rates of decomposition enables the age structure of shoots within a colony to be established with precision in some polar mosses (Longton, 1972b). On this basis, Collins (1976b) showed that the life expectancy of individual *Polytrichum alpestre* shoots on Signy I increased from a young colony where *P. alpestre* was becoming established in a turf of *Chorisodontium aciphyllum* to an older, pure stand of the former (Fig. 8.6). Collins attributed shoot mortality largely to winter kill. However, annual increments were shown to decrease in length and in dry weight with increasing shoot age in boreal and mild-Arctic colonies of *P. alpestre*, in which age structure resembled that in Collins's mature stand. This suggests that the decline in growth, and eventual mortality, might be related to increasing problems of water and mineral nutrient uptake from the basal tomentose region of the shoots to the apices in older, longer individuals (Longton, 1972b, 1979b). Recruitment of

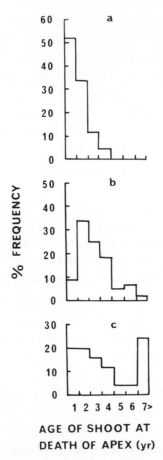

Fig. 8.6. Age structure of *Polytrichum alpestre* in three turfs on Signy I: (a) a shallow turf of *P. alpestre* and *Chorisodontium aciphyllum*; (b) a deeper turf of *P. alpestre* and *C. aciphyllum*; (c) a pure turf of *P. alpestre*. Reproduced from Collins (1976b) by courtesy of The Editor, *Oikos*.

young shoots within the turfs appears to occur principally, if not exclusively, through branching (Fig. 6.3) but the method of colony establishment is unknown.

There is considerable scope for studies on population dynamics in cryptogams, particularly in polar regions where the complicating effects of interference by vascular plants are minimal. Present evidence points to a predominance of asexual propagation, with effective sexual reproduction by spores being most likely in crustose lichens and in annual and other short-lived bryophytes. Sporelings of most species are microscopic and thus difficult to detect, however, and there is no firm evidence whether colony establishment following asexual propagation is supplemented by occasional reproduction by spores either produced locally in some species, or transported by wind from lower latitudes. The few examples discussed above suggest that there may be significant, interspecific differences in the subsequent development and age structure of the populations, but too few studies have yet been undertaken to permit generalisation.

Life history strategies
Strategies in bryophytes

Bryophytes show considerable diversity not only in the frequency of sporophyte production, but also in features such as spore size and colony life expectancy. Thus a range of life history strategies may be recognised, as in the scheme proposed by During (1979) and elaborated upon by Longton & Schuster (1983). The strategies are to some degree correlated with the ecological niches of the species concerned, but the adaptive significance of the various trends remains largely a matter of speculation.

The basic system in bryophytes was visualised by Longton & Schuster (1983) as involving perennial, dioecious gametophytes lacking specialised asexual propagules, commonly producing capsules on long setae with, in mosses, well-developed peristomes, and producing small (under 20 μm diameter) spores. Such a system provides for maximum genetic recombination and high dispersal capability, in cases where the fertilisation problems associated with dioecism are successfully overcome. Derived features would thus include monoecism, production of asexual propagules, immersed capsules with reduced peristomes, and larger spores. The evidence supporting this view is circumstantial, and some authors consider monoecism to be the primitive condition in bryophytes (Wyatt, 1982).

The basic system outlined above has persisted in many species displaying During's perennial stayer life strategy. Some perennial stayers are monoecious, e.g. *Drepanocladus uncinatus*, but the major trend has been towards local rarity (*Pleurozium schreberi, Polytrichum alpestre*), more general rarity, or absence (*Aulacomnium acuminatum*) of sporophytes. Few species produce specialised asexual propagules. Reproductive effort is therefore variable with emphasis on colony development and maintenance by branching, and on outbreeding and the production of small, easily dispersed spores where sexual reproduction occurs. Perennial stayers occupy stable habitats where regular reproduction beyond branching for colony maintenance is not essential to short-term survival. Extreme longevity may be achieved in some colonies, and it is conceivable that contemporary plants represent the founding clones in cold-Antarctic moss banks now several thousand years old (page 87).

Bryophytes exhibiting strategies grouped by During under the heading colonist exploit habitats that occur unpredictably in space and time but persists for several years, for example after fire. Thus the colonies commonly survive for a few years, and where spores are produced they are again typically small and easily dispersed. Some species retain dioecism and high spore output, as in *Ceratodon purpureus*. Others combine dioecism with low fertility and the production of asexual propagules; examples include *Pohlia andrewsii* in the Arctic, which produces axillary bulbils (Shaw, 1981), and leafy liverworts in genera such as *Cephalozia, Lophozia* and *Scapania* where gemmae develop freely on the leaves. Still others, such as *Leptobryum pyriforme*, achieve high fertility associated with monoecism, giving them the advantage that new, fertile colonies can theoretically be produced by the offspring of one spore. Sexual reproduction of such colonies would, however, involve obligate autogamy until other biotypes became established in the population.

A combination of monoecism with prolific production of small spores is also a feature of many lithophytic mosses where colony longevity may perhaps be limited by instability associated with increasing size. A high proportion of the species fruiting in the cold-Antarctic are lithophytes. Reduction in seta length and peristome development are also common trends in such species, e.g. *Schistidium antarctici*, possibly because spores are readily dispersed from exposed habitats, while *Andreaea* spp produce relatively large spores.

Bryophytes showing During's (1979) annual and short-lived shuttle strategies are characteristic of habitats available only for one or a few years, but predictably recurring within a given community. Few produce asexual propagules, but most show high sexual reproductive effort, often associated with monoecism (Longton & Schuster, 1983), and produce relatively large spores (over 25 μm diameter). These features are thought to enable the populations to avoid periods of stress by surviving within a community as spores. Examples include *Pottia austro-georgica* and *Pterygoneurum* cf *ovatum* occurring as pioneers on unstable moraines on Signy I, *Stegonia latifolia* on Alaskan frost boils (Steere, 1965a) and *Riccia sorocarpa* on steep, soil-covered ledges in Greenland (Schuster, 1983). The capsules in *Riccia* spp are immersed in the thalli so that spores are released only when the latter disintegrate.

The principal advantage of larger spores may lie in increased chances of germination and sporeling establishment arising from extensive food reserves, rather than in limited dispersal potential favouring deposition within the parent community, as many small spores also appear to be deposited near the parent colony (page 323). Species of the Splachnaceae which grow on dung and bone in Arctic regions show many of the features typical of shuttle species with additional adaptations promoting local dispersal of spores of suitable habitats by insects (Chapter 2).

Fugitives, which During regards as annuals producing abundant small, easily dispersed spores in transient habitats which do not predictably recur within a given area, are few in number. The most familiar example is *Funaria hygrometrica* which commonly appears early in succession, for example following fire in boreal woodland, and at the Deception I fumaroles.

Bryophyte strategies in relation to those in vascular plants
The perennial stayer is the major strategy, *sensu* During, among polar mosses, being exemplified by species of *Calliergon, Drepanocladus, Pleurozium, Polytrichum alpestre* and many others. It is remarkable how closely these taxa comply with the characteristics of stress-tolerators according to Grime's (1979) triangular theory of life history strategies that was formulated primarily from the standpoint of vascular plants.

Grime visualised stress-tolerators as being subjected to high levels of stress, but low levels of competition and disturbance, as is true in many polar environments where severe climate is combined with low rates of succession (Chapter 3). He considered plants adapted to such conditions to be long-lived perennials, perennating by stress-tolerant, mature organs rather than by dormant buds or seeds. They are evergreen with photosynthetic organs of considerable longevity, and with photosynthesis opportunistic, often uncoupled from vegetative growth and showing strongly developed seasonal acclimation (Chapters 5 and 6). They store photosynthate in vegetative tissues and are characterised by low growth rates (Chapter 6), a feature permitted by unpalatability (Chapter 7) as well as absence of severe competition.

Stress-tolerators were regarded as reproducing sexually only intermittently, with vegetative expansion a major regenerative strategy. *Polytrichum alpestre* and other mosses that produce sporophytes differ from stress-tolerant flowering plants in showing strict seasonality in reproductive development. Many such species fruit freely in favourable climates, and although spore production is apparently unnecessary in the short to medium term, failure in sexual reproduction by the same species in polar regions appears to reflect environmental limitation rather than adaptation and selection.

Grime's second primary strategy, the competitors, are adapted to habitats where there is little disturbance and low levels of stress. *Tortula robusta* shows a response characteristic of such plants in its rapid change in leaf area and biomass in response to fluctuation in the leaf area index of *Acaena magellanica* (Fig. 7.1), but competitors are not well represented in among cryptogams in unproductive polar environments.

Many of the species exhibiting During's shuttle strategies, such as *Pottia austro-georgica* on Signy I, occupy habitats combining climatic stress and regular cryoturbatic disturbance. These species, and also the colonists, conform in many respects with Grime's third primary strategy, i.e. ruderals, in being annuals or short-lived perennials showing early and prolific sexual reproduction. Grime might well place such plants in an intermediate group, the stress-tolerant ruderals of disturbed habitats subject to moderate stress, as he considers that a combination of severe levels of both disturbance and stress represents a situation for which no effective strategy exists.

The three primary strategies are viewed not as discrete entities, but as the extremes of a two-dimensional continuum (Grime, 1979). There may thus be considerable variation in precise patterns of resource exploitation and other features among the plants grouped under each heading. This is evident among stress-tolerant lichens in the contrasting carbon dioxide exchange relationships of chionophytic species and others equipped to undergo net assimilation under cool conditions when much of the tundra is under snow (page 185). Grime has also noted that, in their extreme forms, stress-tolerators and ruderals represent respectively the K- and r-selection extremes of the unidimensional continuum of strategies proposed by MacArthur & Wilson (1967). In K-selected organisms individuals are of long life expectancy and devote a low proportion of captured resources to reproduction, with a consequent limitation on their evolutionary potential, while the converse applies to r-selected forms. The significance of this concept to the evolution of polar cryptogams will become clear in what follows.

Reproductive strategies in lichens

Reproductive strategies in lichens have been discussed by Bowler & Rundel (1975) and Lawrey (1980), who report a degree of divergence between species where the mycobiont produces ascospores following karyogamy and meiosis, and others which produce asexual propagules such as soredia. Both sexual and asexual propagules are formed by some species of *Cladonia* and *Usnea*, however, as in many bryophytes, and this pattern may be more common in temperate than in tropical regions. Sexuality is regarded as the primitive condition. Where pairs of related species are recognised, one sorediate and the other apotheciate, the former is generally the more widespread and abundant but it may be less well equipped to contend with future environmental change.

The differences in some groups are correlated with the ecological niches of the species concerned, the pattern resembling that in bryophytes. Thus Ahti (1982) considers many *Cladonia* spp to show r-selection since they are effective colonisers of disturbed habitats, with a capacity for rapid population growth, early maturation and high fertility. *C. crispata*, occurring relatively early in the post-fire sequence in boreal woodland (page 91), is an example. Other species, such as *C. stellaris*, are considered to show K-selection as they persist for long periods in stable communities and reproduce primarily by vegetative means.

Origin and evolution of polar cryptogamic floras
Age and origin

Reproductive biology and life strategies are likely to exert a profound effect on rates of speciation. The evolution of polar floras must also be considered against the background of a relatively recent, Tertiary origin of tundra environments and repeated disruptions associated with Pleistocene glaciation (Chapter 1).

Among flowering plants, a distinctive flora exists at high latitudes (Löve & Löve, 1974), the majority of species being confined to, or at least most abundant in areas north of the forest boundary, although some species range widely from temperate to mild- and cool-Arctic regions. It is generally agreed that most Arctic flowering plants evolved before the development of Arctic tundra environments, some in alpine habitats, others perhaps in coniferous forests (Löve & Löve, 1974), or in steppe and sparsely wooded areas such as the central Asian highlands where they would become adapted to cool, dry, open situations (Bliss, 1986; Packer, 1974). However, some species are likely to be of more recent, polyploid origin. A contrasting pattern is evident in the southern hemisphere, for most of the vascular plants reported from the Antarctic regions (Fig. 1.2) also grow further north under temperate conditions, perhaps because of the earlier, more complete glaciation of Antarctic than Arctic regions.

The position is also different in cryptogamic floras, particularly bryophytes, in which the proportion of polar endemics is relatively low (Chapter 1). It thus seems even more likely in the case of cryptogams than of higher plants that the majority of species originated before the retreat of forests and other temperate vegetation, their occurrence in the tundra reflecting tolerance of polar environments rather than evolution in response to the selection pressures exerted by such conditions. This view is consistent with the fact that many of the physiological attributes of polar bryophytes and lichens that appear to confer fitness in severe environments are also to be found among temperate species (Chapter 5).

It is also consistent with the notion that bryophytes are evolutionarily conservative because of a prevalence of asexual reproduction restricting genetic recombination and reducing the potential for introgression and allopolyploidy, combined with a possibly high incidence of inbreeding in monoecious taxa. These ideas have been widely discussed (Anderson, 1963; Crum, 1972; Longton, 1976; Longton & Schuster, 1983; Smith, 1978; Wyatt & Anderson, 1984), and are particularly applicable to polar cryptogams due to the generally low levels of fertility.

Incidence of polyploidy

Many higher plants in the Arctic may well be products of rapid speciation, because cytological data indicate clearly that Arctic floras contain a higher proportion of polyploids than those in temperate and tropical regions. The figure for Peary Land is 86%. Interpretation of this pattern has long been a matter of debate. In dicots, polyploidy is most widespread in herbaceous perennials, and its frequency in the Arctic could be merely a reflection of the predominance there of this type of plant.

However, it has been suggested that polyploidy could effectively have paved the way for the exploitation of vacant niches made available as glaciers retreated, by rapidly creating a range of variants, some of which might be better adapted to the new habitats than previously existing forms that had undergone selection in different habitats. Also, many polyploids are apomictic and this could confer a selective advantage by favouring successful reproduction during the colonisation of unstable habitats under severe environmental conditions (Briggs & Walters, 1984; Packer, 1974). A higher incidence of hybridisation, a prerequisite for allopolyploidy, may also be expected between migrating than static populations.

Polyploidy has undoubtedly played an important role in the evolution of mosses (A. J. E. Smith, 1978), but Steere (1954) concluded from a survey of 55 Alaskan species that the distribution of chromosome numbers in Arctic mosses is remarkably similar to that in other areas, both in a general sense (Fig. 8.7) and in particular taxa. Thus four Alaskan species of the Polytrichaceae had $n = 7$, while high number polyploids were most frequent in the Pottiaceae, with a maximum of $n = 50$ in *Pottia obtusifolia*. Alaskan populations of several mosses had chromosome numbers identical to those in temperate material of the same species, e.g. *Ceratodon purpureus* with $n = 13$. Steere also found no evidence of aneuploid variation among the Alaskan material and he reported that meiotic abnormalities were not unusually frequent.

In contrast, some though not all of the endemic Arctic species were found to be polyploid, often at levels uncommon in the genera concerned (Steere, 1954). Thus $n = 42$ was recorded for *Distichium hagenii*, a member of the circumpolar Arctic element, whereas the more widely distributed *D. capillaceum* has $n = 14$ or 28. Even more remarkable was the position in *Bryum*, in which $n = 10$, 11 or 12 in most temperate species. Alaskan material of the cosmopolitan species *B. argenteum* had $n = 11$, but numbers from $n = 20$ (*B. arcticum*, *B. nitidulum*) to $n = 50$ in an unidentified species of *Bryum* were recorded among Arctic alpines.

More recent data suggest that the position with regard to polyploidy may be broadly similar among Antarctic bryophytes. The overall frequency of polyploids is not unusually high, but $n = 20$ has been reported for the apparently endemic *Bryum algens* from both the cold- and frigid-Antarctic (Inoue, 1976; Newton, 1980). Inoue suggested that an earlier report of $n = 20$ in frigid-Antarctic *B. argenteum* may have been based on misidentified material, but $n = 16 + 2m$ for *Cehpaloziella varians* from South Georgia and the South Shetland Is is approximately double the number typical of leafy liverworts.

Polyploidy has also been recorded in South Shetland Is material of bipolar hepatics common in north temperate and Arctic regions, e.g. $n = 18$ (also $n = 8 + 1m$) in *Barbilophozia hatcheri* and $n = 27$ in *Lophozia excisa* (Ochyra, Przywara & Kuta, 1982). Intra-specific variation, both polyploid and aneuploid, is known among Antarctic mosses. Thus $n = 11 + 1m$ and 16 are recorded in *Bartramia patens*, $n = 10$ and 20 in *Drepanocladus uncinatus*, and $n = 10$, 11, 13 and 20 in *Brachythecium austro-salebrosum* (Newton, 1972, 1979, 1980; Kuta, Ochyra & Przywara, 1982). Several morphological variants exist in these three species but

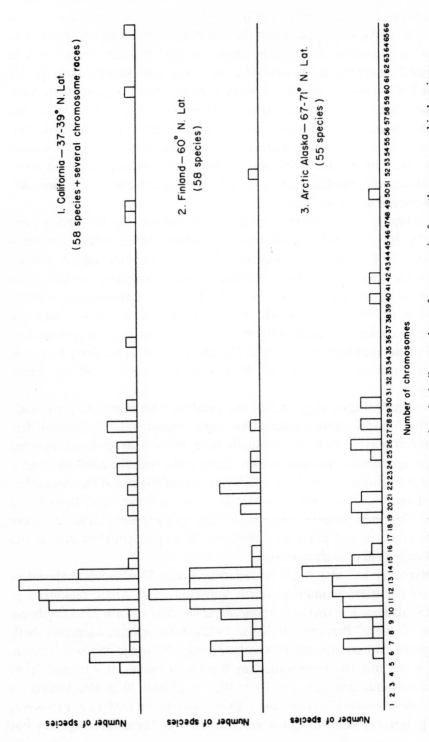

Fig. 8.7. Comparison of gametophytic chromosome numbers in similar numbers of moss species from three geographical regions. Reproduced from W. C. Steere (1954) by courtesy of The Editor, *Botanical Gazette*. Copyright 1954 by the University of Chicago.

it is not clear how closely the morphological and cytological variation is correlated.

One can but speculate on the reasons for the different proportions of polyploids among polar vascular plants and bryophytes. If, as seems likely, polyploidy in bryophytes most commonly arises through apospory or diplospory (A. J. E. Smith, 1978), then its frequency might well be reduced by a generally low level of fertility. Autopolyploidy appears to be prevalent in bryophytes, possibly reflecting relatively large genetic distances and consequent infrequency of interspecific hybridisation in many genera. The incidence of autopolyploidy is less likely to be increased by contact between species during Pleistocene migration than in the case of allopolyploidy.

Further, the possible selective advantage of apomixis associated with polyploidy in vascular plants would not apply to bryophytes in which most fertile species with haploid or ancestral-diploid (A. J. E. Smith, 1978) gametophytes are also capable of vegetative propagation. Finally, if many bryophytic genotypes confer tolerance of a wide range of environmental regimes (Longton & Schuster, 1983) then production of new genetic variants would be less advantageous in the colonisation of new tundra habitats than in groups where the biotypes are less broadly plastic.

Variability in widely ranging species

As the climates at high latitudes deteriorated during the Tertiary one can thus visualise those bryophyte and lichen species which already possessed combinations of morphological and physiological attributes appropriate to one of the newly created tundra habitats persisting to form the basis of modern polar floras. However, this scenario does not imply complete evolutionary stasis during the intervening period. Numerous examples of intra-specific variation in morphological and physiological characters of probably adaptive significance have been reported among polar cryptogams, as discussed in Chapters 5 and 6.

In bryophytes, some of this variation has been shown to have a genetic basis, and the same may well be true of lichens although this remains to be confirmed. The clinal nature of the variation in stature shown by *Polytrichum alpestre* (Figs. 6.13 and 6.17) could best be explained by genetic recombination, especially as this species fruits throughout much of its range and produces small, readily dispersible spores. Selection of mutations, which would always be expressed in haploid ($n = 7$) gametophytes, is an alternative possibility.

In liverworts, several groups such as the section Curtae of *Scapania*, which are relatively stenotypic in temperate regions, show complex species structure and division into many ill-defined taxa in the Arctic (Schuster, 1983). This has been attributed (Schuster, 1966–80) to 'the chaotic history of the Arctic since Tertiary times, with the coupling of intense selection pressures, extinction of biotypes, and remingling of ± imperfectly genetically but once geographically isolated genotypes, plus the demonstrated phenotypic malleability of the taxa'. Such a combination of factors must have applied to many groups of cryptogams.

Origin of Arctic endemics

Most of the species so far investigated from the physiological point of view are wide-ranging, but the polar endemics (Chapter 1) are also of great interest. The traditional view (Steere, 1965a; Thomson, 1972) is that the circumpolar Arctic endemic cryptogams are Tertiary relicts, some of the bryophytes having their closest relationships with southern hemisphere tropical and subtropical floras. More recently, Brassard (1974) has argued persuasively that a number of the bryophytes evolved in the Arctic during the Pleistocene. He considered that relict status was most likely in the case of species in monotypic genera such as *Arnellia fennica* or *Andreaeobryum macrosporum*, a recently described, abundantly fertile taxon, placed in a separate family, and apparently representing a transitional form between the Andreaeales and Bryales (Steere & Murray, 1976). Relict status was also considered likely where the Arctic species show no evidence of a close relationship to their cogeners as in the case of the now sterile *Aulacomnium acuminatum*. However, a more recent origin was suspected where the Arctic species appear closely related either to each other, or to southern species.

Many such taxa occur in groups showing high fertility in the Arctic, often with shuttle or colonist life strategies, and Brassard suspected that hybridisation may have been involved in their evolution. Examples include species of *Funaria* and the Splachnaceae, the polyploid *Bryum* spp, and *Polytrichum hyperboreum* which was seen as a possible hybrid between the colonising species *Polytrichum juniperinum* and *P. piliferum*. Schuster (1983) has pointed out that Arctic endemics in several hepatic genera such as *Radula* and *Anastrophyllum* are monoecious and produce sporophytes whereas related temperate species are dioecious, and he likewise suggested that the Arctic species represent relatively modern derivatives. Many of the species thought to be of recent origin are monoecious: it is conceivable that they show facultative autogamy with the effect of a consequent stabilisation of any adaptive, new gene combinations outweighing a low incidence of genetic recombination. The fertility of some cold-Antarctic endemic lichens (page 315) presents an intriguing parallel with the position among Arctic endemic bryophytes.

Concluding remarks

The broad pattern that is beginning to emerge, albeit tentatively, is that the polar cryptogamic floras comprise, in the main, species with ranges encompassing also forested regions and other areas of relatively favourable climate: they are likely to have evolved before Arctic tundra developed during Tertiary climatic deterioration. They were able to exploit newly created tundra niches by virtue of features that are widely developed among mosses and lichens (page 203), of which the most significant include a high degree of phenotypic plasticity, and poikilohydrous water relations leading to extreme frost resistance and opportunistic responses in carbon dioxide exchange and growth.

Most of the species investigated to date from an ecophysiological standpoint fall into this category. In mosses, for which appropriate techniques are available, this work has demonstrated intra-specific genetic variation of an adaptive, but often relatively minor nature, that may well have evolved in response to local environments. It must be remembered, however, that these species have undergone extensive and repeated migrations in response to Pleistocene glaciation that would undoubtedly break down barriers to reproduction between populations, so that much of the adaptive, intra-specific variation now apparent may represent selection during the relatively brief, post-Wisconsin period. Comparison of populations from formerly glaciated and refugial areas with similar contemporary environments could thus be as rewarding in terms of revealing Arctic adaptations as latitudinal studies of the type so far carried out.

The Arctic endemics represent, if Brassard (1974) is correct, a mixture of ancient species originating in warmer Tertiary environments but surviving at least the Wisconsin glaciation in Arctic refugia, and younger species, some of polyploid origin, that evolved under Arctic conditions. Investigation of environmental responses in both types of taxon would be of the utmost interest, and the same applies to temperate species with disjunct populations in Arctic refugia that are likely to have been isolated throughout at least the most recent glaciation. Glaciation is thought to have become established earlier in the Antarctic than in the Arctic (Chapter 1), but the contemporary floras also appear to be composed largely of post-glacial immigrants, and great interest again centres on the putative endemics, including the newly described endolithic algae and fungi from the frigid-Antarctic (Chapter 2). Hepatics also merit more attention than they have so far received, for in temperate regions they tend to be less tolerant of desiccation and frost than the mosses.

By and large, however, hepatics and endemic mosses are less abundant in tundra vegetation than many of the more widely distributed moss and lichen species. The dominant role of cryptogams in the tundra thus reflects, above all, reduced competition from flowering plants and attributes conferring fitness in polar environments but present among mosses and lichens throughout the world.

References

Aaltonen, K., Pasanen, S. & Aaltonen, H. (1985). Measuring systems for analysing of the CO_2 concentration under snow. *Aquilo, Ser. Bot.*, **23**, 65–8

Abramova, A. L., Savicz-Ljubitskaja, L. I. & Smirnova, Z. N. (1961). *Handbook for the Determination of the Mosses of Arctic Russia*. Leningrad, Komarov Botanical Institute

Addison, P. A. (1977). Studies on evapotranspiration and energy budgets on Truelove Lowland. In *Truelove Lowland, Devon Island, Canada*, ed. L. C. Bliss, pp. 281–300. Edmonton, University of Alberta Press

Addison, P. A. & Bliss, L. C. (1980). Summer climate, microclimate, and energy budget of a polar semidesert on King Christian Island, N.W.T., Canada. *Arctic and Alpine Research*, **12**, 161–70

Ahmadjian, V. (1970). Adaptations of Antarctic terrestrial plants. In *Antarctic Ecology*, ed. M. W. Holdgate, vol. 2, pp. 801–11. London, Academic Press

Ahti, T. (1977). Lichens of the boreal coniferous zone. In *Lichen Ecology*, ed. M. R. D. Seaward, pp. 145–181. London, Academic Press

Ahti, T. (1982). Evolutionary trends in cladoniiform lichens. *Journal of the Hattori Botanical Laboratory*, **52**, 331–41

Ahti, T., Hämet-Ahti, L. & Jalas, J. (1968). Vegetation zones and their sections in northwestern Europe. *Annales Botanici Fennici*, **5**, 169–211

Aleksandrova, V. D. (1980). *The Arctic and Antarctic: their Division into Geobotanical Areas*. Cambridge, Cambridge University Press

Aleksandrova, V. D. (1988). *Vegetation of the Soviet Polar Deserts*. English edition translated by D. Löve. Cambridge, Cambridge University Press

Alexander, V. (1975). Nitrogen fixation by blue-green algae in polar and subpolar regions. In *Nitrogen Fixation by Free-Living Micro-organisms*, ed. W. D. P. Stewart, pp. 175–88. Cambridge, Cambridge University Press

Alexander, V., Billington, M. & Schell, D. M. (1978). Nitrogen fixation in Arctic and alpine tundra. In *Vegetation and Production Ecology of an Alaskan Arctic Tundra*, ed. L. L. Tieszen, pp. 539–58. New York, Springer-Verlag

Alexandrova, V. D. (1970). The vegetation of the tundra zones in the USSR and data about its productivity. In *Productivity and Conservation in Northern Circumpolar Lands*, ed. W. A. Fuller & P G. Kevan, pp. 93–114. Morges, Switzerland, IUCN

Allen, S. E., Grimshaw, H. M. & Holdgate, M. W. (1967). Factors affecting the availability of plant nutrients on an Antarctic island. *Journal of Ecology*, **55**, 381–96

Allen, S. E. & Heal, O. W. (1970). Soils of the maritime Antarctic zone. In *Antarctic Ecology*, ed. M. W. Holdgate, vol. 2, pp. 693–6. London, Academic Press

Alpert, P. & Oechel, W. C. (1982). Bryophyte vegetation and ecology along a topographic gradient in montane tundra in Alaska. *Holarctic Ecology*, **5**, 99–108

Alpert, P. & Oechel, W. C. (1984). Microdistribution and water loss resistances of selected bryophytes in an Alaskan *Eriophorum* tussock tundra. *Holarctic Ecology*, **7**, 111–18

Anderson, I. (1986). A glimpse of the green hills of Antarctica. *New Scientist*, **1515**, 22

Anderson, L. E. (1963). Modern species concepts: mosses. *Bryologist*, **66**, 107–19

Ando, H. (1979). Ecology of terrestrial plants in the Antarctic with particular reference to bryophytes. *Memoirs of National Institute of Polar Research, Special Issue*, **11**, 81–103

Andreev, V. N. (1954). Prirost kormovykh lishainikov i priemy ego regulirovaniia. *Geobotanica*, **9**, 11–74

Andreev, V. N. (1971). Methods of defining above ground phytomass on vast territories of the subarctic. *Reports from the Kevo Subarctic Research Station*, **8**, 3–11

Andrews, J. T. & Barnett, D. M. (1979). Holocene (Neoglacial) moraine and periglacial lake chronology, Barnes Ice Cap, N.W.T., Canada. *Boreas*, **8**, 341–58

Andrews, J. T. & Webber, P. J. (1964). A lichenometrical study of the northwestern margin of the Barnes Ice Cap: a geomorphological technique. *Geographical Bulletin*, **22**, 80–104

Aplin, P. S. & Hill, D. J. (1979). Growth analysis of circular lichen thalli. *Journal of Theoretical Botany*, **78**, 747–63

Arafat, N. M. & Gloschenko, W. A. (1982). The use of bog vegetation as an indicator of atmospheric deposition of arsenic in northern Ontario. *Environmental Pollution*, **B4**, 85–90

Armstrong, R. A. (1975). Studies on the growth rates of lichens. In *Lichenology: Progress and Problems*, ed. D. H. Brown, D. L. Hawksworth & R. H. Bailey, pp. 309–22. London, Academic Press

Armstrong, R. A. (1979). Growth and regeneration of lichen thalli with the centres artificially removed. *Environmental and Experimental Botany*, **19**, 175–8

Armstrong, R. A. (1984). Growth of experimentally reconstructed thalli of the lichen *Parmelia conspersa*. *New Phytologist*, **98**, 497–502

Aro, E.-M. & Valanne, N. (1979). Effect of continuous light on CO_2 fixation and chloroplast structure of the mosses *Pleurozium schreberi* and *Ceratodon purpureus*. *Physiologia Plantarum*, **45**, 460–6

Ascaso, C. (1985). Structural aspects of lichens invading their substrata. In *Surface Physiology of Lichens*, ed. C. Vicenti, D. H. Brown & M. Estrella Legaz, pp. 87–113. Madrid, Universidad Complutense de Madrid

Ascaso, C., Brown, D. H. & Rapsch, S. (1985). Ultrastructural studies of desiccated lichens. In *Lichen Physiology and Cell Biology*, ed. D. H. Brown, pp. 259–74. New York, Plenum Press

Ashton, D. H. & Gill, A. M. (1965). Pattern and process in a Macquarie Island feldmark. *Proceedings of the Royal Society of Victoria*, **79**, 235–45

Atanasiu, L. (1971). Photosynthesis and respiration of three mosses at winter low temperatures. *Bryologist*, **74**, 23–7

Babb, T. A. & Whitfield, D. W. S. (1977). Mineral nutrient cycling and limitation of plant growth in the Truelove Lowland ecosystem. In *Truelove Lowland, Devon Island, Canada: a High Arctic Ecosystem*, ed. L. C. Bliss, pp. 589–606. Edmonton, University of Alberta Press

Bailey, R. H. (1976). Ecological aspects of dispersal and establishment in lichens. In *Lichenology: Progress and Problems*, ed. D. H. Brown, D. L. Hawksworth & R. H. Bailey, pp. 215–47. London, Academic Press

Baker, J. H. (1972). The rate of production and decomposition in *Chorisodontium aciphyllum* (Hook. f. et Wils.) Broth. *British Antarctic Survey Bulletin*, **27**, 123–9

Barashkova, E. A. (1971). Photosynthesis in fruticose lichens *Cladonia alpestris* L. Rabh. and *C. rangiferina* L. Web. in the Taimyr Peninsula. *International Tundra Biome Translation*, **4**, 1–10

Barrett, P. E. & Thomson, J. W. (1975). Lichens from a high Arctic coastal lowland, Devon Island, N.W.T. *Bryologist*, **78**, 160–7

Barry, R. G., Courtin, G. M. & Labine, C. (1981). Tundra climates. In *Tundra Ecosystems: a Comparative Analysis*, ed. L. C. Bliss, O. W. Heal & J. J. Moore, pp. 81–114. Cambridge, Cambridge University Press

Basilier, K., Granhall, U. & Senström, T.-A. (1978). Nitrogen fixation in wet minerotrophic moss communities of a subarctic mire. *Oikos*, **31**, 236–46

Batzli, G. O. & Cole, F. R. (1979). Nutritional ecology of microtine rodents: digestibility of forage. *Journal of Mammalogy*, **60**, 740–50

Batzli, G. O. & Sobaski, S. T. (1980). Distribution, abundance, and foraging patterns of ground squirrels near Atkasook, Alaska. *Arctic and Alpine Research*, **12**, 501–10

Batzli, G. O., White, R. G. & Bunnell, F. L. (1981). Herbivory: a strategy of tundra consumers. In *Tundra Ecosystems: a Comparative Analysis*, ed. L. C. Bliss, O. W. Heal & J. J. Moore, pp. 359–75. Cambridge, Cambridge University Press

Batzli, G. O., White, R. G., MacLean, S. E., Pitelka, F. A. & Collier, B. D. (1980). The herbivore-based trophic system. In *An Arctic Ecosystem: The Coastal Tundra at Barrow, Alaska*, ed. J. Brown, P. C. Miller, L. L. Tieszen & F. L. Bunnell, pp. 335–410. Dowden, Hutchinson & Ross

Bayfield, N. B. (1973). Notes on water relations of *Polytrichum commune* Hedw. *Journal of Bryology*, **7**, 607–17

Bazzaz, F. A., Paolillo, D. J. & Jagels, R. H. (1970). Photosynthesis and respiration of forest and alpine populations of *Polytrichum juniperinum*. *Bryologist*, **73**, 579–85

Beckett, P. J., Boileau, L. J. R., Padovan, D., Richardson, D. H. S. & Nieboer, E. (1982). Lichens and mosses as monitors of industrial activity associated with uranium mining in northern Ontario, Canada – Part 2: Distance dependent uranium and lead accumulation patterns. *Environmental Pollution*, **B4**, 91–107

Bell, K. L. & Bliss, L. C. (1980). Plant reproduction in a high Arctic environment. *Arctic and Alpine Research*, **12**, 1–10

Benninghoff, W. S. (1952). Interaction of vegetation and soil frost phenomena. *Arctic*, **5**, 34–44

Benson, C. S. (1986). Problems of air quality in local arctic and sub-arctic areas, and regional problems of arctic haze. In *Arctic Air Pollution*, ed. B. Stonehouse, pp. 69–84. Cambridge, Cambridge University Press. (7)

Berg, A. (1975). Pigment structure of vascular plants, mosses and lichens at Hardangervidda, Norway. In *Fennoscandian Tundra Ecosystems. I: Plants and Microorganisms*, ed. F. E. Wielgolaski, pp. 216–24. New York, Springer-Verlag.

Beschel, R. E. (1950). Flechten als Altersmasstab rezenter Moränen. *Zeitschrift für Gletscherkunde und Glazialgeologie*, **1**, 152–61

Beschel, R. E. (1961). Dating rock surfaces by lichen growth and its application to glaciology and physiography (lichenometry). In *Geology of the Arctic*, ed. G. O. Raasch, vol. 2, pp. 1044–62. Toronto, University of Toronto Press

Beschel, R. E. & Weidick, A. (1973). Geobotanical and geomorphological reconnaissance in west Greenland, 1961. *Arctic and Alpine Research*, **5**, 311–19

Bigger, C. M. & Oechel, W. C. (1982). Nutrient effect on maximum photosynthesis in Arctic plants. *Holarctic Ecology*, **5**, 158–63

Billington, M. & Alexander, V. (1978). Nitrogen fixation in a black spruce (*Picea mariana* (Mill) B.S.P.) forest in Alaska. In *Environmental Role of Nitrogen-Fixing Blue-Green Algae and Asymbiotic Bacteria*, ed. U. Granhall, pp. 209–15. *Ecological Bulletins (Stockholm)*, 26

Billings, W. D. (1974). Arctic and alpine vegetation: plant adaptation to cold summer climates. In *Arctic and Alpine Environments*, ed. J. D. Ives & R. G. Barry, pp. 403–43. London, Methuen

Billings, W. D. & Peterson, K. M. (1980). Vegetational change and ice-wedge polygons through the thaw-lake cycles in Arctic Alaska. *Arctic and Alpine Research*, **12**, 413–32

Birkenmajer, K. (1980a). Age of the Penguin Island volcano, South Shetland Islands (West Antarctica), by the lichenometric method. *Bulletin de l'Académie Polonaise des Sciences, Série des Sciences de la Terre*, **27**, 69–76

Birkenmajer, K. (1980b). Lichenometric dating of glacier retreat at Admiralty Bay, King George Island (South Shetland Islands, West Antarctica). *Bulletin de l'Académie Polonaise des Sciences, Série des Sciences de la Terre*, **27**, 77–85

Birkenmajer, K. (1981). Lichenometric dating of raised marine beaches at Admiralty Bay, King George Island (South Shetland Islands, West Antarctica). *Bulletin de l'Académie Polonaise des Sciences, Série des Sciences de la Terre*, **29**, 119–27

Black, R. A. & Bliss, L. C. (1978). Recovery sequence of *Picea mariana–Vaccinium uliginosum* forests after burning near Inuvik, Northwest Territories, Canada. *Canadian Journal of Botany*, **56**, 2020–30

Bliss, L. C. (1962). Adaptations of Arctic and alpine plants to environmental conditions. *Arctic*, **15**, 117–44

Bliss, L. C. (1975). Tundra grasslands, herblands and shrublands and the role of herbivores. *Geoscience and Man*, **10**, 51–79

Bliss, L. C. (ed.) (1977). *Truelove Lowland, Devon Island, Canada: a High Arctic Ecosystem*. Edmonton, University of Alberta Press

Bliss, L. C. (1979). Vascular plant vegetation of the southern circumpolar region in relation to Antarctic, alpine and Arctic vegetation. *Canadian Journal of Botany*, **57**, 2167–78

Bliss, L. C. (1981). North American and Scandinavian tundras and polar deserts. In *Tundra Ecosystems: a Comparative Analysis*, ed. L. C. Bliss, O. W. Heal & J. J. Moore, pp. 8–24. Cambridge, Cambridge University Press

Bliss, L. C. (1986). Arctic ecosystems: their structure, function and herbivore carrying capacity. In *Grazing Research at Northern Latitudes*, ed. O. Gudmundsson, pp. 5–25. New York, Plenum Press

Bliss, L. C., Courtin, G. M., Pattie, D. L., Riewe, R. R., Whitfield, D. W. A. & Widden, P. (1973). Arctic tundra ecosystems. *Annual Review of Ecology and Systematics*, **4**, 359–99

Bliss, L. C. & Svoboda, J. (1984). Plant communities and plant production in the western Queen Elizabeth Islands. *Holarctic Ecology*, **7**, 325–44

Bliss, L. C., Svoboda, J. & Bliss, D. I. (1984). Polar deserts, their plant cover and plant production in the Canadian high Arctic. *Holarctic Ecology*, **7**, 305–24

Block, W. (1985). Arthropod interactions in an Antarctic terrestrial community. In *Antarctic Nutrient Cycles and Food Webs*, ed. W. R. Siegfried, P. R. Condy & R. M. Laws, pp. 614–19. Berlin, Springer-Verlag

Blum, O. B. (1973). Water relations. In *The Lichens*, ed. V. Ahmadjian & M. E. Hale, pp. 381–400. New York, Academic Press

Boelcke, O., Moore, D. M. & Roig, F. A. (ed.) (1985). *Transecta Botánica de la Patagonia Austral*. Buenos Aires, Consego Nacional de Investigaciones Científicas y Tecnicas

Boertje, R. (1984). Seasonal diets of the Denali caribou herd, Alaska. *Arctic*, **37**, 161–5

Booth, R. G. & Usher, M. B. (1985). Variation in the physical and chemical environments of a maritime Antarctic moss-turf. *British Antarctic Survey Bulletin*, **67**, 25–40

Bowler, P. A. & Rundel, P. W. (1975). Reproductive strategies in lichens. *Botanical Journal of the Linnean Society*, **70**, 325–40

Brassard, G. R. (1969). *Mielichhoferia elongata*, a copper moss new to North America, found in Arctic Canada. *Nature*, **222**, 384–5

Brassard, G. R. (1971a). The mosses of northern Ellesmere Island, Arctic Canada. I: Ecology and phytogeography, with an analysis for the Queen Elizabeth Islands. *Bryologist*, **74**, 233–81

Brassard, G. R. (1971b). The mosses of northern Ellesmere Island, Arctic Canada. II: Annotated list of the taxa. *Bryologist*, **74**, 282–311

Brassard, G. R. (1974). The evolution of Arctic bryophytes. *Journal of the Hattori Botanical Laboratory*, **38**, 39–48

Brassard, G. R. (1976). The mosses of northern Ellesmere Island, Arctic Canada. III: New or additional records. *Bryologist*, **79**, 480–7

Brassard, G. R. & Longton, R. E. (1970). The flora and vegetation of Van Hauen Pass, northwestern Ellesmere Island. *Canadian Field Naturalist*, **84**, 357–64

Briggs, D. & Walters, S. M. (1984). *Plant Variation and Evolution*, 2nd ed. Cambridge, Cambridge University Press

Broady, P. A. (1981). The ecology of chasmolithic algae at coastal locations of Antarctica. *Phycologia*, **20**, 259–72

Broady, P. A., Given, D., Greenfield, L. & Thompson, K. (1987). The biota and environment of fumaroles on Mt Melbourne, northern Victoria Land. *Polar Biology*, **7**, 97–113

Brodo, I. M. (1973). Substrate ecology. In *The Lichens*, ed. V. Ahmadjian & M. E. Hale, pp. 401–41. New York, Academy Press

Brossard, T., Deruelle, S., Nimis, P. L. & Petit, P. (1984). An interdisciplinary approach to mapping on lichen-dominated systems in high Arctic Environment, Ny Ålesund (Svalbard). *Phytocoenologia*, **12**, 433–53

Brown, D. & Kershaw, K. A. (1985). Electrophoretic and gas exchange patterns of two populations of *Peltigera rufescens*. In *Lichen Physiology and Cell Biology*, ed. D. H. Brown, pp. 111–28. New York, Plenum Press

Brown, D., MacFarlane, J. D. & Kershaw, K. A. (1983). Physiological–environmental interactions in lichens. XVI: A re-examination of resaturation respiration phenomena. *New Phytologist*, **93**, 237–46

Brown, D. H. (1982). Mineral nutrition. In *Bryophyte Ecology*, ed. A. J. E. Smith, pp. 383–444. London, Chapman and Hall

Brown, D. H. (1984). Uptake of mineral elements and their use in pollution monitoring. In *The Experimental Biology of Bryophytes*, ed. A. F. Dyer & J. G. Duckett, pp. 229–55. London, Academic Press

Brown, D. H. & Beckett, R. P. (1984). Uptake and effect of cations on lichen metabolism. *Lichenologist*, **16**, 173–88

Brown, D. H. & Beckett, R. P. (1985a). Minerals and lichens: acquisition, localization and effect. In *Surface Physiology of Lichens*, ed. C. Vicente, D. H. Brown & M. Estrella Legaz, pp. 127–49. Madrid, Universidad Complutense de Madrid

Brown, D. H. & Beckett, R. P. (1985b). Intracellular and extracellular uptake of cadmium by the moss *Rhytidiadelphus squarrosus*. *Annals of Botany*, **55**, 179–88

Brown, D. H. & Buck, G. W. (1979). Desiccation effects and cation distribution in bryophytes. *New Phytologist*, **82**, 115–25

Brown, R. N. Rudmose (1912). The problems of Antarctic plant life. *Report on Scientific Results, Scottish National Antarctic Expedition, S.Y. Scotia, 1902–04.* Vol. 3, *Botany*, no. 1, 3–21

Bryson, R. A. (1966). Air masses, stream lines and the boreal forest. *Geographical Bulletin*, **8**, 228–69

Bubrick, P. & Galun, M. (1986). Spore to spore resynthesis of *Xanthoria parientina*. *Lichenologist*, **18**, 47–9

Bunnell, F. L., MacLean, S. F. & Brown, J. (1975). Barrow, Alaska, U.S.A. In *Structure and Function of Tundra Ecosystems*, ed. T. Rosswall & O. W. Heal, pp. 73–124. *Ecological Bulletins (Stockholm)*, 20

Burger, A. E. (1985). Terrestrial food webs in the sub-Antarctic: island effects. In *Antarctic Nutrient Cycling and Food Webs*, ed. W. R. Siegfried, P. R. Condy & R. M. Laws, pp. 582–91. Berlin, Springer-Verlag

Caldwell, M. M. (1972). Biologically effective solar ultraviolet irradiation in the Arctic. *Arctic and Alpine Research*, **4**, 39–43

Caldwell, M. M., Johnson, D. A. & Fareed, M. (1978). Constraints on tundra productivity: photosynthetic capacity in relation to solar radiation utilization and water stress in Arctic and alpine tundras. In *Vegetation and Production Ecology of an Alaskan Arctic Tundra*, ed. L. L. Tieszen, pp. 323–42. New York, Springer-Verlag

Calkin, P. E. & Ellis, J. M. (1980). A lichenometric dating curve and its application to holocene glacier studies in the central Brooks Range, Alaska. *Arctic and Alpine Research*, **12**, 245–64

Calkin, P. E. & Ellis, J. M. (1981). A cirque-glacier chronology based on emergent lichens and mosses. *Journal of Glaciology*, **27**, 511–15

Callaghan, T. V. & Collins, N. J. (1981). Life cycles, population dynamics, and the growth of tundra plants. In *Tundra Ecosystems: a Comparative Analysis*, ed. L. C. Bliss, O. W. Heal & J. J. Moore, pp. 257–84. Cambridge, Cambridge University Press

Callaghan, T. V., Collins, N. J. & Callaghan, C. H. (1978). Photosynthesis, growth and reproduction of *Hylocomium splendens* and *Polytrichum commune* in Swedish Lapland. *Oikos*, **31**, 73–88

Cameron, R. E. (1972). Pollution and conservation of the Antarctic terrestrial ecosystem. In *Conservation Problems in Antarctica*, ed. B. C. Parker, pp. 267–305. Blacksburg, Virginia Polytechnic Institute and State University

Cameron, R. E., Lacey, G. H., Morelli, F. A. & Marsh, J. B. (1971). Farthest south soil microbiological and ecological investigations. *Antarctic Journal of the United States*, **6**, 105–6

Cameron, R. G. & Troilo, D. (1982). Fly-mediated spore dispersal of *Splachnum ampullaceum* (Musci). *Michigan Botanist*, **21**, 59–65

Carlberg, G. E., Ofstad, E. B., Drangsholt, H. & Steinnes, E. (1983). Atmospheric deposition of organic micropollutants in Norway studied by means of moss and lichen analysis. *Chemosphere*, **12**, 341–56

Carstairs, A. G. & Oechel, W. C. (1978). Effects of several microclimatic factors and nutrients on net carbon dioxide exchange in *Cladonia alpestris* (L.) Rabh. in the subarctic. *Arctic and Alpine Research*, **10**, 81–94

Chapin, F. S. (1983). Direct and indirect effects of temperature on Arctic plants. *Polar Biology*, **2**, 47–52

Chapin, F. S. & Chapin, M. C. (1980). Revegetation of an Arctic disturbed site by native tundra species. *Journal of Applied Ecology*, **17**, 449–56

Chapin, F. S., Johnson, D. A. & McKendrick, J. D. (1980). Seasonal movement of nutrients in plants of differing growth form in an Alaskan tundra ecosystem: implications for herbivory. *Journal of Ecology*, **68**, 189–209

Chernov, Yu. I. (1985). *The Living Tundra*. English edition translated by D. Löve. Cambridge, Cambridge University Press

Christie, P. (1987). Nitrogen in two contrasting Antarctic bryophyte communities. *Journal of Ecology*, **75**, 73–93

Claridge, G. G. C. & Campbell, I. B. (1985). Physical geography – soils. In *Key Environments Antarctica*, ed. W. N. Bonner & D. W. H. Walton, pp. 62–70. Oxford, Pergamon Press

Clarke, G. C. S. & Greene, S. W. (1970). Reproductive performance of two species of *Pohlia* at widely separated stations. *Transactions of the British Bryological Society*, **6**, 114–28

Clarke, G. C. S. & Greene, S. W. (1971). Reproductive performance of two species of *Pohlia* from temperate and sub-Antarctic stations under controlled conditions. *Transactions of the British Bryological Society*, **6**, 278–95

Clarke, G. C. S., Greene, S. W. & Greene, D. M. (1971). Productivity of bryophytes in polar regions. *Annals of Botany*, **35**, 99–108

Clausen, E. (1964). The tolerance of hepatics to desiccation and temperature. *Bryologist*, **67**, 411–17

Clayton-Greene, K. A., Collins, N. J., Green, T. G. A. & Proctor, M. C. F. (1985). Surface wax, structure and function in leaves of Polytrichaceae. *Journal of Bryology*, **13**, 549–62

Clements, F. E. (1916). *Plant Succession*. Washington, Carnegie Institution

Clymo, R. S. (1970). The growth of *Sphagnum*: methods of measurement. *Journal of Ecology*, **58**, 13–49

Collins, N. J. (1973). Productivity of selected bryophytes in the maritime Antarctic. In *Proceedings of the Conference on Primary Production and Production Processes, Tundra Biome*, ed. L. C. Bliss & F. E. Wielgolaski, pp. 177–83. Edmonton, IBP Tundra Biome Steering Committee

Collins, N. J. (1976a). The development of moss-peat banks in relation to changing climate and ice cover on Signy Island in the maritime Antarctic. *British Antarctic Survey Bulletin*, **43**, 85–102

Collins, N. J. (1976b). Growth and population dynamics of the moss *Polytrichum alpestre* in the maritime Antarctic. *Oikos*, **27**, 389–401

Collins, N. J. (1977). The growth of mosses in two contrasting communities in the maritime Antarctic: measurement and prediction of net annual production. In *Adaptations within Antarctic Ecosystems*, ed. G. A. Llano, pp. 921–33. Washington, Smithsonian Institution

Collins, N. J., Baker, J. H. & Tilbrook, P. J. (1975). Signy Island, maritime Antarctic. In *Structure and Function of Tundra Ecosystems*, ed. T. Rosswall & O. W. Heal, pp. 345–74. *Ecological Bulletin (Stockholm)*, 20

Collins, N. J. & Callaghan, T. V. (1980). Predicted patterns of photosynthetic production in maritime Antarctic mosses. *Annals of Botany*, **45**, 601–20

Collins, N. J. & Oechel, W. C. (1974). The pattern of growth and translocation of photosynthate in a tundra moss, *Polytrichum alpinum*. *Canadian Journal of Botany*, **52**, 355–63

Corner, R. W. M. & Smith, R. I. Lewis (1973). Botanical evidence of ice recession in the Argentine Islands. *British Antarctic Survey Bulletin*, **35**, 83–6

Costin, A. B. & Moore, D. M. (1960). The effects of rabbit grazing on the grasslands of Macquarie Island. *Journal of Ecology*, **48**, 729–32

Courtin, G. M. & Mayo, J. M. (1975). Arctic and alpine plant water relations. In *Physiological Adaptations to the Environment*, ed. F. J. Vernberg, pp. 201–24. New York, Intext Educational Publishers

Cowles, S. (1982). Preliminary results investigating the effects of lichen ground cover on the growth of black spruce. *Naturaliste Canadien*, **109**, 573–81

Cox, C. B. & Moore, P. D. (1980). *Biogeography: an Ecological and Evolutionary Approach*, 3rd ed. Oxford, Blackwell

Coxson, D. S., Brown, D. & Kershaw, K. A. (1983). The interaction between CO_2 diffusion and the degree of thallus dehydration in lichens: some further comments. *New Phytologist*, **93**, 247–60

Crittenden, P. D. (1983). The role of lichens in the nitrogen economy of subarctic woodlands: nitrogen loss from the nitrogen-fixing lichen *Stereocaulon paschale* during rainfall. In *Nitrogen as an Ecological Factor*, ed. J. A. Lee, S. McNeill & I. H. Rorison, pp. 43–68. Oxford, Blackwell

Crum, H. (1972). The geographic origins of the mosses of North America's eastern deciduous forest. *Journal of the Hattori Botanical Laboratory*, **35**, 269–98

Curl, H., Hardy, J. T. & Ellermeier, R. (1972). Spectral absorption of solar radiation in alpine snowfields. *Ecology*, **53**, 1189–94

Daniëls, F. J. A. (1975). Vegetation of the Angmagssalik District, Southeast Greenland. III: Epilithic macrolichen communities. *Meddelelser om Grønland*, **198(3)**, 1–32

Daniëls, F. J. A. (1982). Vegetation of the Angmagssalik District, Southeast Greenland, IV: Shrub, dwarf shrub and terricolous lichens. *Meddelelser om Grønland, Bioscience*, **10**, 1–78

Daniëls, F. J. A. (1985). Floristic relationship between plant communities of corresponding habitats in southeast Greenland and alpine Scandinavia. *Vegetatio*, **59**, 145–50

Davey, A. (1983). Effects of abiotic factors on nitrogen fixation by bluegreen algae in Antarctica. *Polar Biology*, **2**, 95–100

Davey, A. & Marchant, H. J. (1983). Seasonal variation in nitrogen fixation by *Nostoc commune* Vaucher at the Vestfold Hills, Antarctica. *Phycologia*, **22**, 377–85

Davies, L. & Greene, S. W. (1976). Notes sur la végétation de l'Ile de la Possession (Archipel Crozet). *Comité National Français des Recherches Antarctiques*, **41**, 1–20

Davis, R. C. (1981). Structure and function of two Antarctic terrestrial moss communities. *Ecological Monographs*, **51**, 125–43

Davis, R. C. (1983). Prediction of net primary production in two Antarctic mosses by two models of net CO_2 fixation. *British Antarctic Survey Bulletin*, **59**, 47–61

Davis, R. C. & Harrison, P. M. (1981). Prediction of photosynthesis in maritime Antarctic mosses. *Comité National Français des Recherches Antarctiques*, **51**, 241–7

Dawson, H. J., Hrutfiord, B. F. & Ugolini, F. C. (1984). Mobility of lichen compounds from *Cladonia mitis* in Arctic soil. *Soil Science*, **138**, 40–5

Derwent, R. G. & Nodop, K. (1986). Long-range transport and deposition of acidic nitrogen species in north-west Europe. *Nature*, **324**, 356–8

Dilks, T. J. K. & Proctor, M. C. F. (1975). Comparative experiments on temperature responses in bryophytes: assimilation, respiration and freezing damage. *Journal of Bryology*, **8**, 317–36

Dilks, T. J. K. & Proctor, M. C. F. (1976). Effects of intermittent desiccation on bryophytes. *Journal of Bryology*, **9**, 249–64

Dilks, T. J. K. & Proctor, M. C. F. (1979). Photosynthesis, respiration and water content in bryophytes. *New Phytologist*, **82**, 97–114

Dodge, C. W. (1973). *Lichen Flora of the Antarctic Continent and Adjacent Islands*. Canaan, New Hampshire, Pheonix Publishing

Dowding, P., Chapin, F. S., Wielgolaski, F. E. & Kilfeather, P. (1981). Nutrients in tundra ecosystems. In *Tundra Ecosystems: a Comparative Analysis*, ed. L. C. Bliss, O. W. Heal & J. J. Moore, pp. 647–83. Cambridge, Cambridge University Press

Du Rietz, G. E. (1940). Problems of bipolar plant distribution. *Acta Phytogeographica Suecica*, **13**, 215–82

Dubreuil, M.-A. & Moore, T. R. (1982). A laboratory study of postfire nutrient redistribution in subarctic spruce–lichen woodlands. *Canadian Journal of Botany*, **60**, 2511–17

Dunbar, M. J. (1973). Stability and fragility in polar ecosystems. *Arctic*, **26**, 179–85

During, H. J. (1979). Life strategies of bryophytes: a preliminary review. *Lindbergia*, **5**, 2–18

Eckardt, F. E., Heerfordt, L., Jorgensen, H. M. & Vaag, P. (1982). Photosynthetic production in Greenland as related to climate, plant cover and grazing pressure. *Photosynthetica*, **16**, 71–100

Edward, N. & Miller, P. C. (1977). Validation of a model of the effect of tundra vegetation on soil temperature. *Arctic and Alpine Research*, **9**, 89–104

Edwards, J. A. (1979). An experimental introduction of vascular plants from South Georgia to the maritime Antarctic. *British Antarctic Survey Bulletin*, **49**, 73–80

Edwards, J. A. & Greene, D. M. (1973). The survival of Falkland Islands transplants at South Georgia and Signy Island, South Orkney Islands. *British Antarctic Survey Bulletin*, **33 & 34**, 33–45

Elliot, D. H. (1985). Physical geography – geological evolution. In *Key Environments Antarctica*, ed. W. N. Bonner & D. W. H. Walton, pp. 39–61. Oxford, Pergamon Press

Engel, J. J. (1978). A taxonomic and phytogeographic study of Brunswick Peninsula (Strait of Magellan) Hepaticae and Anthocerotae. *Fieldiana Botany*, **41**, 1–319

Engel, J. J. & Schuster, R. M. (1973). On some tidal zone Hepaticae from south Chile, with comments on marine dispersal. *Bulletin of the Torrey Botanical Club*, **100**, 29–35

Engelskjøn, T. (1981). Terrestrial vegetation of Bouvetøya: a preliminary account. *Skrifter om Norsk Polarinstitutt*, **175**, 17–28

Eschrich, W. & Steiner, M. (1967). Autoradiographische Untersuchungen zum Stofftransport bei *Polytrichum commune*. *Planta*, **74**, 330–49

Everett, K. R., Vassiljevskaya, V. D., Brown, J. & Walker, B. D. (1981). Tundra and analogous soils. In *Tundra Ecosystems: a Comparative Analysis*, ed. L. C. Bliss, O. W. Heal & J. J. Moore, pp. 139–79. Cambridge, Cambridge University Press

Farrar, J. F. (1976). Ecological physiology of the lichen *Hypogymnia physodes*. II: Effects of wetting and drying cycles and the concept of physiological buffering. *New Phytologist*, **77**, 105–13

Farrar, J. F. (1978). Ecological physiology of the lichen *Hypogymnia physodes*. IV: Carbon allocation at low temperatures. *New Phytologist*, **81**, 65–9

Farrar, J. F. & Smith, D. C. (1976). Physiological ecology of the lichen *Hypogymnia physodes*. III: The importance of the rewetting phase. *New Phytologist*, **77**, 115–25

Fenton, J. H. C. (1980). The rate of peat accumulation in Antarctic moss banks. *Journal of Ecology*, **68**, 211–28

Fenton, J. H. C. (1982a). The formation of vertical edges on Antarctic moss peat banks. *Arctic and Alpine Research*, **14**, 21–6

Fenton, J. H. C. (1982b). Vegetation re-exposed after burial by ice and its relationship to changing climate in the South Orkney Islands. *British Antarctic Survey Bulletin*, **51**, 247–55

Fenton, J. H. C. (1983). Concentric fungal rings in Antarctic moss communities. *Transactions of the British Mycological Society*, **80**, 415–20

Filson, R. B. (1966). The lichens and mosses of MacRobertson Land. *Australian National Antarctic Expeditions Scientific Reports*, B(II) Botany, **82**, 1–169

Filson, R. B. (1982). Lichens of continental Antarctica. *Journal of the Hattori Botanical Laboratory*, **53**, 357–60

Folkeson, L. (1984). Deterioration of moss and lichen vegetation in a forest polluted by heavy metals. *Ambio*, **13**, 37–9

Foster, D. R. (1985). Vegetation development following fire in *Picea mariana* (black spruce) – *Pleurozium* forests of south-eastern Labrador, Canada. *Journal of Ecology*, **73**, 517–34

Frankland, J. C. (1974). Decomposition of lower plants. In *Biology of Plant Litter Decomposition*, ed. C. H. Dickinson & G. J. F. Pugh, vol. 1, pp. 3–36. London, Academic Press

Freedman, W. & Hutchinson, T. C. (1976). Physical and biological effects of experimental crude oil spills on low Arctic tundra in the vicinity of Tuktoyaktuk, N.W.T., Canada. *Canadian Journal of Botany*, **54**, 2219–30

French, D. D. & Smith, V. R. (1985). A comparison of northern and southern hemisphere tundras and related ecosystems. *Polar Biology*, **5**, 5–21

Fridriksson, S. (1975). *Surtsey: Evolution of Life on a Volcanic Island*. London, Butterworth

Friedmann, E. I. (1982). Endolithic microorganisms in the Antarctic cold desert. *Science*, **215**, 1045–53

Friedmann, E. I. & Kibler, A. P. (1980). Nitrogen economy of endolithic microbial communities in hot and cold deserts. *Microbial Ecology*, **6**, 95–108

Furmańczyk, K. & Ochyra, R. (1982). Plant communities of the Admiralty Bay region (King George Island, South Shetland Islands, Antarctic). I: Jasnarzewski Gardens. *Polskie Badania Polarne*, **3**, 25–39

Gaare, E. & Skogland, T. (1975). Wild reindeer food habits and range use at Hardangervidda. In *Fennoscandian Tundra Ecosystems. 2: Animals and Systems Analysis*, ed. F. E. Wielgolaski, pp. 195–205. New York, Springer-Verlag

Gannutz, T. P. (1969). Effects of environmental extremes on lichens. In *Colloque sur les Lichéns et la Symbiose Lichenique*, ed. R. G. Werner, pp. 169–79. Paris, Societé Botanique de France

Gannutz, T. P. (1970). Photosynthesis and respiration of plants in the Antarctic Peninsula area. *Antarctic Journal of the United States*, **5**, 49–52

Gannutz, T. P. (1971). Ecodynamics of lichen communities in Antarctica. In *Research in the Antarctic*, ed. L. O. Quam, pp. 213–326. Washington, AAAS

Garnier, B. J. & Ohmura, A. (1968). A method of calculating the direct shortwave income on slopes. *Journal of Applied Meteorology*, **7**, 796–800

Gates, D. M. (1962). *Energy Exchange in the Biosphere*. New York, Harper & Row

Gauslaa, Y. (1984). Heat resistance and energy budget in different Scandinavian plants. *Holarctic Ecology*, **7**, 1–78

Gellerman, J. L., Anderson, W. H., Richardson, D. G. & Schlenk, H. (1975). Distribution of arachidonic and eicosapentaenoic acids in the lipids of mosses. *Biochimica et Biophysica Acta*, **338**, 277–90

Gerson, U. (1982). Bryophytes and invertebrates. In *Bryophyte Ecology*, ed. A. J. E. Smith, pp. 291–332. London, Chapman & Hall

Gerson, U. & Seaward, M. R. D. (1977). Lichen–invertebrate associations. In *Lichen Ecology*, ed. M. R. D. Seaward, pp. 69–119. London, Academic Press

Gimingham, C. H. & Birse, E. M. (1957). Ecological studies on growth-form in bryophytes. I: Correlations between growth-form and habitat. *Journal of Ecology*, **45**, 433–45

Gimingham, C. H. & Smith, R. I. Lewis (1970). Bryophyte and lichen communities in the maritime Antarctic. In *Antarctic Ecology*, ed. M. W. Holdgate, vol. 2, pp. 752–85. London, Academic Press

Gimingham, C. H. & Smith, R. I. Lewis (1971). Growth form and water relations of mosses in the maritime Antarctic. *British Antarctic Survey Bulletin*, **25**, 1–21

Godley, E. J. (1960). The botany of southern Chile in relation to New Zealand and the Subantarctic. *Proceedings of the Royal Society*, **B152**, 457–75

Grace, B., Gillespie, T. J. & Puckett, K. J. (1985). Sulphur dioxide threshold concentration values for *Cladonia rangiferina* in the Mackenzie Valley, NWT. *Canadian Journal of Botany*, **63**, 806–12

Granhall, U. & Lid-Torsvik, V. (1975). Nitrogen fixation by bacteria and free-living blue-green algae in tundra areas. In *Fennoscandian Tundra Ecosystems. 1: Plants and Microorganisms*, ed. F. E. Wielgolaski, pp. 305–15. New York, Springer-Verlag

Granhall, U. & Lindberg, T. (1978). Nitrogen fixation in some coniferous forest ecosystems. In *Environmental Role of Nitrogen-Fixing Blue-Green Algae and Asymbiotic Bacteria*, ed. U. Granhall, pp. 178–92. *Ecological Bulletins (Stockholm)*, 26

Greene, D. M. (1986). *A Conspectus of the mosses of Antarctica, The Falkland Islands and Southern South America*. Cambridge, British Antarctic Survey

Greene, S. W. (1960). The maturation cycle, or the stages of development of gametangia and capsules in mosses. *Transactions of the British Bryological Society*, **3**, 736–45

Greene, S. W. (1964). Plants of the land. In *Antarctic Research*, ed. R. Priestley, R. J. Adie & G. de Q. Robin, pp. 240–53. London, Butterworths

Greene, S. W. (1967). Bryophyte distribution. In *Terrestrial Life of Antarctica*, ed. V. C. Bushnell, pp. 11–13. New York, American Geographical Society

Greene, S. W. (1968). Studies in Antarctic bryology. I: A basic checklist for mosses. *Revue Bryologique et Lichénologique*, **36**, 132–8

Greene, S. W., Greene, D. M., Brown, P. D. & Pacey, J. M. (1970). Antarctic moss flora. I: The genera *Andreaea*, *Pohlia*, *Polytrichum*, *Psilopilum* and *Sarconeurum*. *British Antarctic Survey Scientific Reports*, **64**, 1–118

Greene, S. W., Hässel de Menéndez, G. G. & Matteri, C. M. (1985). La contribucion de les briofitas en la vegetacion de la transecta. In *Transecta Botánica de la Patagonia Austral*, ed. O. Boelcke, D. M. Moore & F. A. Roig, pp. 557–91. Buenos Aires, Consego Nacional de Investigaciones Cientificas y Tecnicas

Greene, S. W. & Walton, D. W. H. (1975). An annotated check list of the sub-Antarctic and Antarctic vascular flora. *Polar Record*, **17**, 473–84

Greenfield, L. B. & Wilson, G. J. (1981). *University of Canterbury Antarctic Research Unit, Expedition 19*. Christchurch, University of Canterbury

Gremmen, N. J. M. (1982). *The Vegetation of the Subantarctic Islands Marion and Prince Edward*. The Hague, Junk

Greig-Smith, P. (1979). Pattern in vegetation. *Journal of Ecology*, **67**, 755–79

Gressitt, J. L. & Shoup, J. (1967). Ecological notes on free-living mites in north Victoria Land. In *Entomology of Antarctica*, ed. J. L. Gressitt, pp. 307–20. Washington, American Geophysical Union

Grime, J. P. (1979). *Plant Strategies and Vegetation Processes*. Chichester, Wiley

Grolle, R. (1972). The hepatics of the South Sandwich Islands and South Georgia. *British Antarctic Survey Bulletin*, **28**, 83–95

Haag, R. W. & Bliss, L. C. (1974). Energy budget changes following surface disturbance to upland tundra. *Journal of Applied Ecology*, **11**, 355–74

Hale, M. E. (1973). Growth. In *The Lichens*, ed. V. Ahmadjian & M. E. Hale, pp. 473–92. New York, Academic Press

Hale, M. E. (1981). Pseudocyphellae and pored epicortex in the Parmeliaceae: their delimitation and evolutionary significance. *Lichenologist*, **13**, 1–10

Hale, M. E. (1983). *The Biology of Lichens*, 3rd ed. London, Arnold

Hale, M. E. (1987). Epilithic lichens in the Beacon sandstone formation, Victoria Land, Antarctica. *Lichenologist*, **19**, 269–287

Halliday, G., Kliim-Nielsen, L. & Smart, I. H. M. (1974). Studies of the flora of the north Blosseville Kyst and on the hot springs of Greenland. *Meddelelser om Grønland*, **199(2)**, 1–49

Hanson, W. C. (1982). ^{137}Cs concentrations in northern Alaskan eskimos, 1962–79: effects of ecological, cultural and political factors. *Health Physics*, **42**, 433–47

Hare, F. K. (1950). Climate and zonal divisions of the boreal forest formations in eastern Canada. *Geographical Review*, **40**, 615–35

Hare, F. K. & Hay, J. E. (1974). The climate of Canada and Alaska. In *Climates of North America*, ed. R. A. Bryson & F. K. Hare, pp. 49–192. Amsterdam, Elsevier

Hässel de Menéndez, G. G. & Solari, S. S. (1975). Bryophyta: Hepaticopsida Calobryales Jungermanniales Balantiopsaceae. *Flora Criptogámica de Tierra del Fuego*, **15(1)**. Buenos Aires, Fundación para la Educación, la Ciencia y la Cultura

Haukioja, E. (1981). Invertebrate herbivory at tundra sites. In *Tundra Ecosystems: a Comparative Analysis*, ed. L. C. Bliss, O. W. Heal & J. J. Moore, pp. 547–55. Cambridge, Cambridge University Press

Hawksworth, D. L. & Hill, D. J. (1984). *The Lichen-Forming Fungi*. Glasgow, Blackie

Hawksworth, D. L. & Rose, F. (1976). *Lichens as Pollution Monitors*. London, Arnold

Headland, R. K. (1984). *The Island of South Georgia*. Cambridge, Cambridge University Press

Heal, O. W., Flanagan, P. W., French, D. D. & MacLean, S. F. (1981). Decomposition and accumulation of organic matter. In *Tundra Ecosystems: a Comparative Analysis*, ed. L. C. Bliss, O. W. Heal & J. J. Moore, pp. 587–633. Cambridge, Cambridge University Press

Hébant, C. (1977). *The Conducting Tissues of Bryophytes*. Vaduz, Cramer

Heikkilä, H. & Kallio, P. (1969). On the problem of subarctic basidiolichens, II. *Reports from the Kevo Subarctic Research Station*, **4**, 90–7

Heilbronn, T. D. & Walton, D. W. H. (1984). Plant colonization of actively sorted stone stripes in the subantarctic. *Arctic and Alpine Research*, **16**, 161–72

Henssenn, A. (1986). *Edwardiella mirabilis*, a holocarpous lichen from Marion Island. *Lichenologist*, **18**, 51–6

Hesselbo, A. (1918). The bryophytes of Iceland. In *The Botany of Iceland*, ed. L. K. Rosenvinge & E. Warming, vol. 1, pp. 395–676. Copenhagen, Frimodt

Hicklenton, P. R. & Oechel, W. C. (1976). Physiological aspects of the ecology of *Dicranum fuscescens* in the subarctic. I: Acclimation and acclimation potential of CO_2 exchange in relation to habitat, light, and temperature. *Canadian Journal of Botany*, **54**, 1104–19

Hicklenton, P. R. & Oechel, W. C. (1977a). The influence of light intensity and temperature on the field carbon dioxide exchange in *Dicranum fuscescens* in the subarctic. *Arctic and Alpine Research*, **9**, 407–19

Hicklenton, P. R. & Oechel, W. C. (1977b). Physiological aspects of the ecology of *Dicranum fuscescens* in the subarctic. II: Seasonal patterns of organic nutrient content. *Canadian Journal of Botany*, **55**, 2168–77

Hill, J. D. (1984). Studies on the growth of lichens. I: Lobe formation and the maintenance of circularity in crustose species. *Lichenologist*, **16**, 273–8

Hoddinott, J. & Bain, J. (1979). The influence of simulated canopy light on the growth of six acrocarpous moss species. *Canadian Journal of Botany*, **57**, 1236–42

Hoffman, G. R. & Gates, D. M. (1970). An energy budget approach to the study of water loss in cryptogams. *Bulletin of the Torrey Botanical Club*, **97**, 361–6

Holdgate, M. W. (1964a). Terrestrial ecology in the maritime Antarctic. In *Biologie Antarctique*, ed. R. Carrick, M. W. Holdgate & J. Prevost, pp. 181–94. Paris, Hermann

Holdgate, M. W. (1964b). An experimental introduction of plants to the Antarctic. *British Antarctic Survey Bulletin*, **3**, 13–16

Holdgate, M. W., Allen, S. E. & Chambers, M. J. G. (1967). A preliminary investigation of the soils of Signy Island, South Orkney Islands. *British Antarctic Survey Bulletin*, **12**, 53–71

Holmen, K. (1955). Notes on the bryophyte vegetation of Peary Land, north Greenland. *Mitteilungen der Thuringischen Botanischen Gesellschaft*, **1**, 96–106

Holmen, K. (1960). The mosses of Peary Land, north Greenland. *Meddelelser om Grønland*, **163(2)**, 1–96

Holpainen, T. & Kärenlampi, L. (1984). Injuries to lichen ultrastructure caused by sulphur dioxide fumigations. *New Phytologist*, **98**, 285–94

Hooker, T. N. (1977). The Growth and Physiology of Antarctic Lichens. Ph.D. Thesis, University of Bristol

Hooker, T. N. (1980a). Lobe growth and marginal zonation in crustose lichens. *Lichenologist*, **12**, 313–23

Hooker, T. N. (1980b). Factors affecting the growth of Antarctic crustose lichens. *British Antarctic Survey Bulletin*, **50**, 1–19

Hooker, T. N. (1980c). Growth and production of *Cladonia rangiferina* and *Sphaerophorus globosus* on Signy Island, South Orkney Islands. *British Antarctic Survey Bulletin*, **50**, 27–34

Hooker, T. N. (1980d). Growth and production of *Usnea antarctica* and *U. fasciata* on Signy Island, South Orkney Islands. *British Antarctic Survey Bulletin*, **50**, 35–49

Hooker, T. N. & Brown, D. H. (1977). A photographic method for accurately measuring the growth of crustose and foliose saxicolous lichens. *Lichenologist*, **9**, 65–75

Horikawa, Y. & Ando, H. (1967). The mosses of Ongul Islands and adjoining coastal areas of the Antarctic continent. *Japanese Antarctic Research Expeditions Scientific Reports, Special Issue*, **1**, 245–52

Hudson, M. A. & Brustkern, P. (1965). Resistance of young and mature leaves of *Mnium undulatum* (L.) to frost. *Planta*, **66**, 135–55

Hughes, J. G. (1982). Penetration by rhizoids of the moss *Tortula muralis* Hedw. into well cemented oolitic limestone. *International Biodeterioration Bulletin*, **18**, 43–6

Huneck, S., Sainsbury, M., Rickard, T. M. A. & Smith, R. I. Lewis (1984). Ecological and chemical investigations of lichens from South Georgia and the maritime Antarctic. *Journal of the Hattori Botanical Laboratory*, **56**, 461–80

Huntley, B. J. (1971). Vegetation. In *Marion and Prince Edward Islands*, ed. J. M. van Zinderen Bakker, J. M. Winterbottom & R. A. Dyer, pp. 98–160. Capetown, Balkema

Hustich, I. (1951). The lichen woodlands in Labrador and their importance as winter pastures for domesticated reindeer. *Acta Geographica*, **12(1)**, 1–48

Hustich, I. (1975). The next phase in northern ecological research. *Proceedings of the Circumpolar Conference on Northern Ecology*, Ottawa 1975, **5**, 1–10

Hustich, I. (1979). Ecological concepts and biogeographical zonation in the north: the need for a generally accepted terminology. *Holarctic Ecology*, **2**, 208–17

Hutchison-Benson, E., Svoboda, J. & Taylor, H. W. (1985). The latitudinal inventory of ^{137}Cs in vegetation and topsoil in northern Canada, 1980. *Canadian Journal of Botany*, **63**, 748–91

Huttunen, S., Karhu, M. & Kallio, S. (1981). The effect of air pollution on transplanted mosses. *Silva Fennica*, **15**, 495–504

Imshaug, H. A. (1972). Need for the conservation of terrestrial vegetation in the Subantarctic. In *Conservation Problems in Antarctica*, ed. B. C. Parker, pp. 229–38. Blacksburg, Virginia Polytechnic Institute and State University

Ingold, C. T. (1965). *Spore Liberation*. Oxford, Clarendon Press

Innes, J. L. (1982). Lichenometric use of an aggregated *Rhizocarpon* 'species'. *Boreas*, **11**, 53–7

Innes, J. L. (1986). Influence of sampling design on lichen size–frequency distribution and its effect on derived lichenometric indices. *Arctic and Alpine Research*, **18**, 201–8

Ino, Y. (1983). Estimation of primary production in moss community on East Ongul Island, Antarctica. *Antarctic Record*, **80**, 30–8

Ino, Y. (1985). Comparative study of the effects of temperature on net photosynthesis and respiration in lichens from the Antarctic and subalpine zones in Japan. *Botanical Magazine, Tokyo*, **98**, 41–53

Ino, Y. & Nakatsubo, T. (1986). Distribution of carbon, nitrogen and phosphorous in a moss community-soil system developed on a cold desert in Antarctica. *Ecological Research*, **1**, 59–69

Inoue, S. (1976). Chromosome studies on five species of Antarctic mosses. *Kumamoto Journal of Science, Biology*, **13**, 1–5

Ives, J. D. (1974). Permafrost. In *Arctic and Alpine Environments*, ed. J. D. Ives & R. G. Barry, pp. 159–94. London, Methuen

Ives, J. D. & Barry, R. G. (1974). *Arctic and Alpine Environments*. London, Methuen

Jacobsen, E., Hove, K., Bjarghov, R. S. & Skjenneberg, S. (1981). Supplementary feeding of female reindeer on a lichen diet during the last part of pregnancy. *Acta Agriculturae Scandinavica*, **31**, 81–6

Jahns, H. M. (1973). Anatomy, morphology and development. In *The Lichens*, ed. V. Ahmadjian & M. E. Hale, pp. 3–58. New York, Academic Press

Janetschek, H. (1970). Environments and ecology of terrestrial arthropods in the high Antarctic. In *Antarctic Ecology*, ed. M. W. Holdgate, vol. 2, pp. 871–85. London, Academic Press

Jenkin, J. F. (1975). Macquarie Island, Subantarctic. In *Structure and Function of Tundra Ecosystems*, ed. T. Rosswall & O. W. Heal, pp. 375–97. *Ecological Bulletins (Stockholm)*, 20

Jochimsen, M. (1973). Does the size of lichen thalli really constitute a valid measure for dating glacial deposits? *Arctic and Alpine Research*, **5**, 417–24

Jóhannsson, B. (1983). A list of Icelandic bryophyte species. *Acta Naturalia Islandica*, **30**, 1–29

Johnson, E. A. (1981). Vegetation organization and dynamics of lichen woodland communities in the Northwest Territories, Canada. *Ecology*, **62**, 202–15

Jones, D. & Wilson, M. J. (1985). Chemical activity of lichens on mineral surfaces – a review. *International Biodeterioration Bulletin*, **21**, 99–104

Jones, H. G. (1980). *Plants and Microclimate*. Cambridge, Cambridge University Press

Kallio, P. (1975). Kevo, Finland. In *Structure and Function of Tundra Ecosystems*, ed. T. Rosswall & O. W. Heal, pp. 193–223. *Ecological Bulletins (Stockholm)*, 20

Kallio, P. & Heinonen, S. (1971). Influence of short-term low temperature on net photosynthesis in some subarctic lichens. *Reports from the Kevo Subarctic Research Station*, **8**, 63–72

Kallio, P. & Heinonen, S. (1973). Ecology of *Rhacomitrium lanuginosum* (Hedw.) Brid. *Reports from the Kevo Subarctic Research Station*, **10**, 43–54

Kallio, P. & Heinonen, S. (1975). CO_2 exchange and growth of *Rhacomitrium lanuginosum* and *Dicranum elongatum*. In *Fennoscandian Tundra Ecosystems. Part 1. Plants and Microorganisms*, ed. F. E. Wielgolaski, pp. 138–48. New York, Springer-Verlag

Kallio, P. & Kallio, S. (1978). Adaptation of nitrogen fixation to temperature in the *Peltigera aphthosa* group. In *Environmental Role of Nitrogen-Fixing Blue-Green Algae and Asymbiotic Bacteria*, ed. U. Granhall, pp. 225–33. *Ecological Bulletins (Stockholm)*, 26

Kallio, P. & Kärenlampi, L. (1975). Photosynthesis in mosses and lichens. In *Photosynthesis and Productivity in Different Environments*, ed. J. P. Cooper, pp. 393–423. Cambridge, Cambridge University Press

Kallio, P. & Saarnio, E. (1986). The effect on mosses of transplantation to different latitudes. *Journal of Bryology*, **14**, 159–78

Kallio, P. & Valanne, N. (1975). On the effects of continuous light on photosynthesis in mosses. In *Fennoscandian Tundra Ecosystems. Part 1.*

Plants and Microorganisms, ed. F. E. Wielgolaski, pp. 149–62. New York, Springer-Verlag

Kallio, S. (1973). The ecology of nitrogen fixation in *Stereocaulon paschale*. *Reports from the Kevo Subarctic Research Station*, **10**, 34–42

Kallio, S. & Varheenma, T. (1974). On the effect of air pollution on nitrogen fixation in lichens. *Reports from the Kevo Subarctic Research Station*, **11**, 42–6

Kanda, H. (1979). Regenerative development in culture of Antarctic plants of *Ceratodon purpureus* (Hedw.) Brid. *Memoirs of the National Institute of Polar Research, Special Issue*, **11**, 58–69

Kanda, H. (1981a). Flora and vegetation of mosses in ice-free areas of Soya Coast and Prince Olav Coast, East Antarctica. *Hikobia, Supplement*, **1**, 91–100

Kanda, H. (1981b). Two moss species of the genus *Pottia* collected in the vicinity of Syowa Station, East Antarctica. *Antarctic Record*, **71**, 96–108

Kappen, L. (1973). Responses to extreme environments. In *The Lichens*, ed. V. Ahmadjian & M. E. Hale, pp. 310–80. New York, Academic Press

Kappen, L. (1983). Ecology and physiology of the Antarctic fruticose lichen *Usnea sulphurea* (Koenig) Th. Fries. *Polar Biology*, **1**, 249–55

Kappen, L. (1984). Ecological aspects of exploitation of the non-living resources of the Antarctic continent. In *Antarctic Challenge*, ed. R. Wolfrum, pp. 211–17. Berlin, Duncker & Humblot

Kappen, L. (1985a). Vegetation and ecology of ice-free areas of northern Victoria Land, Antarctica. I: The lichen vegetation of Birthday Ridge and an inland mountain. *Polar Biology*, **4**, 213–25

Kappen, L. (1985b). Vegetation and ecology of ice-free areas of northern Victoria Land, Antarctica. II: Ecological conditions in typical microhabitats of lichens at Birthday Ridge. *Polar Biology*, **4**, 227–36

Kappen, L. (1985c). Water relations and net photosynthesis of *Usnea*. A comparison between *Usnea fasciata* (maritime Antarctic) and *Usnea sulphurea* (continental Antarctic). In *Lichen Physiology and Cell Biology*, ed. D. H. Brown, pp. 41–6. London, Plenum Press

Kappen, L. & Friedmann, E. I. (1983a). Kryptoendolithische Flichten als Beispiel einer Anpassung an extrem trochen-kalte Klimabedingungen. *Verhandlungen der Gesellschaft für Ökologie*, **10**, 517–19

Kappen, L. & Friedmann, E. I. (1983b). Ecophysiology of lichens in the dry valleys of southern Victoria Land, Antarctica. II: CO_2 gas exchange in cryptoendolithic lichens. *Polar Biology*, **1**, 227–32

Kärenlampi, L. (1970a). Distribution of chlorophyll in the lichen *Cladonia alpestris*. *Reports from the Kevo Subarctic Research Station*, **7**, 1–8

Kärenlampi, L. (1970b). Morphological analysis of the growth and productivity of the lichen *Cladonia alpestris*. *Reports from the Kevo Research Station*, **7**, 9–15

Kärenlampi, L. (1971a). Studies on the relative growth rate of some fruticose lichens. *Reports from the Kevo Subarctic Research Station*, **7**, 33–9

Kärenlampi, L. (1971b). On methods for measuring and calculating the energy flow through lichens. *Reports from the Kevo Subarctic Research Station*, **7**, 40–6

Kärenlampi, L. (1973). Biomass and estimated yearly net production of the ground vegetation at Kevo. In *Proceedings of the Conference on Primary*

Production and Production Processes, Tundra Biome, ed. L. C. Bliss & F. E. Wielgolaski, pp. 111–14. Edmonton, IBP Tundra Biome Steering Committee

Kärenlampi, L. & Pelkonen, M. (1971). Studies on the morphological variation of the lichen *Cladonia uncialis*. *Reports from the Kevo Subarctic Research Station*, **7**, 47–56

Karunen, P. (1981). The role of neutral lipids in the physiology and ecology of subarctic *Dicranum elongatum*. *Canadian Journal of Botany*, **59**, 1902–9

Karunen, P. & Kallio, P. (1976). Seasonal variation in the total lipid content of subarctic *Dicranum elongatum*. *Reports from the Kevo Subarctic Research Station*, **13**, 63–70

Karunen, P. & Liljenberg, C. (1981). Temperature induced changes in the ultrastructure and lipid composition in green and senescent leaves of *Dicranum elongatum*. *Physiologia Plantarum*, **53**, 48–54

Karunen, P. & Mikola, H. (1980). Distribution of triglycerides in *Dicranum elongatum*. *Phytochemistry*, **19**, 319–21

Karunen, P. & Salin, M. (1982). Seasonal changes in lipids of photosynthetically active and senescent parts of *Sphagnum fuscum*. *Lindbergia*, **8**, 35–44

Kaspar, M., Simmons, G. M., Parker, B. C., Seaburg, K. B., Wharton, R. A. & Smith, R. I. Lewis (1982). *Bryum* Hedw. collected from Lake Vanda, Antarctica. *Bryologist*, **85**, 424–30

Kay, P. A. (1979). Multivariate statistical estimates of Holocene vegetation and climatic change, forest-tundra transition zone, NWT, Canada. *Quaternary Research*, **11**, 125–40

Keizer, P. J., Van Tooren, B. F. & During, H. J. (1985). Effects of bryophytes on seedling emergence and establishment of short-lived forbs in chalk grassland. *Journal of Ecology*, **73**, 493–504

Kershaw, K. A. (1974). Studies on lichen-dominated systems. X: The sedge meadows of the coastal raised beaches. *Canadian Journal of Botany*, **52**, 1947–72

Kershaw, K. A. (1975a). Studies on lichen-dominated systems. XII: The ecological significance of thallus color. *Canadian Journal of Botany*, **53**, 660–7

Kershaw, K. A. (1975b). Studies on lichen-dominated systems. XIV: The comparative ecology of *Alectoria nitidula* and *Cladina alpestris*. *Canadian Journal of Botany*, **53**, 2608–13

Kershaw, K. A. (1977a). Studies on lichen-dominated systems. XX: An examination of some aspects of the northern boreal lichen woodlands in Canada. *Canadian Journal of Botany*, **55**, 393–410

Kershaw, K. A. (1977b). Physiological–environmental interactions in lichens. II: The pattern of net photosynthetic acclimation in *Peltigera canina* (L.) Wild var. *praetextata* (Floerke in Somm.) Hue, and *P. polydactyla* (Neck.) Hoffm. *New Phytologist*, **79**, 377–90

Kershaw, K. A. (1978). The role of lichens in boreal tundra transition areas. *Bryologist*, **81**, 294–306

Kershaw, K. A. (1983). The thermal operating-environment of a lichen. *Lichenologist*, **15**, 191–207

Kershaw, K. A. (1985). *Physiological Ecology of Lichens*. Cambridge, Cambridge University Press

Kershaw, K. A. & Field, G. F. (1975). Studies on lichen-dominated systems. XV: The temperature and humidity profiles in a *Cladina alpestris* mat. *Canadian Journal of Botany*, **53**, 2614–20

Kershaw, K. A. & Harris, G. P. (1971). A technique for measuring the light profile in a lichen canopy. *Canadian Journal of Botany*, **49**, 609–11

Kershaw, K. A. & MacFarlane, J. D. (1980). Physiological–environmental interactions in lichens. X: Light as an ecological factor. *New Phytologist*, **84**, 687–702

Kershaw, K. A. & MacFarlane, J. D. (1982). Physiological–environmental interactions in lichens. XIII: Seasonal constancy of nitrogenase activity, net photosynthesis and respiration, in *Collema furfuraceum* (Am.) DR. *New Phytologist*, **90**, 723–34

Kershaw, K. A., MacFarlane, J. D., Webber, M. R. & Fovargue, A. (1983). Phenotypic differences in the seasonal pattern of net photosynthesis in *Cladonia stellaris*. *Canadian Journal of Botany*, **61**, 2169–80

Kershaw, K. A. & Rouse, W. R. (1971). Studies on lichen-dominated systems. II: The growth pattern of *Cladonia alpestris* and *C. rangiferina*. *Canadian Journal of Botany*, **49**, 1401–10

Kershaw, K. A. & Rouse, W. R. (1973). Studies on lichen-dominated systems. V: A primary survey of a raised-beach system in northwestern Ontario. *Canadian Journal of Botany*, **51**, 1280–1307

Kershaw, K. A. & Smith, M. M. (1978). Studies on lichen-dominated systems. XXI: The control of seasonal rates of net photosynthesis by moisture, light and temperature in *Stereocaulon paschale*. *Canadian Journal of Botany*, **56**, 2825–30

Kershaw, K. A. & Watson, S. (1983). The control of seasonal rates of net photosynthesis by moisture, light and temperature in *Parmelia disjuncta* Erichs. *Bryologist*, **86**, 31–43

Kightly, S. P. J. & Smith, R. I. Lewis (1976). The influence of reindeer on the vegetation of South Georgia. I: Long-term effects of unrestricted grazing and the establishment of exclosure experiments in various plant communities. *British Antarctic Survey Bulletin*, **44**, 57–76

Klein, D. R. (1968). The introduction, increase, and crash of reindeer on St Matthew Island. *Journal of Wildlife Management*, **32**, 350–67

Klein, D. R. (1982). Fire, lichens, and caribou. *Journal of Range Management*, **35**, 390–5

Koerner, R. M. (1980). The problem of lichen-free zones in Arctic Canada. *Arctic and Alpine Research*, **12**, 87–94

Komarkova, V. & Webber, P. J. (1980). Two low Arctic vegetation maps near Atkasook, Alaska. *Arctic and Alpine Research*, **12**, 447–72

Koponen, A. (1978). The peristome and spores in Splachnaceae and their evolutionary significance. In *Congrès International de Bryologie*, ed. C. Suire, pp. 535–67. Vaduz, Cramer

Koponen, A. & Koponen, T. (1978). Evidence of entomophily in Splachnaceae (Bryophyta). In *Congrès International de Bryologie*, ed. C. Suire, pp. 569–77. Vaduz, Cramer

Köppen, W. (1918). Klassifikation der Klimate nach Temperatur, Niederschlag und Jahreslauf. *Petermanns Geographische Mitteilungen*, **64**, 193–203

Krebs, J. S. & Barry, R. G. (1970). The Arctic front and the tundra/taiga boundary in Eurasia. *Geographical Review*, **60**, 548–54

Kuc, M. (1970). Additions to the Arctic moss flora. V: The role of mosses in plant succession and the development of peat on Fitzwilliam Owen Island (Western Canadian Arctic). *Revue Bryologique et Lichénologique*, **37**, 931–9

Kuta, E., Ochyra, R. & Przywara, L. (1982). Karyological studies on Antarctic mosses. I. *Bryologist*, **85**, 131–8

Kvillner, E. & Sonesson, M. (1980). Plant distribution and environment of a subarctic mire. In *Ecology of a Subarctic Mire*, ed. M. Sonesson, pp. 97–111. *Ecological Bulletins (Stockholm)*, 30

Lamb, I. M. (1968). Antarctic lichens. II: The genera *Buellia* and *Rinodina*. *British Antarctic Survey Scientific Reports*, **61**, 1–129

Lamb, I. M. (1970). Antarctic terrestrial plants and their ecology. In *Antarctic Ecology*, ed. M. W. Holdgate, vol. 2, pp. 733–751. London, Academic Press

Lange, B. (1973). The *Sphagnum* flora of hot springs in Iceland. *Lindbergia*, **2**, 81–93

Lange, O. L. (1953). Hitze- und Trochenresistenz der Flechten in Beziehung zu ihrer Verbreitung. *Flora (Jena)*, **140**, 39–97

Lange, O. L. & Kappen, L. (1972). Photosynthesis of lichens from Antarctica. In *Antarctic Terrestrial Biology*, ed. G. A. Llano, pp. 83–95. Washington, American Geophysical Union

Lange, O. L. & Tenhunen, J. D. (1981). Moisture content and CO_2 exchange of lichens. II: Depression of net photosynthesis in *Ramalina maciformis* at high water content is caused by increased thallus carbon dioxide diffusion resistance. *Oecologia*, **51**, 426–9

Larsen, J. A. (1972). The vegetation of northern Keewatin. *Canadian Field Naturalist*, **86**, 45–72

Larsen, J. A. (1974). Ecology of the northern forest border. In *Arctic and Alpine Environments*, ed. J. D. Ives & R. G. Barry, pp. 341–69. London, Methuen

Larson, D. W. (1978). Patterns of lichen photosynthesis and respiration following prolonged frozen storage. *Canadian Journal of Botany*, **56**, 2119–23

Larson, D. W. (1979). Lichen water relations under drying conditions. *New Phytologist*, **82**, 713–31

Larson, D. W. (1981). Differential wetting in some lichens and mosses: the role of morphology. *Bryologist*, **84**, 1–15

Larson, D. W. (1982). Environmental stress and *Umbilicaria* lichens: the effect of high temperature pretreatments. *Oecologia*, **55**, 102–7

Larson, D. W. (1983a). Environmental stress in *Umbilicaria* lichens: the effects of subzero temperature pretreatment. *Oecologia*, **55**, 268–78

Larson, D. W. (1983b). The pattern of production within individual *Umbilicaria* lichen thalli. *New Phytologist*, **94**, 409–19

Larson, D. W. (1984). Thallus size as a complicating factor in the physiological ecology of lichens. *New Phytologist*, **97**, 87–97

Larson, D. W. & Kershaw, K. A. (1975a). Measurement of CO_2 exchange in lichens: a new method. *Canadian Journal of Botany*, **53**, 1535–41

Larson, D. W. & Kershaw, K. A. (1975b). Studies on lichen-dominated systems. XIII: Seasonal and geographical variation of net CO_2 exchange of *Alectoria ochroleuca*. *Canadian Journal of Botany*, **53**, 2598–607

Larson, D. W. & Kershaw, K. A. (1972c). Studies on lichen-dominated systems. XVI: Comparative patterns of net CO_2 exchange in *Cetraria nivalis* and *Alectoria ochroleuca* collected from a raised-beach ridge. *Canadian Journal of Botany*, **53**, 2884–92

Larson, D. W. & Kershaw, K. A. (1975d). Studies on lichen-dominated systems. XI: Lichen-heath and winter snow cover. *Canadian Journal of Botany*, **53**, 621–6

Larson, D. W. & Kershaw, K. A. (1976). Studies on lichen-dominated systems. XVIII: Morphological control of evaporation in lichens. *Canadian Journal of Botany*, **54**, 2061–73

Lawrey, J. D. (1980). Sexual and asexual reproductive patterns in *Parmotrema* (Parmeliaceae) that correlate with latitude. *Bryologist*, **83**, 344–50

Lawrey, J. D. (1983). Lichen herbivore preference: a test of two hypotheses. *American Journal of Botany*, **70**, 1188–94

Leader-Williams, N., Scott, T. A. & Pratt, R. M. (1981). Forage selection by introduced reindeer on South Georgia and its consequences for the vegetation. *Journal of Applied Ecology*, **18**, 83–106

LeBlanc, F., Rao, D. N. & Comeau, G. (1972). The epiphytic vegetation of *Populus balsamifera* and its significance as an air pollution indicator in Sudbury, Ontario. *Canadian Journal of Botany*, **50**, 519–28

Lechowicz, M. J. (1978). Carbon dioxide exchange in *Cladina* lichens from subarctic and temperate habitats. *Oecologia*, **32**, 225–37

Lechowicz, M. J. (1981a). The effects of climatic pattern on lichen productivity: *Cetraria cucullata* (Bell.) Ach. in the Arctic tundra of northern Alaska. *Oecologia*, **50**, 210–16

Lechowicz, M. J. (1981b). Adaptation and the fundamental niche: evidence from lichens. In *The Fungal Community: its Organization and Role in the Ecosystem*, ed. D. T. Wicklow & G. Carroll, pp. 89–108. New York, Dekker

Lechowicz, M. J. (1982a). Ecological trends in lichen photosynthesis. *Oecologia*, **53**, 330–6

Lechowicz, M. J. (1982b). The effects of simulated acid precipitation on photosynthesis in the caribou lichen *Cladina stellaris* (Opiz.) Brodo. *Water, Air, and Soil Pollution*, **18**, 421–30

Lechowicz, M. J. (1983). Age dependence of photosynthesis in the caribou lichen *Cladina stellaris*. *Plant Physiology*, **71**, 893–5

Levitt, J. (1980). *Responses of Plants to Environmental Stress*, 2nd ed. vol. 1: *Chilling, Freezing and High Temperature Stresses*. New York, Academic Press

Light, J. J. & Heywood, R. B. (1975). Is the vegetation of continental Antarctica predominantly aquatic? *Nature*, **256**, 199–200

Lightowlers, P. J. (1983). Taxonomic studies in Antarctic bryophytes with particular reference to the genus *Tortula*. Ph.D. Thesis, University of Edinburgh

Lightowlers, P. J. (1985). A synoptic flora of South Georgia: *Tortula*. *British Antarctic Survey Bulletin*, **67**, 41–77

Lindsay, D. C. (1971). Vegetation of the South Shetland Islands. *British Antarctic Survey Bulletin*, **25**, 59–83

Lindsay, D. C. (1973a). Estimates of lichen growth rates in the maritime Antarctic. *Arctic and Alpine Research*, **5**, 341–6

Lindsay, D. C. (1973b). Probable introductions of lichens to South Georgia. *British Antarctic Survey Bulletin*, **33 & 34**, 169–72

Lindsay, D. C. (1974). The macrolichens of South Georgia. *British Antarctic Survey Scientific Reports*, **89**, 1–91

Lindsay, D. C. (1975). Growth rates of *Cladonia rangiferina* (L.) Web. on South Georgia. *British Antarctic Survey Bulletin*, **40**, 49–53

Lindsay, D. C. (1977). Lichens of cold deserts. In *Lichen Ecology*, ed. M. R. D. Seaward, pp. 183–209. London, Academic Press

Lindsay, D. C. (1978). The role of lichens in Antarctic ecosystems. *Bryologist*, **81**, 268–76

Lindsay, D. C. & Brook, D. (1971). Lichens from the Theron Mountains. *British Antarctic Survey Bulletin*, **25**, 95–8

Link, S. O. & Nash, T. H. (1984). An analysis of an Arctic lichen community with resect to slope on the siliceous rocks of Anaktuvuk Pass, Alaska. *Bryologist*, **87**, 162–6

Llano, G. A. (1965). The flora of Antarctica. In *Antarctica*, ed. T. Hatherton, pp. 331–50. London, Methuen

Lock, W. W., Andrews, J. T. & Webber, P. J. (1979). *A Manual for Lichenometry. British Geomorphological Research Group, Technical Bulletin*, 26

Longton, R. E. (1966a). Alien vascular plants on Deception Island, South Shetland Islands. *British Antarctic Survey Bulletin*, **9**, 55–60

Longton, R. E. (1966b). Botanical studies in the Antarctic during the 1963–64 and 1964–65 seasons. *British Antarctic Survey Bulletin*, **10**, 85–95

Longton, R. E. (1967). Vegetation in the maritime Antarctic. *Philosophical Transactions of the Royal Society*, **B252**, 213–35

Longton, R. E. (1970). Growth and productivity of the moss *Polytrichum alpestre* Hoppe in Antarctic regions. In *Antarctic Ecology*, ed. M. W. Holdgate, vol. 2, pp. 818–37. London, Academic Press

Longton, R. E. (1972a). Reproduction of Antarctic mosses in the genera *Polytrichum* and *Psilopilum* with particular reference to temperature. *British Antarctic Survey Bulletin*, **27**, 51–96

Longton, R. E. (1972b). Growth and reproduction in northern and southern hemisphere populations of the peat forming moss *Polytrichum alpestre* Hoppe with reference to the estimation of productivity. *Proceedings of the Fourth International Peat Congress, Helsinki*, **1**, 259–75

Longton, R. E. (1973a). A classification of terrestrial vegetation near McMurdo Sound, continental Antarctica. *Canadian Journal of Botany*, **51**, 2339–46

Longton, R. E. (1973b). The occurrence of radial infection patterns in colonies of polar bryophytes. *British Antarctic Survey Bulletin*, **32**, 41–9

Longton, R. E. (1974a). Microclimate and biomass in communities of the *Bryum* association on Ross Island, continental Antarctica. *Bryologist*, **77**, 109–27

Longton, R. E. (1974b). Genecological differentiation in bryophytes. *Journal of the Hattori Botanical Laboratory*, **38**, 49–65

Longton, R. E. (1976). Reproductive biology and evolutionary potential in bryophytes. *Journal of the Hattori Botanical Laboratory*, **41**, 205–23

Longton, R. E. (1979a). Vegetation ecology and classification in the maritime Antarctic zone. *Canadian Journal of Botany*, **57**, 2264–78

Longton, R. E. (1979b). Studies on growth, reproduction and population ecology in relation to microclimate in the bipolar moss *Polytrichum alpestre* Hoppe. *Bryologist*, **82**, 325–67

Longton, R. E. (1980). Physiological ecology of mosses. In *The Mosses of North America*, ed. R. J. Taylor & A. E. Leviton, pp. 77–113. San Francisco, Pacific Division AAAS

Longton, R. E. (1981). Inter-population variation in morphology and physiology in the cosmopolitan moss *Bryum argenteum* Hedw. *Journal of Bryology*, **11**, 501–20

Longton, R. E. (1982). The biosystematic approach to bryology. *Journal of the Hattori Botanical Laboratory*, **53**, 1–19

Longton, R. E. (1984). The role of bryophytes in terrestrial ecosystems. *Journal of the Hattori Botanical Laboratory*, **55**, 147–63

Longton, R. E. (1985a). Terrestrial habitats – Vegetation. In *Key Environments Antarctica*, ed. W. N. Bonner & D. W. H. Walton, pp. 73–105. Oxford, Pergamon Press

Longton, R. E. (1985b). Reproductive biology and susceptibility to air pollution in *Pleurozium schreberi* (Brid.) Mitt. (Musci) with particular reference to Manitoba, Canada. *Monographs in Systematic Botany from the Missouri Botanical Garden*, **11**, 51–69

Longton, R. E. & Greene, S. W. (1967). The growth and reproduction of *Polytrichum alpestre* Hoppe on South Georgia. *Philosophical Transactions of the Royal Society*, **B252**, 295–322

Longton, R. E. & Holdgate, M. W. (1967). Temperature relationships of Antarctic vegetation. *Philosophical Transactions of the Royal Society*, **B252**, 237–50

Longton, R. E. & Holdgate, M. W. (1979). The South Sandwich Islands, IV. Botany. *British Antarctic Survey Scientific Reports*, **94**, 1–53

Longton, R. E. & MacIver, M. A. (1977). Climatic relationships in Antarctic and Northern Hemisphere populations of a cosmopolitan moss, *Bryum argenteum* Hedw. In *Adaptations within Antarctic Ecosystems*, ed. G. A. Llano, pp. 899–919. Washington, Smithsonian Institution

Longton, R. E. & Schuster, R. M. (1983). Reproductive biology. In *New Manual of Bryology*, ed. R. M. Schuster, vol. 1, pp. 386–462. Nichinan, Hattori Botanical Laboratory

Lösch, R., Kappen, L. & Wolf, A. (1983). Productivity and temperature biology of two snowbed bryophytes. *Polar Biology*, **1**, 243–8

Löve, A. & Löve, D. (1974). Origin and evolution of Arctic and alpine floras. In *Arctic and Alpine Environments*, ed. J. D. Ives & R. G. Barry, pp. 571–603. London, Methuen

Lydolph, P. E. (1977). *Climates of the Soviet Union*. Amsterdam, Elsevier

MacArthur, R. H. & Wilson, E. D. (1967). *The Theory of Island Biogeography*. Princeton, Princeton University Press

MacFarlane, J. D. & Kershaw, K. A. (1980). Physiological–environmental interactions in lichens. IX: Thermal stress and lichen ecology. *New Phytologist*, **84**, 669–85

MacFarlane, J. D. & Kershaw, K. A. (1982). Physiological–environmental interactions in lichens. XIV: The environmental control of glucose movement from alga to fungus in *Peltigera polydactyla*, *P. rufescens* and *Collema furfuraceum*. *New Phytologist*, **91**, 93–101

MacFarlane, J. D. & Kershaw, K. A. (1985). Some aspects of carbohydrate metabolism in lichens. In *Lichen Physiology and Cell Biology*, ed. D. H. Brown, pp. 1–8. New York, Plenum Press

MacKenzie, D. (1986). The rad-dosed reindeer. *New Scientist*, **1539**, 37–40

MacLean, S. F. (1980). The detritus-based trophic system. In *An Arctic Ecosystem: The Coastal Tundra at Barrow, Alaska*, ed. J. Brown, P. C. Miller, L. L. Tieszen & F. L. Bunnell, pp. 411–57. Dowden, Hutchinson & Ross

Magomedova, M. A. (1980). Succession of communities of lithophilic lichens in the highlands of northern Ural. *Ékologiya*, **3**, 29–38

Maikawa, E. & Kershaw, K. A. (1976). Studies on lichen-dominated systems. XIX: The postfire recovery sequence of black spruce–lichen woodland in the Abitau Lake region, N.W.T. *Canadian Journal of Botany*, **54**, 2679–87

Markova, K. K., Bardin, V. I., Lebedev, V. C., Orlov, A. I. & Suetora, I. A. (1970). *The Geography of Antarctica*. Jerusalem, Israel Program for Scientific Translation

Mårtensson, O. (1956). Bryophytes of the Torneträsk area of northern Swedish Lapland. II: Musci. *Kunglia Svenska Vetenskakademiens Avhandlinger i Naturskyddsarden*, **14**, 1–321

Matsuda, T. (1968). Ecological study of the moss community and microorganisms in the vicinity of Syowa Station, Antarctica. *Japanese Antarctica Research Expedition Scientific Reports*, **E29**, 1–58

Matteri, C. M. (1982). Patagonian bryophytes. 6: Fruiting *Sarconeurum glaciale* (C. Muell.) Card. et Bryhn newly found in southern Patagonia. *Lindbergia*, **8**, 105–9

Matteri, C. M. (1986). Los Musci (Bryophyta) de las Islas Malvinas, su habitat y distribución. *Nova Hedwigia*, **43**, 159–89

Mattsson, S. & Lidén, K. (1975). ^{137}Cs in carpets of the forest moss *Pleurozium schreberi*, 1961–73. *Oikos*, **26**, 323–7

Matveyeva, N. V., Parinkina, O. M. & Churnov, Yu. I. (1975). Maria Pronchistsheva Bay, USSR. In *Structure and Function of Tundra Ecosystems*, ed. T. Rosswall & O. W. Heal, pp. 61–72. *Ecological Bulletins (Stockholm)*, 20

Matveyeva, N. V., Polozova, T. G., Blagodatskykh, L. S. & Dorogostaiskaya, E. V. (1974). A brief essay on the vegetation in the vicinity of the Taimyr Biogeocoenological Station. *International Tundra Biome Translation*

McKay, C. P. & Friedemann, E. I. (1985). The cryptoendolithic microbial environment in the Antarctic cold desert: temperature variations in nature. *Polar Biology*, **4**, 19–25

Miles, C. J. & Longton, R. E. (1987). The life history of the moss, *Atrichum undulatum* (Hedw.) P. Beauv. *Symposia Biologica Hungarica*, **35**, 193–207

Miller, A. (1976). The climate of Chile. In *Climates of Central and South America*, ed. W. Schwerdtfeger, pp. 113–45. Amsterdam, Elsevier

Miller, G. H. (1973). Variations in lichen growth from direct measurement: preliminary curves for *Alectoria miniscula* from eastern Baffin Island, N.W.T., Canada. *Arctic and Alpine Research*, **5**, 333–9

Miller, N. G. (1980). Mosses as paleoecological indicators of lateglacial terrestrial environments: some North American studies. *Bulletin of the Torrey Botanical Club*, **107**, 373–91

Miller, N. B. (1985). Fossil evidence of the dispersal and establishment of mosses as gametophyte fragments. *Monographs in Systematic Botany from the Missouri Botanical Garden*, **11**, 71–8

Miller, N. G. & Alpert, P. (1984). Plant associations and edaphic features of a high Arctic mesotrophic setting. *Arctic and Alpine Research*, **16**, 11–24

Miller, N. G. & Ambrose, L. J. H. (1976). Growth in culture of wind-blown bryophyte gametophyte fragments from Arctic Canada. *Bryologist*, **79**, 55–63

Miller, P. C. (1982). Environmental and vegetational variation across a snow accumulation area in montane tundra in central Alaska. *Holarctic Ecology*, **5**, 85–98

Miller, P. C., Miller, P. M., Blake-Jacobson, M., Chapin, F. S., Everett, K. R., Hilbert, D. W., Kummerow, J., Linkins, A. E., Marion, G. M., Oechel, W. C., Roberts, S. W. & Stuart, L. (1984). Plant-soil processes in *Eriophorum vaginatum* tussock tundra in Alaska: a systems modelling approach. *Ecological Monographs*, **54**, 361–405

Miller, P. C., Oechel, W. C., Stoner, W. A. & Sveinbjörnsson, B. (1978). Simulation of CO_2 exchange and water relations of four Arctic bryophytes at Point Barrow, Alaska. *Photosynthetica*, **12**, 7–20

Milner, C. & Hughes, R. E. (1968). *Methods for the Measurement of the Primary Production of Grassland*. Oxford, Blackwell

Mogensen, G. S. (1973). A revision of the moss genus *Cinclidium* Sw. (Mniaceae Mitt.). *Lindbergia*, **2**, 49–80

Mogensen, G. S. (ed.) (1985). Illustrated moss flora of North America and Greenland. 1: Polytrichaceae. *Meddelelser om Grønland, Bioscience*, **17**, 1–57

Mogensen, G. S. & Lewinsky, J. (1982). Distribution maps of bryophytes in Greenland 9. *Lindbergia*, **8**, 189–92

Monteith, J. L. (1973). *Principles of Environmental Physics*. London, Arnold

Mooney, H. A. (1976). Some contributions of physiological ecology to plant population biology. *Systematic Botany*, **1**, 269–83

Moore, C. J., Luff, S. E. & Hallam, N. D. (1982). Fine structure and physiology of the desiccation-tolerant mosses, *Barbula torquata* Tayl. and *Triquetrella papillata* (Hook. F. and Wils.) Broth., during desiccation and rehydration. *Botanical Gazette*, **143**, 358–67

Moore, D. M. (1968). The vascular flora of the Falkland Islands. *British Antarctic Survey Scientific Reports*, **60**, 1–202

Moore, D. M. (1979). Southern oceanic wet-heathlands (including magellanic moorland). In *Heathlands and Related Shrublands of the World. A. Descriptive Studies*. ed. R. L. Specht, pp. 489–97. Amsterdam, Elsevier

Moser, T. J. & Nash, T. H. (1978). Photosynthetic patterns of *Cetraria cucullata* (Bell.) Ach. at Anaktuvuk Pass, Alaska. *Oecologeia*, **34**, 37–43

Moser, T. J., Nash, T. H. & Clark, W. D. (1980). Effects of long-term sulfur dioxide fumigation on Arctic caribou forage lichens. *Canadian Journal of Botany*, **58**, 2235–40

Moser, T. J., Nash, T. H. & Link, S. O. (1983). Diurnal gross photosynthetic patterns and potential CO_2 assimilation in *Cladonia stellaris* and *Cladonia rangiferina*. *Canadian Journal of Botany*, **61**, 642–55

Moser, T. J., Nash, T. H. & Olafsen, A. G. (1983). Photosynthetic recovery in Arctic caribou forage lichens following a long-term field sulfur dioxide fumigation. *Canadian Journal of Botany*, **61**, 367–70

Moser, T. J., Nash, T. H. & Thomson, J. W. (1979). Lichens of Anaktuvuk Pass, Alaska, with emphasis on the impact of caribou grazing. *Bryologist*, **82**, 393–408

Muc, M. (1977). Ecology and primary production of sedge-moss meadow communities, Truelove Lowland. In *Truelove Lowland, Devon Island,*

Canada, A High Arctic Ecosystem, ed. L. C. Bliss, pp. 157–84. Edmonton, University of Alberta Press

Muc, M. & Bliss, L. C. (1977). Plant communities of Truelove Lowland. In *Truelove Lowland, Devon Island, Canada, A High Arctic Ecosystem*, ed. L. C. Bliss, pp. 143–54. Edmonton, University of Alberta Press

Muller, C. H. (1952). Plant succession in Arctic heath and tundra in northern Scandinavia. *Bulletin of the Torrey Botanical Club*, **79**, 296–309

Murray, D. F., Murray, B. M., Yertsev, B. A. & Howenstein, R. (1983). Biogeographic significance of steppe vegetation in subarctic Alaska. In *Permafrost: Fourth International Conference Proceedings*, pp. 883–8. Washington, National Academy Press

Nakanishi, S. (1977). Ecological studies of the moss and lichen communities in the ice-free areas near Syowa Station, Antarctica. *Antarctic Record*, **59**, 68–96

Nakatsubo, T. & Ino, Y. (1986). Nitrogen cycling in an Antarctic ecosystem. 1: Biological nitrogen fixation in the vicinity of Syowa Station. *Memoirs of the National Institute of Polar Research*, **E37**, 1–10

Nash, T. H., Moser, T. J. & Link, S. D. (1980). Nonrandom variation of gas exchange within Arctic lichens. *Canadian Journal of Botany*, **58**, 1181–6

Newton, M. E. (1972). Chromosome studies in some South Georgian bryophytes. *British Antarctic Survey Bulletin*, **30**, 41–9

Newton, M. E. (1979). A taxonomic assessment of *Brachythecium* on South Georgia. *British Antarctic Survey Bulletin*, **48**, 119–32

Newton, M. E. (1980). Chromosome studies on some Antarctic and sub-Antarctic bryophytes. *British Antarctic Survey Bulletin*, **50**, 77–86

Nieboer, E. & Richardson, D. H. S. (1980). The replacement of the nondescript term 'heavy metals' by a biologically and chemically significant classification of metal ions. *Environmental Pollution*, **B1**, 3–26

Nieboer, E., Richardson, D. H. S. & Tomassini, F. D. (1978). Mineral uptake and release by lichens. An overview. *Bryologist*, **81**, 226–46

Nörr, M. (1974). Hitzeresistenz bei Moosen. *Flora (Jena)*, **163**, 388–97

Ochyra, R., Przywara, L. & Kuta, E. (1982). Karyological studies on some Antarctic liverworts. *Journal of Bryology*, **12**, 259–63

Oechel, W. C. (1976). Seasonal patterns of temperature response of CO_2 flux and acclimation in arctic mosses growing *in situ*. *Photosynthetica*, **10**, 447–56

Oechel, W. C. & Collins, N. J. (1976). Comparative CO_2 exchange patterns in mosses from two tundra habitats at Barrow, Alaska. *Canadian Journal of Botany*, **54**, 1355–69

Oechel, W. C. & Sveinbjörnsson, B. (1978). Primary production processes in Arctic bryophytes at Barrow, Alaska. In *Vegetation Production and Production Ecology of an Alaskan Arctic Tundra*, ed. L. L. Tieszen, pp. 269–98. New York, Springer-Verlag

Oechel, W. C. & Van Cleve, K. (1986). The role of bryophytes in nutrient cycling in the taiga. In *Forest Ecosystems in the Alaskan Taiga*, ed. K. Van Cleve, F. S. Chapin, P. W. Flanagan, L. A. Viereck & C. T. Dyrness, pp. 121–37. New York, Springer-Verlag

Oksanen, L. (1978). Lichen grounds of Finnmarksvidda, northern Norway, in relation to summer and winter grazing by reindeer. *Reports from the Kevo Subarctic Research Station*, **14**, 64–71

Oksanen, L. & Oksanen, T. (1981). Lemmings (*Lemmus lemmus*) and grey-sided voles (*Clethrionomys rufocanus*) in interaction with their resources and predators on Finnmarksvidda, northern Norway. *Reports from the Kevo Subarctic Research Station*, **17**, 7–31

Oliver, M. J. & Bewley, J. D. (1984). Desiccation and ultrastructure in bryophytes. *Advances in Bryology*, **2**, 91–132

Oosting, H. J. & Anderson, L. E. (1939). Plant succession on granite rock in eastern North Carolina. *Botanical Gazette*, **100**, 750–68

Packer, J. G. (1974). Differentiation and dispersal in alpine floras. *Arctic and Alpine Research*, **6**, 117–28

Pakarinen, P. (1978). Production and nutrient ecology of three *Sphagnum* species in southern Finnish raised bogs. *Annales Botanici Fennici*, **15**, 15–26

Pakarinen, P. (1981). Nutrient and trace element content and retention in reindeer lichen carpets of Finnish ombrotrophic bogs. *Annales Botanici Fennici*, **18**, 265–74

Pakarinen, P. & Tolonen, K. (1976). Regional survey of heavy metals in peat mosses (*Sphagnum*). *Ambio*, **5**, 38–40

Pakarinen, P. & Vitt, D. H. (1974). The major organic components and caloric contents of high Arctic bryophytes. *Canadian Journal of Botany*, **52**, 1151–61

Pegau, R. E. (1968). Growth rates of important reindeer forage lichens on the Seward Peninsula, Alaska. *Arctic*, **21**, 255–9

Person, S. J., Pegau, R. E., White, R. G. & Luick, J. R. (1980). *In vitro* and nylon bag digestibilities of reindeer and caribou forages. *Journal of Wildlife Management*, **44**, 613–22

Peterson, K. M. & Billings, W. D. (1980). Tundra vegetational patterns and succession in relation to microtopography near Atkasook, Alaska. *Arctic and Alpine Research*, **12**, 473–82

Philippi, G. (1973). *Moosflora und Moosvegetation des Freeman-Sund-Gebietes (Sudost-Spitzbergen)*. Wiesbaden, Franz Steiner Verlag

Phillpot, H. R. (1985). Physical geography – climate. In *Key Environments Antarctica*, ed. W. N. Bonner & D. W. H. Walton, pp. 23–38. Oxford, Pergamon Press

Pickard, J. & Seppelt, R. D. (1984). Phytogeography of Antarctica. *Journal of Biogeography*, **11**, 83–102

Pierce, W. G. & Kershaw, K. A. (1976). Studies on lichen-dominated systems. XVII: The colonization of young raised beaches in NW Ontario. *Canadian Journal of Botany*, **54**, 1672–83

Pisano, E. (1983). The magellanic tundra complex. In *Ecosystems of the World 4B. Mires: Swamp, Bog, Fen and Moor*, ed. A. J. P. Gore, pp. 295–329. Amsterdam, Elsevier

Polunin, N. (1935). The vegetation of Akpatok Island. Part II. *Journal of Ecology*, **23**, 161–209

Polunin, N. (1936). Plant succession in Norwegian Lapland. *Journal of Ecology*, **24**, 372–91

Polunin, N. (1951). The real Arctic: suggestions for its delimitation, subdivision and characterization. *Journal of Ecology*, **39**, 308–15

Polunin, N. (1955). Long-distance plant dispersal in the north polar regions. *Nature*, **176**, 22–4

Polunin, N. (1959). *Circumpolar Arctic Flora*. Oxford, Oxford University Press

Porsild, A. E. (1951). Plant life in the Arctic. *Canadian Geographical Journal*, **42**, 120–145

Porsild, A. E. (1958). Geographical distribution of some elements in the flora of Canada. *Geographical Bulletin*, **11**, 57–77

Pratt, R. M. & Smith, R. I. Lewis (1982). Seasonal trends in chemical composition of reindeer forage plants on South Georgia. *Polar Biology*, **1**, 13–22

Priddle, J. (1980). The production ecology of benthic plants in some Antarctic lakes. II: Laboratory physiology studies. *Journal of Ecology*, **68**, 155–66

Prins, H. H. Th. (1982). Why are mosses eaten in cold environments only? *Oikos*, **38**, 374–80

Proctor, M. C. F. (1977). The growth curve of the crustose lichen *Buellia canescens* (Dicks.) De Not. *New Phytologist*, **79**, 659–63

Proctor, M. C. F. (1979). Structure and eco-physiological adaptation in bryophytes. In *Bryophyte Systematics*, ed. G. C. S. Clarke & J. G. Duckett, pp. 479–509. London, Academic Press

Proctor, M. C. F. (1980). Diffusion resistance in bryophytes. In *Plants and their Atmospheric Environment*, ed. J. Grace, E. D. Ford & P. G. Jarvis, pp. 219–29. Oxford, Blackwell

Proctor, M. C. F. (1981). Physiological ecology of bryophytes. *Advances in Bryology*, **1**, 79–166

Proctor, M. C. F. (1982). Physiological ecology: water relations, light and temperature responses, carbon balance. in *Bryophyte Ecology*, ed. A. J. E. Smith, pp. 333–81. London, Chapman & Hall

Proctor, M. C. F. (1983). Size and growth rates of thalli of the lichen *Rhizocarpon geographicum* on the moraines of the Glacier de Valsory, Valais, Switzerland. *Lichenologist*, **15**, 249–62

Pruitt, W. O. (1978). *Boreal Ecology*. London, Arnold

Pryor, M. E. (1962). Some environmental features of Hallett Station, Antarctica, with special reference to soil arthropods. *Pacific Insects*, **4**, 681–728

Puckett, K. J. & Finegan, E. J. (1980). An analysis of the element content of lichens from the Northwest Territories, Canada. *Canadian Journal of Botany*, **58**, 2073–89

Pyatt, F. B. (1973). Lichen propagules. In *The Lichens*, ed. V. Ahmadjian & M. E. Hale, pp. 117–45. New York, Academic Press

Pyysalo, H., Koponen, A. & Koponen, T. (1983). Studies on entomophily in Splachnaceae (Musci). II: Volatile compounds in the hypophysis. *Annales Botanici Fennici*, **20**, 335–8

Racine, C. H. (1981). Tundra fire effects on soils and three plant communities along a hill-slope gradient in the Seward Peninsula, Alaska. *Arctic*, **34**, 71–84

Rao, D. N. (1982). Responses of bryophytes to air pollution. In *Bryophyte Ecology*, ed. A. J. E. Smith, pp. 445–71. London, Chapman & Hall

Rao, D. N., Robitaille, G. & LeBlanc, F. (1977). Influence of heavy metal pollution on lichens and bryophytes. *Journal of the Hattori Botanical Laboratory*, **42**, 213–39

Rastorfer, J. R. (1970). Effects of light intensity and temperature on photosynthesis and respiration of two East Antarctic mosses, *Bryum argenteum* and *Bryum antarcticum*. *Bryologist*, **73**, 544–56

Rastorfer, J. R. (1971a). Effects of temperature on carbon dioxide compensation points of the moss *Drepanocladus uncinatus*. *Antarctic Journal of the United States*, **6**, 162–3

Rastorfer, J. R. (1971b). Vegetative regeneration and sporophyte development of *Bryum antarcticum* in an artificial environment. *Journal of the Hattori Botanical Laboratory*, **34**, 391–7

Rastorfer, J. R. (1972). Comparative physiology of four west Antarctic mosses. In *Antarctic Terrestrial Biology*, ed. G. A. Llano, pp. 143–61. Washington, American Geophysical Union

Rastorfer, J. R. (1974). Element content of three Alaskan-Arctic mosses. *Ohio Journal of Science*, **74**, 55–9

Rastorfer, J. R. (1976). Caloric values of three Alaskan-Arctic mosses. *Bryologist*, **79**, 76–8

Rastorfer, J. R. (1978). Composition and bryomass of the moss layers of two wet-tundra-meadow communities near Barrow, Alaska. In *Vegetation and Production Ecology of an Alaskan Arctic Tundra*, ed. L. L. Tieszen, pp. 169–83. New York, Springer-Verlag

Reimers, E. (1977). Population dynamics of two populations of reindeer in Svalbard. *Arctic and Alpine Research*, **9**, 369–81

Richards, P. W. & Smith, A. J. E. (1975). A progress report on *Campylopus introflexus* (Hedw.) Brid. and *C. polytrichoides* De Not. in Britain. *Journal of Bryology*, **8**, 293–8

Richardson, D. H. S. & Finegan, E. (1977). Studies on the lichens of Truelove Lowland. In *Truelove Lowland, Devon Island, Canada: A High Arctic Ecosystem*, ed. L. C. Bliss, pp. 245–62. Edmonton, University of Alberta Press

Richardson, D. H. S. & Nieboer, E. (1981). Lichens and pollution monitoring. *Endeavour*, **NS5**, 127–33

Richardson, D. H. S. & Young, C. M. (1977). Lichens and vertebrates. In *Lichen Ecology*, ed. M. R. D. Seaward, pp. 121–44. London, Academic Press

Rinne, R. J. K. & Barclay-Estrup, P. (1980). Heavy metals in a feather moss, *Pleurozium schreberi* and its soils in northwest Ontario, Canada. *Oikos*, **34**, 59–67

Ritchie, J. C. (1956). The native plants of Churchill, Manitoba. *Canadian Journal of Botany*, **34**, 269–320

Robinson, H. E. (1972). Observations on the origin and taxonomy of the Antarctic moss flora. In *Antarctic Terrestrial Biology*, ed. G. A. Llano, pp. 163–77. Washington, American Geophysical Union

Rodgers, G. A. & Henriksson, E. (1976). Associations between the blue-green algae *Anabaena variabilis* and *Nostoc muscorum* and the moss *Funaria hygrometrica* with reference to the colonization of Surtsey. *Acta Botanica Islandica*, **4**, 10–15

Rosenberg, N. J. (1974). *Microclimate: The Biological Environment*. New York, Wiley

Rosswall, T., Flower-Ellis, J. G. K., Johansson, L. G., Jonsson, S., Rydén, B. E. & Sonesson, M. (1975). Stordalen (Abisco), Sweden. In *Structure and Function of Tundra Ecosystems*, ed. T. Rosswall & O. W. Heal, pp. 265–94. *Ecological Bulletins (Stockholm)*, **20**

Rosswall, T. & Granhall, U. (1980). Nitrogen cycling in a subarctic ombrotrophic mire. In *Ecology of a Subarctic Mire*, ed. M. Sonesson, pp. 209–34. *Ecological Bulletins (Stockholm)*, 30

Rosswall, T., Veum, A. K. & Kärenlampi, L. (1975). Plant litter decomposition at Fennoscandian tundra sites. In *Fennoscandian Tundra Ecosystems. 1: Plants and Microorganisms*, ed. F. E. Wielgolaski, pp. 268–78. New York, Springer-Verlag

Rouse, W. R. (1976). Microclimatic changes accompanying burning in subarctic lichen woodland. *Arctic and Alpine Research*, **8**, 357–76

Rouse, W. R. & Kershaw, K. A. (1971). The effects of burning on the heat and water regimes of lichen-dominated subarctic surfaces. *Arctic and Alpine Research*, **3**, 291–304

Rudolph, E. D. (1966). Terrestrial vegetation in Antarctica, past and present studies. In *Antarctic Soils and Soil Forming Processes*, ed. J. C. F. Tedrow, pp. 109–24. Washington, American Geophysical Union

Rudolph, E. D. (1967a). Climate. In *Terrestrial Life of Antarctica*, ed. V. C. Bushnell, pp. 5–6. New York, American Geographical Society

Rudolph, E. D. (1967b). Lichen distribution. In *Terrestrial Life of Antarctica*, ed. V. C. Bushnell, pp. 9–11. New York, American Geophysical Union

Rudolph, E. D. (1970). Local dissemination of plant propagules in Antarctica. In *Antarctica Ecology*, ed. M. W. Holdgate, vol. 2, pp. 812–17. London, Academic Press

Rudolph, E. D. (1971). Ecology of land plants in Antarctica. In *Research in the Antarctica*, ed. L. O. Quam, pp. 191–211. Washington, AAAS

Rühling, A. & Tyler, G. (1984). Recent changes in the deposition of heavy metals in northern Europe. *Water, Soil, and Air Pollution*, **22**, 173–80

Rundel, P. W. (1978). The ecological role of secondary lichen substances. *Biochemical Systematics and Ecology*, **6**, 157–70

Russell, R. S. & Wellington, P. S. (1940). Physiological and ecological studies on an Arctic vegetation. I: The vegetation of Jan Mayen Island. *Journal of Ecology*, **28**, 153–79

Russell, S. (1984). Growth measurement in bryophytes: a case study. *Journal of the Hattori Botanical Laboratory*, **55**, 147–57

Russell, S. (1985). Bryophyte productivity at Marion Island. In *Antarctic Nutrient Cycles and Food Webs*, ed. W. R. Siegfried, P. R. Condy & R. M. Laws, pp. 200–3. Berlin, Springer-Verlag

Sage, B. (1985). Conservation and exploitation. In *Key Environments Antarctica*, ed. W. N. Bonner & D. W. H. Walton, pp. 351–69. London, Pergamon

Savicz-Ljubitskaja, L. I. & Smirnova, Z. N. (1961). On the modes of reproduction of *Sarconeurum glaciale* (Hook. fil. et Wils.) Card. et Bryhn, an endemic moss of the Antarctica. *Revue Bryologique et Lichénologique*, **NS30**, 216–22

Scheirer, D. C. (1983). Leaf parenchyma with transfer cell-like characteristics in the moss, *Polytrichum commune* Hedw. *American Journal of Botany*, **70**, 897–992

Schljakov, R. N. (1982). *The Hepaticae of the Northern U.S.S.R. Vol. 5. Hepaticae: Lophocoleaceae – Ricciaceae*. Leningrad, Nauka

Schofield, E. (1972). Preserving the scientific value of cold desert ecosystems: past and present practices and a rationale for the future. In *Conservation Problems in Antarctica*, ed. B. C. Parker, pp. 193–225. Blacksberg, Virginia Polytechnic Institute and State University

Schofield, E. & Ahmadjian, V. (1972). Field observations and laboratory studies of some Antarctic cold desert cryptogams. In *Antarctic Terrestrial Biology*, ed. G. A. Llano, pp. 97–142. Washington, American Geophysical Union

Schofield, E. & Hamilton, W. L. (1970). Probable damage to tundra biota through sulphur dioxide destruction of lichens. *Biological Conservation*, **2**, 278–80

Schofield, W. B. (1972). Bryology in Arctic and boreal North America and Greenland. *Canadian Journal of Botany*, **50**, 1111–33

Scholander, P. F., Flagg, W., Walters, V. & Irving, L. (1952). Respiration in some arctic and tropical lichens in relation to temperature. *American Journal of Botany*, **39**, 707–13

Schuster, R. M. (1959). Introduction. In *The Terrestrial Cryptogams of Northern Ellesmere Island. National Museums of Canada Bulletin*, **164**, 1–14

Schuster, R. M. (1966–80). *The Hepaticae and Anthocerotae of North America*. Vols. 1–4. New York, Columbia University Press

Schuster, R. M. (1979). On the persistence and dispersal of transantarctic Hepaticae. *Canadian Journal of Botany*, **57**, 2179–225

Schuster, R. M. (1983). Phytogeography of the Bryophyta. In *New Manual of Bryology*, ed. R. M. Schuster, pp. 463–626. Nichinan, Hattori Botanical Laboratory

Schwerdtfeger, W. (1971). The climate of the Antarctic. In *Climates of the Polar Regions*, ed. S. Orvig, pp. 253–355. Amsterdam, Elsevier

Scoggan, H. J. (1978–79). *The Flora of Canada*. Parts 1–4. Ottawa, National Museums of Canada.

Scott, M. G. & Larson, D. W. (1985). The effect of winter field conditions on the distribution of two species of *Umbilicaria*. I: CO_2 exchange in reciprocally transplanted thalli. *New Phytologist*, **101**, 89–101

Scotter, G. W. (1962). Productivity of arboreal lichens and their possible importance to barren-ground caribou (*Rangifer arcticus*). *Arch. Soc. Zool. Bot. Fenn. Vanamo*, **16**, 155–61

Scotter, G. W. (1963). Growth rates of *Cladonia alpestris*, *C. mitis* and *C. rangiferina* in the Talston River region, N.W.T. *Canadian Journal of Botany*, **41**, 1199–202

Scotter, G. W. (1965). Chemical composition of forage lichens from northern Saskatchewan as related to use by barren-ground caribou. *Canadian Journal of Plant Science*, **45**, 245–50

Scotter, G. W. (1970). Wildfires in relation to the habitat of barren-ground caribou in the taiga of northern Canada. *Proceedings of the Tenth Annual Tall Timbers Fire Conference*, Fredericton, New Brunswick, pp. 85–105

Scotter, G. W. (1972). Chemical composition of forage plants from the Reindeer Preserve, North West Territories. *Arctic*, **25**, 21–7

See, M. G. & Bliss, L. C. (1980). Alpine lichen-dominated communities in Alberta and Yukon. *Canadian Journal of Botany*, **58**, 2148–70

Seki, T. (1974). A moss flora of Provincia de Aisén, Chile. *Journal of Science of the Hiroshima University*, Series B, Division 2 (Botany), **15**, 9–101

Selkirk, P. M. (1981). Protonemal gammae on *Amblystegium serpens* (Hedw.) B.S.G. from Macquarie Island. *Journal of Bryology*, **11**, 719–21

Selkirk, P. M. (1984). Vegetative reproduction and dispersal of bryophytes on subantarctic Macquarie Island and in Antarctica. *Journal of the Hattori Botanical Laboratory*, **55**, 105–11

Selkirk, P. M., Costin, A. B., Seppelt, R. D. & Scott, J. J. (1983). Rabbits, vegetation and erosion on Macquarie Island. *Proceedings of the Linnean Society of New South Wales*, **106**, 337–46

Sendstad, E. (1981). Soil ecology of a lichen heath at Spitzbergen, Svalbard: effects of artificial removal of the lichen plant cover. *Journal of Range Management*, **34**, 442–5

Seppelt, R. D. (1977). Studies on the bryoflora of Macquarie Island. I: Introduction and checklist of species. *Bryologist*, **80**, 167–70

Seppelt, R. D. (1978). Studies on the bryoflora of Macquarie Island. II: *Ulota phyllantha*. *New Zealand Journal of Botany*, **16**, 21–3

Seppelt, R. D. (1980). A synoptic moss flora of Macquarie Island. *Australian Department of Science and the Environment, Antarctic Division Technical Memorandum*, **93**, 1–8

Seppelt, R. D. (1983a). *Cephaloziella exiliflora* (Tayl.) Steph. from the Windmill Islands, continental Antarctica. *Lindbergia*, **9**, 27–8

Seppelt, R. D. (1983b). The status of *Bryum korotkevicziae*. *Lindbergia*, **9**, 21–6

Seppelt, R. D. (1984). The bryoflora of the Vestfold Hills and Ingrid Christensen Coast, Antarctica. *Australian National Antarctic Expeditions Research Notes*, **20**, 1–31

Seppelt, R. D. & Ashton, D. H. (1978). Studies on the ecology of the vegetation of Mawson Station, Antarctica. *Australian Journal of Ecology*, **3**, 373–88

Seppelt, R. D. & Selkirk, P. M. (1984). Effects of submersion on morphology and the implications of induced environmental modification on the taxonomic interpretation of selected Antarctic moss species. *Journal of the Hattori Botanical Laboratory*, **55**, 273–9

Shaw, A. J. (1981). *Pohlia andrewsii* and *P. tundrae*, two new Arctic-alpine propaguliferous species from North America. *Bryologist*, **84**, 65–74

Sheard, J. W. & Geale, D. W. (1983). Vegetation studies at Polar Bear Pass, Bathurst Island, N.W.T. II: Vegetation–environment relationships. *Canadian Journal of Botany*, **61**, 1637–46

Sigal, L. (1984). Lichen research and regulatory decisions. *Bryologist*, **87**, 185–92

Siple, P. A. (1938). The Second Byrd Antarctic Expedition – Botany. I: Ecology and geographical distribution. *Annals of the Missouri Botanical Garden*, **25**, 467–514

Skogland, T. (1984). Wild reindeer foraging-niche organization. *Holarctic Ecology*, **7**, 345–79

Skottsberg, C. (1960). Remarks on the plant geography of the southern cold temperate zone. *Proceedings of the Royal Society*, **B152**, 447–57

Skre, O., Berg, A. & Wielgolaski, F. E. (1975). Organic compounds in alpine plants. In *Fennoscandian Tundra Ecosystems. 1: Plants and Microorganisms*, ed. F. E. Wielgolaski, pp. 339–50. New York, Springer-Verlag

Skre, O., Oechel, W. C. & Miller, P. M. (1983a). Moss leaf water content and solar radiation at the moss surface in a mature black spruce forest in central Alaska. *Canadian Journal of Forestry Research*, **13**, 860–8

Skre, O., Oechel, W. C. & Miller, P. M. (1983b). Patterns of translocation of carbon in four common moss species in a black spruce (*Picea mariana*) dominated forest in interior Alaska. *Canadian Journal of Forestry Research*, **13**, 869–78

Slocum, R. D., Ahmadjian, V. & Hildreth, K. C. (1980). Zoosporogenesis in *Trebouxia gelatinosa*: ultrastructure potential for zoospore release and implications for the lichen association. *Lichenologist*, **12**, 173–87

Smith, A. J. E. (1978). Cytogenetics, biosystematics and evolution in the Bryophyta. *Advances in Botanical Research*, **6**, 195–276

Smith, A. J. E. (1982). Epiphytes and epiliths. In *Bryophyte Ecology*, ed. A. J. E. Smith, pp. 191–227. London, Chapman & Hall

Smith, D. C. & Drew, E. A. (1965). Studies on the physiology of lichens. V: Translocation from the algal layer to the medulla in *Peltigera polydactyla*. *New Phytologist*, **64**, 195–200

Smith, D. C. & Molesworth, S. (1973). Lichen physiology. XIII: Effects of rewetting dry lichens. *New Phytologist*, **72**, 525–33

Smith, M. W. (1975). Microclimatic influences on ground temperature and permafrost distribution, McKenzie Delta, Northwest Territories. *Canadian Journal of Earth Science*, **12**, 1421–38

Smith, R. I. Lewis (1972). Vegetation of the South Orkney Islands with particular reference to Signy Island. *British Antarctic Survey Scientific Reports*, **68**, 1–124

Smith, R. I. Lewis (1978). Summer and winter concentrations of sodium, potassium and calcium in some maritime Antarctic cryptogams. *Journal of Ecology*, **66**, 891–909

Smith, R. I. Lewis (1981). Types of peat and peat-forming vegetation on South Georgia. *British Antarctic Survey Bulletin*, **53**, 119–39

Smith, R. I. Lewis (1982a). Plant succession and re-exposed moss banks on a deglaciated headland in Arthur Harbour, Anvers Island. *British Antarctic Survey Bulletin*, **51**, 193–9

Smith, R. I. Lewis (1982b). Growth and production in South Georgian bryophytes. *Comité National Français des Recherches Antarctiques*, **51**, 229–39

Smith, R. I. Lewis (1984a). Terrestrial plant biology of the sub-Antarctic and Antarctic. In *Antarctic Ecology*, ed. R. M. Laws, vol. 1, pp. 61–162. London, Academic Press

Smith, R. I. Lewis (1984b). Colonization by bryophytes following recent volcanic activity on an Antarctic island. *Journal of the Hattori Botanical Laboratory*, **56**, 53–63

Smith, R. I. Lewis (1984c). Colonization and recovery by cryptogams following recent volcanic activity at Deception Island, South Shetland Islands. *British Antarctic Survey Bulletin*, **62**, 25–51

Smith, R. I. Lewis (1985a). Nutrient cycling in relation to biological productivity in Antarctic and sub-Antarctic terrestrial and freshwater ecosystems. In *Antarctic Nutrient Cycles and Food Webs*, ed. W. R. Siegfried, P. R. Condy & R. M. Laws, pp. 138–55. Berlin, Springer-Verlag

Smith, R. I. Lewis (1985b). A unique community of pioneer mosses dominated by *Pterygoneurum* cf *ovatum* in the Antarctic. *Journal of Bryology*, **13**, 509–14

Smith, R. I. Lewis (1986). Plant ecological studies in the fellfield ecosystem near Casey Station, Australian Antarctic Territory, 1985–86. *British Antarctic Survey Bulletin*, **72**, 81–91

Smith, R. I. Lewis & Corner, R. W. M. (1973). Vegetation of the Arthur Harbour-Argentine Islands region of the Antarctic Peninsula. *British Antarctic Survey Bulletin*, **33 & 34**, 89–122

Smith, R. I. Lewis & Gimingham, C. H. (1976). Classification of cryptogamic communities in the maritime Antarctic. *British Antarctic Survey Bulletin*, **43**, 25–47

Smith, R. I. Lewis & Walton, D. W. H. (1973). Calorific values of South Georgian plants. *British Antarctic Survey Bulletin*, **36**, 123–7

Smith, R. I. Lewis & Walton, D. W. H. (1975). South Georgia, Subantarctic. In *Structure and Function of Tundra Ecosystems*, ed. T. Rosswall & O. W. Heal, pp. 399–423. *Ecological Bulletins (Stockholm)*, 20

Smith, V. R. (1977). Notes on the feeding of *Ectomnorrhinus similis* Waterhouse (Curculionidae) adults on Marion Island. *Oecologia*, **29**, 269–73

Smith, V. R. & Russell, S. (1982). Acetylene reduction by bryophyte–cyanobacteria associations on a subantarctic island. *Polar Biology*, **1**, 153–7

Snelgar, W. P., Green, T. G. A. & Wilkins, A. L. (1981). Carbon dioxide exchange in lichens: resistance to CO_2 uptake at different thallus water contents. *New Phytologist*, **88**, 353–61

Sohlberg, E. H. & Bliss, L. C. (1984). Microscale pattern of vascular plant distribution in two high Arctic plant communities. *Canadian Journal of Botany*, **62**, 2033–42

Sonesson, M. (1978). Studies on the mire vegetation in the Torneträsk area, northern Sweden. IV: Some habitat conditions of the poor mires. *Botaniska Notiser*, **123**, 67–111

Sonesson, M. (ed.) (1980). *Ecology of a Subarctic Mire. Ecological Bulletins (Stockholm)*, 30

Sonesson, M. & Bergman, H. (1980). Area-harvesting as a method of estimating phytomass changes in a tundra mire. In *Ecology of a Subarctic Mire*, ed. M. Sonesson, pp. 127–37. *Ecological Bulletins (Stockholm)*, 30

Sonesson, M. & Johansson, S. (1973). Bryophyte growth, Stordalen, 1973. In *Technical Report 16, Swedish Tundra Biome Project*, ed. M. Sonesson, pp. 17–27. Stockholm, International Biological Program

Sonesson, M., Persson, S., Basilier, K. & Stenström, T.-A. (1980). Growth of *Sphagnum riparium* Ångstr. in relation to some environmental factors in the Stordalen mire. In *Ecology of a Subarctic Mire*, ed. M. Sonesson, pp. 191–207. *Ecological Bulletins (Stockholm)*, 30

Sørensen, T. (1941). Temperature relations and phenology of the northeast Greenland flowering plants. *Meddelelser om Grønland*, **125(9)**, 1–305

Spatt, P. D. & Miller, M. C. (1981). Growth conditions and vitality of *Sphagnum* in a tundra community along the Alaska pipeline haul road. *Arctic*, **34**, 48–54

Staaland, H., Brattbakk, I., Ekern, K. & Kildemo, K. (1983). Chemical composition of reindeer forage plants in Svalbard and Norway. *Holarctic Ecology*, **6**, 109–22

Staaland, H., Jacobsen, E. & White, R. G. (1979). Comparison of the digestive tract in Svalbard and Norwegian reindeer. *Arctic and Alpine Research*, **11**, 457–66

Staaland, H., Jacobsen, E. & White, R. G. (1984). The effect of mineral supplements on nutrient concentrations and pool sizes in the alimentary tract of reindeer fed on lichens and on concentrates during winter. *Canadian Journal of Zoology*, **62**, 1232–41

Steere, W. C. (1954). Chromosome number and behavior in Arctic mosses. *Botanical Gazette*, **116**, 93–133

Steere, W. C. (1965a). The boreal bryophyte flora as affected by Quaternary glaciation. In *The Quaternary of the United States*, ed. H. E. Wright & G. Frey, pp. 485–96. Princeton, Princeton University Press

Steere, W. C. (1965b). Antarctic bryophyta. *BioScience*, **15**, 283–5

Steere, W. C. (1973). Observations on the genus *Aplodon* (Musci: Splachnaceae). *Bryologist*, **76**, 347–55

Steere, W. C. (1978a). *The Mosses of Arctic Alaska*. Vaduz, Cramer

Steere, W. C. (1978b). Floristics, phytogeography and ecology of Arctic Alaskan bryophytes. In *Vegetation and Production Ecology of an Alaskan Arctic Tundra*, ed. L. L. Tieszen, pp. 141–67. New York, Springer-Verlag

Steere, W. C. & Inoue, H. (1978). The Hepaticae of Arctic Alaska. *Journal of the Hattori Botanical Laboratory*, **44**, 251–345

Steere, W. C. & Murray, B. M. (1976). *Andreaeobryum macrosporum*, a new genus and species of Musci from northern Alaska and Canada. *Phytologia*, **33**, 407–10

Stoner, W. A., Miller, P. & Miller, P. C. (1982). Seasonal dynamics of standing crops of biomass and nutrients in a subarctic tundra vegetation. *Holarctic Ecology*, **5**, 172–9

Stoner, W. A., Miller, P. C. & Oechel, W. C. (1978). Simulation of the effects of the tundra vascular plant canopy on the productivity of four plant species. In *Vegetation and Production Ecology of an Alaskan Arctic Tundra*, ed. L. L. Tieszen, pp. 371–87. New York, Springer-Verlag

Sveinbjörnsson, B. & Oechel, W. C. (1981). Controls on CO_2 exchange in two *Polytrichum* moss species. 2: The implications of belowground plant parts on the whole-plant carbon balance. *Oikos*, **36**, 348–54

Sveinbjörnsson, B. & Oechel, W. C. (1983). The effect of temperature preconditioning on the temperature sensitivity of net CO_2 flux in geographically diverse populations of the moss *Polytrichum commune* Hedw. *Ecology*, **64**, 1100–8

Svensson, B. H. & Rosswall, T. (1980). Energy flow through a subarctic mire at Stordalen. In *Ecology of a Subarctic Mire*, ed. M. Sonesson, pp. 283–301. *Ecological Bulletins (Stockholm)*, 30

Svoboda, J. & Taylor, H. W. (1979). Persistence of cesium-137 in Arctic lichens, *Dryas integrifolia* and lake sediments. *Arctic and Alpine Research*, **11**, 95–108

Syers, J. K. & Iskandar, I. K. (1973). Pedogenetic significance of lichens. In *The Lichens*, ed. V. Ahmadjian & M. E. Hale, pp. 225–48. New York, Academic Press

Tansley, A. G. (1935). The use and abuse of vegetational concepts and terms. *Ecology*, **16**, 284–307

Taylor, H. W., Hutchison-Benson, E. & Svoboda, J. (1985). Search for latitudinal trends in the effective half-life of fallout ^{137}Cs in vegetation of the Canadian Arctic. *Canadian Journal of Botany*, **63**, 792–6

Tedrow, J. C. F. (1977). *Soils of the Polar Landscapes*. New Brunswick, Rutgers University Press

Tegler, B. & Kershaw, K. A. (1980). Studies on lichen-dominated systems. XXIII: The control of seasonal rates of net photosynthesis by moisture, light, and temperature in *Cladonia rangiferina*. *Canadian Journal of Botany*, **58**, 1851–8

Tegler, B. & Kershaw, K. A. (1981). Physiological–environmental interactions in lichens. XII: The seasonal variation of the heat stress response of *Cladonia rangiferina*. *New Phytologist*, **87**, 395–401

Thomas, D. C. & Edmonds, J. (1983). Rumen content and habitat selection of Peary caribou in winter, Canadian Arctic Archipelago. *Arctic and Alpine Research*, **15**, 97–105

Thomas, D. C. & Kroeger, P. (1980). *In vitro* digestibilities of plants in rumen fluids of Peary caribou. *Arctic*, **33**, 757–67

Thomas, D. C. & Kroeger, P. (1981). Digestibility of plants in ruminal fluids of barren-ground caribou. *Arctic*, **34**, 321–4

Thompson, D. C. & McCourt, K. H.(1981). Seasonal diets of the porcupine caribou herd. *American Midland Naturalist*, **105**, 70–6

Thomson, J. W. (1972). Distribution patterns of American Arctic lichens. *Canadian Journal of Botany*, **50**, 1135–56

Thomson, J. W. (1979). *Lichens of the Alaskan Arctic Slope*. Toronto, University of Toronto Press

Thomson, J. W. (1982). Lichen vegetation and ecological patterns in the high Arctic. *Journal of the Hattori Botanical Laboratory*, **53**, 361–4

Thomson, J. W. (1984). *American Arctic Lichens* I. *The Macrolichens*. New York, Columbia University Press

Tieszen, L. L. (1974). Photosynthetic competence of the subnivean vegetation of an Arctic tundra. *Arctic and Alpine Research*, **6**, 253–6

Tieszen, L. L. (ed.) (1978a). *Vegetation and Production Ecology of an Alaskan Arctic Tundra*. New York, Springer-Verlag

Tieszen, L. L. (1978b). Photosynthesis in the principal Barrow, Alaska, species: a summary of field and laboratory responses. In *Vegetation and Production Ecology of an Alaskan Arctic Tundra*, ed. L. L. Tieszen, pp. 241–68. New York, Springer-Verlag

Tieszen, L. L. (1978c). Summary. In *Vegetation and Production Ecology of an Alaskan Arctic Tundra*, ed. L. L. Tieszen, pp. 621–45. New York, Springer-Verlag

Tieszen, L. L. & Johnson, P. C. (1968). Pigment structure of some Arctic tundra communities. *Ecology*, **49**, 370–3

Tilbrook, P. J. (1970). The terrestrial environment and invertebrate fauna of the maritime Antarctic. In *Antarctic Ecology*, ed. M. W. Holdgate, vol. 2, pp. 886–96. London, Academic Press

Tiskov, A. A. (1985). Trophic connections of vertebrate animals and bryophytes. *Acta Academiae Agriensis*, **NS17**, *Suppl.* **1**, 18

Topham, P. B. (1977). Colonization, growth, succession and competition. In *Lichen Ecology*, ed. M. R. D. Seaward, pp. 31–68. London, Academic Press

Troll, C. (1960). The relationship between the climates, ecology and plant geography of the southern cold temperate zone and of the tropical high mountains. *Proceedings of the Royal Society*, **B152**, 529–32

Tschermak-Woess, E. & Friedmann, E. I. (1984). *Hemichloris antarctica*, gen. et sp. nov. (Chlorococcales, Chlorophyta), a cryptoenolithic alga from Antarctica. *Phycologia*, **23**, 443–54

Ugolini, F. C. (1970). Antarctic soils and their ecology. In *Antarctic Ecology*, ed. M. W. Holdgate, vol. 2, pp. 673–92. London, Academic Press

Ugolini, F. C. (1977). The protoranker soils and the evolution of an ecosystem at Kar Plateau, Antarctica. In *Adaptations within Antarctic Ecosystems*, ed. G. A. Llano, pp. 1091–110. Washington, Smithsonian Institution

Ugolini, F. C. & Edmonds, R. L. (1983). Soil biology. In *Pedogenesis and Soil Taxonomy. I: Concepts and Interactions*, ed. L. P. Wilding, N. E. Smeck & G. F. Hall, pp. 193–231. Amsterdam, Elsevier

Ulrich, A. & Gersper, P. L. (1978). Plant nutrient limitations of tundra plant growth. In *Vegetation and Production Ecology of an Alaskan Arctic Tundra*, ed. L. L. Tieszen, pp. 457–81. New York, Springer-Verlag

Usher, M. B. (1983). Pattern in the simple moss-turf communities of the sub-Antarctic and maritime Antarctic. *Journal of Ecology*, **71**, 945–58

Usher, M. B. & Edwards, M. (1986). The selection of conservation areas in Antarctica: an example using the arthropod fauna of Antarctic islands. *Environmental Conservation*, **13**, 115–22

Valanne, N. (1984). Photosynthesis and photosynthetic products in mosses. In *The Experimental Biology of Bryophytes*, ed. A. F. Dyer & J. G. Duckett, pp. 257–73. London, Academic Press

Van Cleve, K. & Alexander, V. (1981). Nitrogen cycling in tundra and boreal ecosystems. In *Terrestrial Nitrogen Cycles*, ed. F. F. Clark & T. Rosswall, pp. 375–404. *Ecological Bulletins (Stockholm)*, 33

Vassiljevskaja, V. D., Ivanov, V. V., Bogatyrev, L. G., Pospelova, E. B., Shalaeva, N. M. & Grishina, L. A. (1975). Agapa, USSR. In *Structure and Function of Tundra Ecosystems*, ed. T. Rosswall & O. W. Heal, pp. 141–58. *Ecological Bulletins (Stockholm)*, 20

Vitt, D. H. (1975). A key and annotated synopsis of the mosses of the northern lowlands of Devon Island, NWT, Canada. *Canadian Journal of Botany*, **53**, 2158–97

Vitt, D. H. & Pakarinen, P. (1977). The bryophyte vegetation, production, and organic compounds of Truelove Lowland. In *Truelove Lowland, Devon Island, Canada: A High Arctic Ecosystem*, ed. L. C. Bliss, pp. 225–44. Edmonton, University of Alberta Press

Vogel, M., Remmert, H. & Smith, R. I. Lewis (1984). Introduced reindeer and their effects on the vegetation and the epigeic invertebrate fauna on South Georgia (subantarctic). *Oecologia*, **62**, 102–9

Vowinkel, E. & Orvig, S. (1970). The climate of the North Polar Basin. In *Climates of the Polar Regions*, ed. S. Orvig, pp. 129–252. Amsterdam, Elsevier

Wace, N. M. (1960). The botany of the southern oceanic islands. *Proceedings of the Royal Society*, **B152**, 475–90

Wace, N. M. (1965). Vascular plants. In *Biogeography and Ecology in Antarctica*, ed. P. van Oye & J. van Mieghem, pp. 201–66. The Hague, Junk

Walton, D. W. H. (1973). Changes in standing crop and dry matter production in an *Acaena* community on South Georgia. In *Proceedings of the Conference on Primary Production and Production Processes, Tundra Biome*, ed. L. C. Bliss & F. E. Wielgolaski, pp. 185–90. Edmonton, IBP Tundra Biome Steering Committee

Walton, D. W. H. (1975). European weeds and other alien species in the sub-Antarctic. *Weed Research*, **15**, 271–82

Walton, D. W. H. (1977). Radiation and soil temperatures 1972–74: Signy Island terrestrial sites. *British Antarctic Survey Data Reports*, **1**, 1–51

Walton, D. W. H. (1982). The Signy Island terrestrial reference sites. XV: Micro-climate monitoring, 1972–74. *British Antarctic Survey Bulletin*, **55**, 111–26

Walton, D. W. H. (1984). The terrestrial environment. In *Antarctic Ecology*, ed. R. M. Laws, vol. 1, pp. 1–60. London, Academic Press

Walton, D. W. H. (1985a). A preliminary study of the action of crustose lichens on rock surfaces in Antarctica. In *Antarctic Nutrient Cycles and Food Webs*, ed. W. R. Siegfried, P. R. Condy & R. M. Laws, pp. 180–5. Berlin, Springer-Verlag

Walton, D. W. H. (1985b). Cellulose decomposition and its relationships to nutrient cycling at South Georgia. In *Antarctic Nutrient Cycles and Food Webs*, ed. W. R. Siegfried, P. R. Condy & R. M. Laws, pp. 191–9. Berlin, Springer-Verlag

Walton, D. W. H. & Smith, R. I. Lewis (1979). The chemical composition of South Georgian vegetation. *British Antarctic Survey Bulletin*, **49**, 117–35

Webb, R. (1973). Reproductive behaviour of mosses on Signy Island, South Orkney Islands. *British Antarctic Survey Bulletin*, **36**, 61–77

Webber, P. J. (1978). Spatial and temporal variation of the vegetation and its production, Barrow, Alaska. In *Vegetation and Production Ecology of an Alaskan Arctic Tundra*, pp. 37–112. New York, Springer-Verlag

Webber, P. A. & Andrews, J. T. (1973). Lichenometry: a commentary. *Arctic and Alpine Research*, **5**, 295–302

Weber, M. G. & Van Cleve, K. (1984). Nitrogen transformations in feather moss and forest floor layers of interior Alaska black spruce ecosystems. *Canadian Journal of Forestry Research*, **14**, 278–90

Wein, R. W. & Bliss, L. C. (1973). Experimental crude oil spills on Arctic plant communities. *Journal of Applied Ecology*, **10**, 671–82

Wein, R. W. & Bliss, L. C. (1974). Primary production in Arctic cottongrass tussock tundra communities. *Arctic and Alpine Research*, **6**, 261–74

Welch, H. E. & Kalff, J. (1974). Benthic photosynthesis and respiration in Char Lake. *Journal of the Fisheries Research Board of Canada*, **31**, 609–20

Weller, G. & Holmgren, G. (1974). The microclimate of the Arctic tundra. *Journal of Applied Meteorology*, **13**, 854–62

West, S. D. (1982). Dynamics of colonization and abundance in central Alaskan populations of the northern red-backed vole, *Clethrionomys rutilis*. *Journal of Mammalogy*, **63**, 128–43

Westman, L. (1973). Notes on the taxonomy and ecology of an Arctic lichen: *Lecanora symmicta* var. *sorediosa* Westm. *Lichenologist*, **5**, 457–60

Wetmore, C. M. (1982). Lichen decomposition in a black spruce bog. *Lichenologist*, **14**, 267–71

Weyant, W. S. (1966). The Antarctic climate. In *Antarctic Soils and Soil Forming Processes*, ed. J. C. F. Tedrow, pp. 47–59. Washington, American Geophysical Union

White, R. G., Bunnell, F. L., Gaare, E., Skogland, T. & Hubert, B. (1981). Ungulates on Arctic ranges. In *Tundra Ecosystems: A Comparative Analysis*, ed. L. C. Bliss, O. W. Heal & J. J. Moore, pp. 397–483

White, R. G., Jacobsen, E. & Staaland, H. (1984). Secretion and absorption of nutrients in the alimentary tract of reindeer fed lichens and concentrates during winter. *Canadian Journal of Zoology*, **62**, 2364–76

White, R. G. & Trudell, J. (1980). Habitat preference and forage consumption by reindeer and caribou near Atkasook, Alaska. *Arctic and Alpine Research*, **12**, 511–29

Whittaker, R. H. (1953). A consideration of the climax theory: the climax as a population and pattern. *Ecological Monographs*, **23**, 41–78

Wielgolaski, F. E. (1972). Vegetation types and plant biomass in tundra. *Arctic and Alpine Research*, **4**, 291–305

Wielgolaski, F. E. (ed.) (1975a). *Fennoscandian Tundra Ecosystems. I: Plants and Microorganisms*. New York, Springer-Verlag

Wielgolaski, F. E. (1975b). Primary production in tundra. In *Photosynthesis and Productivity in Different Environments*, ed. J. P. Cooper, pp. 75–106. Cambridge, Cambridge University Press

Wielgolaski, F. E. (1975c). Functioning of Fennoscandian tundra ecosystems. In *Fennoscandian Tundra Ecosystems. 2: Animals and Systems Analysis*, ed. F. E. Wielgolaski, pp. 300–26. New York, Springer-Verlag

Wielgolaski, F. E. & Kjelvik, S. (1975). Production of plants (vascular plants and cryptogams) in alpine tundra, Hardangervidda. In *Primary Production and Production Processes, Tundra Biome*, ed. L. C. Bliss & F. E. Wielgolaski, pp. 75–86. Edmonton, IBP Tundra Biome Steering Committee

Williams, M. E. & Rudolph, E. D. (1974). The role of lichens and associated fungi in the chemical weathering of rock. *Mycologia*, **64**, 648–60

Williams, M. E., Rudolph, E. D., Schofield, E. A. & Prasher, D. C. (1978). The role of lichens in the structure, productivity, and mineral cycling of the wet coastal Alaskan tundra. In *Vegetation and Production Ecology of an Alaskan Arctic Tundra*, ed. L. L. Tieszen, pp. 185–206. New York, Springer-Verlag

Wilson, A. T. (1970). The McMurdo dry valleys. In *Antarctic Ecology*, ed. M. W. Holdgate, vol. 1, pp. 21–30. London, Academic Press

Wilson, J. Warren (1951). Observations of concentric 'fairy rings' in Arctic moss mat. *Journal of Ecology*, **39**, 407–16

Winner, W. E. & Bewley, J. D. (1978a). Contrasts between bryophyte and vascular plant synecological responses in an SO_2-stressed white spruce association in central Alberta. *Oecologia*, **33**, 311–25

Winner, W. E. & Bewley, J. D. (1978b). Terrestrial mosses as bioindicators of SO_2 pollution stress. *Oecologia*, **33**, 221–30

Winner, W. E. & Koch, G. W. (1982). Water relations and SO_2 resistance of mosses. *Journal of the Hattori Botanical Laboratory*, **52**, 431–40

Wise, K. A. J., Fearon, C. E. & Wilkes, O. R. (1964). Entomological investigations in Antarctica, 1962–63 season. *Pacific Insects*, **6**, 541–70

Wise, K. A. J. & Gressitt, J. L. (1965). Far southern animals and plants. *Nature*, **207**, 101–2

Woodwin, S., Press, M. C. & Lee, J. A. (1985). Nitrate reductase activity in *Sphagnum fuscum* in relation to wet deposition of nitrate from the atmosphere. *New Phytologist*, **99**, 381–8

Worsley, P. & Ward, M. R. (1974). Plant colonization of recent 'annual' moraine ridges at Austre Ostinbreen, North Norway. *Arctic and Alpine Research*, **6**, 217–30

Wyatt, R. (1982). Population ecology of bryophytes. *Journal of the Hattori Botanical Laboratory*, **52**, 179–98

Wyatt, R. & Anderson, L. E. (1984). Breeding systems in bryophytes. In *The Experimental Biology of Bryophytes*, ed. A. F. Dyer & J. G. Duckett, pp. 39–64. London, Academic Press

Wynn-Williams, D. D. (1980). Seasonal fluctuations in microbial activity in Antarctic moss peat. *Biological Journal of the Linnean Society*, **14**, 11–28

Wynn-Williams, D. D. (1984). Comparative respirometry of peat decomposition on a latitudinal transect in the maritime Antarctic. *Polar Biology*, **3**, 171–81

Yarranton, G. A. (1975). Population growth in *Cladonia stellaris* (Opiz.) Pouz. and Vezda. *New Phytologist*, **75**, 99–110

Yarrington, M. R. & Wynn-Williams, D. D. (1985). Methanogenesis and the anaerobic micro-biology of a wet moss community at Signy Island. In *Antarctic Nutrient Cycles and Food Webs*, ed. W. R. Siegfried, P. R. Condy & R. M. Laws, pp. 229–33. Berlin, Springer-Verlag

Zanten, B. O. van (1978). Experimental studies on trans-oceanic long-range dispersal of moss spores in the Southern Hemisphere. *Journal of the Hattori Botanical Laboratory*, **44**, 455–82

Zanten, B. O. van & Pócs, T. (1981). Distribution and dispersal of bryophytes. *Advances in Bryology*, **1**, 479–562

Zoltai, P. & Tarnocai, C. (1975). Perennially frozen peat-lands in the western Arctic and subarctic of Canada. *Canadian Journal of Earth Science*, **12**, 28–43

Index of generic and specific names

Acaena, 15, 19, 57
 magellanica, 58, 260–2, 286, 288, 334
Acarospora, 52, 86
 macrocyclos, 241
Acrophyllum, 55
Agrostis, 53
Alectoria, 44, 48–9, 55, 127, 143, 159, 283
 chalbeiformis, 98
 cryptochlorophaea, 100
 jubata, 266
 miniscula, 228
 nigricans, 92, 100, 306
 nitidula, 125
 ochroleuca, 84–6, 125–6, 130, 145–6, 157, 174–5, 177, 185, 200, 206–7
Alnus, 14, 55
Alopecurus, 57
Amblystegium serpens, 314
Amphidium, 60
Anabaena variabilis, 298
Anastrophyllum, 340
Andreaea, 35, 38, 40–1, 47, 50, 54, 59–62, 87, 97–8, 103, 233, 249, 283, 332
 acutifolia, 233–4
 alpina, 267
 depressinervis, 76
 gainii, 98
 rupestris, 48, 62
Andreaeobryum macrosporum, 340
Andromeda polifolia, 97
Anthelia, 98
 juratzkana, 62, 148, 155
Aplodon wormskjoldii, 312
Archidium alternifolium, 69, 327
Arctagrostis, 14
 latifolia, 53
Arctostaphylos rubra, 96
Arnellia fennica, 321, 340
Aspicilia caesiocinerea, 85
Astelia, 14
Aulocomnium, 55, 293–4
 acuminatum, 271, 331, 340
Azorella, 15
 selago, 103

Barbilophozia, 44, 54
 hatcheri, 337

Bartramia, 62
 patens, 98, 104, 311, 337
Berberis, 55
Betula, 14, 54, 259
 nana, 53, 289
 tortuosa, 289
Blepharidophyllum, 53
 densifolium, 63, 267, 270
Botrydina, 36
Brachythecium, 35, 55, 89–90, 100, 213, 233
 austro-salebrosum, 44–6, 234–5, 267, 270, 337
 rutabulum, 58, 267, 270, 278, 292
 subplicatum, 267, 270
Breutelia, 55
Bryoria, 127, 143, 159
 nitidula, 76, 130, 140, 145, 157, 177, 184–5
 trichodes, 181
Bryum, 23, 30, 33, 35, 41, 48, 50, 62, 64, 89, 100, 215, 340
 algens, 45, 50, 104, 248, 257, 319, 321, 337
 antarcticum, 30, 50, 78, 150, 216, 256–7
 arcticum, 337
 argenteum, 50, 58, 64, 78, 95, 112, 132, 134–7, 150, 160–2, 167, 169, 181, 200, 202, 216, 246, 251–2, 256–7, 314, 321, 324, 328–9, 337
 cryophilum, 62, 268, 271
 inconnexum, 233–4
 nitidulum, 337
 pseudotriquetrum, 58, 258
Buellia, 27, 38–9, 44, 49, 52, 315
 canescens, 224–5
 frigida, 70–1, 78, 96, 146, 165–6, 172, 228
 grimmiae, 50
 latemarginata, 241

Calliergidium, 214, 233, 258
 austro-stramineum, 44, 234–5, 242, 267
Calliergon, 35, 53, 63, 65, 98–9, 139, 177, 213, 258, 279, 281, 333
 giganteum, 271
 sarmentosum, 43, 64, 73, 100,

124, 143, 148, 152, 155, 163, 173, 179, 181–2
Caloplaca, 36, 85–6
 cirrochroa, 241
Campylium, 64, 99
 stellatum, 48, 271, 276
Campylopus, 44, 57, 69, 95
 canescens, 326
 introflexus, 58, 326–7
 pyriformis, 327
Candelariella, 49
Carex, 14
 aquatilis, 53, 97, 100, 276
 bigelowii, 289
 rotundifolia, 97
Cassiope, 14
 tetragona, 55, 265
Cephalozia, 35
 badia, 44
Cephaloziella, 35, 41, 44, 48, 50, 53, 95
 varians, 44, 98, 337
Cerastium, 48
Ceratodon, 30, 41, 89–90, 94
 purpureus, 93, 114–15, 124, 246, 304, 328, 332, 337
Cetraria, 36, 48, 53–7, 60–1, 282–3, 303
 cucullata, 146, 194–5, 268
 islandica, 96, 306
 laevigata, 84–6
 nivalis, 96–7, 121, 125–6, 144–5, 174–5, 185, 306
Cheliotrichum, 14, 55
Chorisodontium, 233, 258
 aciphyllum, 42–3, 47, 58, 87, 101, 104, 139–40, 201, 234, 242, 267, 270, 292, 329–30
Cinclidium, 53
 arcticum, 28
Cladonia, 36–7, 41, 44, 53–7, 60–1, 82, 86, 93, 219, 232, 238, 240, 246, 256, 263, 266, 273, 275, 282–6, 303–4, 315, 323–4
 amaurocraea, 275–6
 arbuscula, 84–5, 116, 283
 coccifera, 128
 crispata, 91, 335
 cryptochlorophaea, 100
 evansii, 206
 mitis, 83, 92

Cladonia (*contd.*)
  rangiferina, 84, 91–2, 112, 116,
    145, 157, 159, 222, 227,
    239–40, 260, 268, 283
  stellaris, 56, 76, 91–2, 112, 116,
    125–30, 132, 140, 144–5, 157,
    159, 174–5, 177, 184–5, 206–7,
    220–2, 239, 246, 248, 289, 297,
    299, 304–5, 329, 335
  sylvatica, 239
  uncialis, 233–4
Clethrionomys rutilis, 279
Climacium, 35
  dendroides, 57
Coccomyxa, 36
Collema, 36, 48, 57
  ceraniscum, 48
  furfuraceum, 144, 152, 157, 159
Colobanthus, 15
  quitensis, 37
Coriscium, 36
Cornicularua aculeata, 98
  divergens, 99–100
Cortadaria, 15
  pilosa, 57
Coscinodon cribrosus, 77
Cotula, 15, 57
Crassula moschata, 57–8
Cryptochila grandiflora, 45, 69, 95
Cyrtomnium hymenophylloides, 60

Dactylina, 53, 55
  arctica, 100
Deschampsia antarctica, 37, 82, 86,
    270
Desmatodon leucostoma, 64
Dicranella, 64
Dicranoloma, 57
Dicranoweisia, 38, 41, 60
  crispula, 62
  grimmiaceae, 234
Dicranum, 35, 54–5, 143, 181, 279
  angustum, 149, 152, 155, 182
  elongatum, 53, 97, 99–100, 124,
    148–9, 152, 155–6, 182, 198,
    203, 211–12, 218–19, 231, 236
  fuscescens, 148, 156, 160,
    167–70, 181–5, 198, 209,
    218–19, 269
Dicrostonyx groenlandicus, 279
Distichium, 64
  capillaceum, 337
  hagenii, 337
Ditrichum, 55, 59–60, 104
  flexicaule, 48, 57
  strictum, 104, 267, 270
Draba, 15, 48
Drepanocladus, 35, 53, 63, 65, 83,
    89, 91, 139, 213–14, 258, 281,
    333
  brevifolius, 100
  schulzei, 97, 236, 289
  uncinatus, 37, 43–5, 58, 76, 82,
    86, 102, 143, 149, 157–9, 160,
    163–4, 167, 174, 181, 201,
    234–5, 331
Dryas, 15, 19, 57, 59
  integrifolia, 306
Dupontia fisheri, 53, 100, 276, 280–1

Edwardiella mirabilis, 325
Empetrum, 14, 55
  nigrum, 289
Encalypta, 62
Eriophorum, 14, 53, 97
  angustifolium, 100, 281
  russeolum, 100
  vaginatum, 52, 289
Evernia, 54, 283

Festuca, 15, 60
  contracta, 261
Fissidens, 60
Fontinalis, 63–4
Frullania, 55
Funaria, 340
  arctica, 279
  hygrometrica, 298, 326–8, 333
  polaris, 64, 314

Gaimardia, 14
Gasparrinia elegans, 85
Gleocapsa, 325
Grimmia, 35, 60, 124
  lawiana, 30, 104, 257
  pulvinata, 319
Gymnomitrion corraloides, 59

Haematomma ventosum, 84
Hebe, 55
Herzogobryum, 41
Himantormia, 260
  lugubris, 38, 98
Holodontium, 59
Hygrohypnum, 63
Hygrolembidium, 41
Hylocomium, 35, 55, 57
  splendens, 54, 91, 115, 172, 208,
    213–18, 271, 295, 303
Hypnum, 55, 59
  revolutum, 57
Hypogymnia, 59

Isopterygium, 60

Jamesoniella colorata, 63
Juncus, 14, 53
  trifidus, 289

Kobresia, 14, 15

Larix, 14, 54
  laricina, 4
Lecanora, 48, 229
  melanopthalma, 146, 172, 200
  polytropa, 84
  tophroeceta, 248
Lecidea, 23, 36, 39, 48–9, 52, 59,
    91
  flavocaerulescens, 84
  granulosa, 92
  pantherina, 84–5
Ledum, 14, 55
Lemmus amurensis, 279
  lemmus, 279
  sibiricus, 279–81
Lepraria, 36, 41
Leptobryum pyriforme, 326, 328
Leptogium, 46, 60

Leskeela, 60
Lobaria, 53
Lophozia, 48, 55, 319
  excisa, 337
Luzula, 14
  campestris, 103
  confusa, 100

Macromitrium, 55
Marchantia, 35
  berteroana, 45, 95, 292, 300
  polymorpha, 42, 93
Marsippospermum, 53
Marsupella, 64
Masonhalea richardsonii, 100, 104,
    275–6
Mastodia, 36
  tesselata, 46, 325
Meesia, 53
  triquetra, 235, 246–7, 271
Mielichhoferia, 77
  eckloni, 58
Muelleriella crassifolia, 38
Myurella, 60

Nephroma arcticum, 37, 121–2,
    143–4, 151, 198–9, 297
Neuropogon acromelanus, 146, 172
  sulphureus, 48–9, 128, 146–7,
    181–2, 207, 257
Nostoc, 37, 46, 48, 50, 297
Nothofagus, 14, 20
  betuloides, 54

Ochrolechia, 41, 44, 48, 59, 229
  frigida, 41, 98
Omphalina, 36
Omphalodiscus decussatus, 71, 147
Oncophorus wahlenbergii, 100
Orthodontium lineare, 58
Orthothecium chryseum, 48
Ovibos moschatus, 5, 278
Oxyria, 15, 58

Pachyglossa, 41
Pannaria, 98
Papaver, 48
  radicata, 265
Parmelia, 59–60, 64, 143
  conspersa, 122, 227, 229
  disjuncta, 122, 145, 157, 159,
    178, 185
  olivacea, 144, 198–9
  saxatilis, 89
Pedicularis, 53, 57–8
Peltigera, 36, 53–4, 91, 273, 285
  aphthosa, 177, 180, 202–3, 206,
    276
  canina, 100
  polydactyla, 227
  rufescens, 208
Pernettya, 14, 55
Pertusaria, 48, 283
Petasites frigidus, 276
Philonotis fontana, 58, 62
  scabrifolia, 58
Physcia, 64
  caesia, 241

Picea, 14, 54, 56
  glauca, 4, 296
  mariana, 4, 93, 266, 296
Pinus banksiana, 266
  sylvestris, 289
Pilgerodendron uvifera, 54
Placopsis contortuplicata, 89
Placynthium nigrum, 85–6
Pleurophyllum, 15
  hookeri, 261
Pleurozium, 35, 55, 93, 115, 333
  schreberi, 54, 91–2, 143, 172,
    198, 203, 208, 213, 218, 248,
    295, 303, 305, 314, 321, 331
Poa, 15, 58
  annua, 68
  arctica, 53, 100
  flabellata, 286
  foliosa, 261, 300
Pohlia, 60, 89–90
  andrewsii, 332
  bulbifera, 328
  cruda, 319
  nutans, 41–4, 89, 95–8, 318
  wahlenbergii, 213–14, 233, 235,
    267–8, 292–4
Polyblastia, 57
  bryophila, 59
Polygonum, 58
Polytrichum, 35, 55, 57, 93, 167,
  171, 213–14, 233, 254, 279,
  293–4
  alpestre, 42–4, 47, 58, 87, 91,
    104, 132–3, 137–40, 143, 150,
    154, 156–60, 164–7, 174, 181,
    214–17, 230, 234–8, 242–6,
    249–51, 258, 266–7, 270, 273,
    288, 290, 292, 294, 314, 316,
    321–3, 329–33, 339
  alpinum, 42, 44, 47–8, 53, 72–6,
    89–90, 94–5, 98–101, 132–3,
    140, 149, 153, 155, 157, 159,
    163, 173–4, 179, 181–2, 214,
    217, 235, 263, 311
  commune, 55, 115, 148–9, 172,
    176–7, 208–9, 214, 217–18,
    271, 276
  hyperboreum, 276, 340
  juniperinum, 92, 100, 305–6, 340
  piliferum, 91, 340
  sexangulare, 148, 155
Potentilla, 15
Pottia, 35, 316
  austro-georgica, 329, 332, 334
  heimii, 30, 314
  obtusifolia, 337
Prasiola, 46, 325
  crispa, 44–7, 50, 78, 248
Pseudephebe, 49, 241
  miniscula, 226, 232

  pubescens, 60, 122, 127–8
Pseudotrebouxia, 312
Psilopilum cavifolium, 100
Psoroma, 36, 41
Pterygoneurum, 41, 316, 329, 332
Ptilidium, 54–5
Ptilium crista-castrensis, 92

Racomitrium, 35, 41, 57, 59, 100
  canescens, 101, 328
  crispulum, 102–3
  fasciculare, 62
  lanuginosum, 48, 60–2, 99, 143,
    148–9, 174–7, 198, 203, 208,
    213, 230, 234, 248, 271, 328
Radula, 340
Ramalina, 36
  terebrata, 38
Rangifer tarandus, 280–6, 307
Ranunculus, 15, 58
Rhizocarpon, 36, 39, 48, 240
  flavum, 78
  geographicum, 60, 84, 86, 224–7,
    240–1
  jemptlandicum, 241
  tinei, 241
Rhizoplaca, 49
Rhytidiadelphus, 57
Riccardia, 53
  pinguis, 53
Riccia sorocarpa, 332
Rinodina, 27, 44, 49
Rostkovia, 14, 53
  magellanica, 288
Rubus chamaemorus, 289

Salix, 14, 53–7, 259
  arctica, 19, 97
  rotundifolia, 99–100
Sarconeurum glaciale, 50, 315
Saxifraga, 15, 48, 57–8
Scapania, 63, 319, 339
Schistidium, 38, 41, 63
  antarctici, 30, 73, 132, 179,
    319–20, 332
Scorpidium, 63, 83
Seligeria, 60
  polaris, 53
  pusilla, 27
Silene, 15
Solarina, 59
Spermophilus parrayi, 279
Sphaerophorus, 59
  globosus, 98, 240, 260
Sphagnum, 18, 20, 22, 35, 53, 70,
  83, 92, 97, 99, 135–6, 177, 213,
  238, 287, 292, 296–7
  balticum, 97, 289
  fimbriatum, 233, 235, 267, 270
  fuscum, 97, 219, 236, 289, 305

  lindbergii, 236, 289
  palustre, 69
  riparium, 233, 236, 247–8
  subsecundum, 115, 218
Stegonia latifolia, 332
Stereocaulon, 4, 54–7, 60–1, 82,
  282, 285
  paschale, 76, 91, 93, 128, 144–5,
    154, 157, 163, 173–4, 297–8
  rivularum, 48
  vesuvianum, 328
Stilbocarpa polaris, 300

Thamnolia, 53, 55, 283
  subobscura, 59
  vermicularis, 268–8, 276
Timmia austriaca, 75
Tomenthypnum, 59
  nitens, 55, 271
Tortula, 41, 53, 58, 124, 179
  excelsa, 45
  robusta, 58, 231, 240, 260,
    267–70, 273, 287–8, 292, 294,
    334
Trapelia coarctata, 328
Trebouxia, 36, 51, 313, 325
Triandrophyllum subtrifidum, 45
Tritomaria, 48

Ulota, 55
  phyllantha, 327
Umbilicaria, 36, 48–9, 60, 170, 219,
  229, 260
  aprina, 147
  arctica, 121–2
  caroliniana, 27
  deusta, 201
  hypoborea, 241
  lyngei, 194
  proboscidea, 84, 241
  vellea, 198–201
Usnea, 36, 44, 49, 54, 60, 97–8, 256,
  260, 278, 283, 315, 335
  antarctica, 38, 40, 76, 86, 329
  fasciata, 38, 40, 240, 260, 315
  picata, 257
  sphacelata, 132

Vaccinium, 14, 55
  uliginosum, 55, 97
Verrucaria, 38, 60, 86
Voitia hypoborea, 314

Xanthoparmelia centrifuga, 84
Xanthoria, 37, 64
  candelaria, 328
  elegans, 37, 60, 78, 96, 122, 147,
    229, 241, 245
  mawsoni, 160, 172, 200
  parietina, 325

Subject index

absorptance, 114, 119, 121
acclimation, 156–9, 167–8, 204–5, 209–10, 333
accumulation phase, 219, 230, 266
acetylene reduction, 296–7
acid deposition, 304–5
adaptation, vii, 63, 68, 81, 141, 203–11, 252, 285, 318, 334
age structure, 329–31
albedo, 10, 93, 111, 113, 116–17, 119, 122, 190
alder, 55
allelopathy, 299
allogenic control, 79–80, 88
altitude, 7, 13, 38, 42, 62, 69
Amblystegiaceae, 70
ammonia, 13, 78, 248, 295–6, 298
amphi-Atlantic distribution, 25
amphi-Beringian distribution, 25
Andreaeales, 340
aneuploidy, 337
Antarctic convergence, 17
Antarctic cryptogam tundra formation, 33, 37–52
Antarctic herb tundra formation, 33, 37
Antarctic zone, 16–18
antheridia, 310, 316–17
rarity of, 319–22
api, 139
apical cell, 213, 244
aplanospores, 313
apomixis, 336, 339
apothecia, 37, 52, 228, 312, 315–16, 323, 329, 335
appressed lichen subformation, 33, 37–8, 49
arachidonic acid, 279
archegonia, 244, 311, 316–17, 319–22
Arctic front, 20
Arctic haze, 300
areolae, 111, 226–8, 251
aridity, 11–12, 22, 52, 195, 289–90
arsenic, 308
Ascomycetes, 36, 101–2, 312
ascospores, 312–13, 315, 323, 334
asexual reproduction, 223, 312–13, 329–30, 336
ash content, 268, 270–1, 273
aspect, 69, 75, 111, 137, 241

associations, 32
atmospheric pollution, 300, 302–8
autogamy, 332, 340
autogenic control, 79–80, 86, 88, 96, 99
autotrophic microorganism subformation, 33, 46, 51

bacteria, 49, 51, 82, 278, 296–7
bacteriostatic agent, 285, 287
Barents Province, 47
barrens, 15, 18, 59, 265–6
Basidiolichens, 36
basidiomycetes, 36, 101
Beer's law, 108, 111
benthic bryophytes, 64–5
Beringian land bridge, 2, 4
bicentric distribution, 28, 30, 31
biogenic control, 79
biomass (see also phytomass), 195, 222, 277–8, 328
biotic influence, 46–7, 78–9
bipolar distribution, 25, 28, 30–1, 326–7
birch, 54, 55, 82, 88, 217, 274
bird perch, 64, 79
birds, 38, 64, 77, 79, 248, 278, 289, 324, 328
borderline metals, 292, 303–4, 308
boreal forest, 16, 55, 74, 139, 172, 203, 237–8, 244–5, 269, 297, 299, 316, 318, 321
boundary layer, 120, 122, 125
Bowen ratio, 126
branching pattern, 34–5, 50, 213–16, 220–1, 230
Braun-Blanquet approach, 32
Bryales, 340
bryophage, 279
bryophyte carpet and mat subformation, 33, 44, 47, 49
bulbil, 251, 332

C₃ pathway, 151
caesium-137, 305–8
calcicole, 77
calcifuge, 77
calcium, 77, 248, 273, 280, 288, 292–6
Canadian Province, 47
canopy, 186, 191–2, 223

influence on solar radiation, 114–16
as moss growth form (defined), 35
capillarity, 72, 170, 231, 291, 295
capsule, 279, 310–12, 319, 321–2, 331
carbohydrate, 198, 218, 227, 273, 284
carbon dioxide (CO_2), 81
dark fixation, 189
carbonic acid, 81
caribou, 266, 280–6, 301, 307
carpet (defined), 35
cation exchange capacity, 287, 303
cation leakage, 198
cephalodia, 37
Chernobyl accident, 307
chionophobes, 121
chionophytes, 61, 76, 122, 334
chlorophyll
$a:b$ ratio, 153
content, 143, 152, 203, 205, 207, 260, 302
Chlorophyta, 52
circular form, 223, 229, 232
circumpolar Arctic element, 25–6, 251, 314, 337
climate, 7–12, 79–80, 104, 191
climax, 79–80, 82, 87, 91–2, 301
cline, 237, 339
cloud, 8, 10, 48, 50, 70, 111, 116, 118, 185
CO_2 concentration, 141–2, 162, 179
subivean, 190
CO_2 exchange (see also net assimilation), 131, 140, 210–11, 213, 246, 334, 341
after short-term freezing, 198–200
prediction of, 191–2
$^{14}CO_2$ uptake, 188–9, 214, 218
cold-Arctic/Antarctic (defined), 22
Coleoptera, 278
collateral cultivation, 207–9, 250–1
collembola, 278
colonisation, 13, 26, 58, 79–94, 336
colonist strategy, 332, 340
compact mat (defined), 35
competition, 51, 54, 79, 86, 88, 94, 223, 229, 258, 325, 333

competitors, 334
conservation, 306–9
continental Antarctica, 17
continental drift, 1, 4, 326
continuous hydration, 194
continuous illumination, 7, 106,
    108, 153, 167, 186, 192
    as a stress, 202–3
convective heat flux (*H*), 117,
    126–7, 130, 181
    factors controlling, 120–3, 125,
        134
cool-Arctic/Antarctic (defined), 21
cool temperate zone, 16–18
copper, 77, 292, 304
coprophilous species, 63
cosmopolitan distribution, 25, 28,
    30
Cretaceous, 2
crude fibre, 273, 288, 290
crustose lichen (defined), 36
cryoturbatic disturbance, 41, 52,
    77, 82, 104, 316, 334
cryptogamic fellfield, 41, 62
cushion bog, 14, 19–20, 52–4
cuticle, 72, 171–2
cyanobacteria, 37, 48–51, 53, 81,
    103, 170, 258, 296–8, 302, 328

dead organic matter (DOM),
    286–7, 290, 295–6
decomposition, 87, 220, 231, 253,
    260, 286–90, 294, 297, 305,
    307, 329
degeneration phase, 220
desiccation, 74, 172, 177, 180, 202,
    245, 302
    during spore dispersal, 325–7
    resistance, 172, 192–3, 203–4,
        342
detritus pathway, 275, 278, 286
dew, 62, 118, 170, 189, 195
Dicranaceae, 57
digestibility, 275, 280–1, 284–5
dioecism, 310, 314, 319, 321, 331–2
Diptera, 63
disjunct distributions, 25, 327
dispersal, 31, 310, 315, 331
    barrier, 19, 68, 300, 327
    long distance, 325–7
DOM – *see* dead organic matter
dry meadow, 18, 259, 274
dry valley region, 23, 51, 70, 257
dwarf shrub (defined), 19
dwarf shrub heath, 14, 18, 20, 55,
    255, 263, 265

ecotype, 208
ectohydric species, 72, 171–2, 174,
    177
edaphic factors, 12–13, 77–8, 299,
    327
elaters, 323
electrolyte leakage, 201
electron transport, 153
emittance, 109, 119
endemics, 25, 30–1, 64, 312, 315,
    325, 336–7, 341

endohydric species, 72, 171–2, 174,
    217, 291
endoliths, 18, 23, 36, 51–2, 96, 162,
    256, 297, 341
energy budget, 7, 117–20, 141, 191
    of *Cladonia stellaris*, 126–30
    of crustose and foliose lichens,
        120–3
    of fruticose lichens and mosses,
        123–30
    of plant communities, 130–1
energy content, 268–73
energy conversion efficiency, 188
energy flow, 275, 277, 291, 298
energy yield, 280–81
engulfment, 229
environmental gradient, 19, 80, 90,
    95
enzyme inhibition, 304
Eocene, 2
ephemerals, 316
epiliths (*see also* lithophytes), 49,
    52, 60, 70, 96, 171, 183, 189,
    256–7, 282–3
epiphytes, 41, 54–5, 59, 63–4, 157,
    266, 269, 283, 291, 296–8
erosion, 58, 75, 81, 87, 94, 101–4,
    289–90
esker, 7
establishment, 31, 87–90, 196, 251,
    310, 324–30, 332
esterase banding, 208
evaporation, 11, 117, 119, 123, 125,
    127, 130, 176, 179, 196, 206
evolution, vi, 68, 210, 310, 335–42
exchange sites, 292, 295, 303
exclosure experiments, 280, 286
extinction coefficient, 108, 113

fairy rings, 101–2
fall critical period, 139–40
fatty acids, 269, 279–80
feeding inhibitors, 275, 285
fellfield, 15, 18, 20, 58–9, 102, 255,
    265–7
ferns, 17, 57
fertilisation, 311, 316, 319, 321,
    323, 331
    range, 322
fire, 54, 80, 91–3, 222, 300–1,
    332–3, 335
floras
    Antarctic, 26–31, 309
    Arctic, 23–6
    Beringian, 26
    origin and evolution of, 335–42
    southern hemisphere, 325–7
floristic element, 23, 25, 28, 30
fluoride, 308
fog, 48, 170, 189, 195
foliose lichen (defined), 36
forest
    boundary, 19–20, 335
    line, 3, 21
forest-tundra ecotone, 16, 20, 54
fossils, 5
freeze-dehydration, 197
freeze-thaw cycles, 75, 81, 138–40,
    196, 316

freezing
    extracellular, 154, 196–7
    intracellular, 196
    prolonged, 200–1
    short-term, 198–200
    of spores, 325–7
frigid-Antarctic (defined), 22
frost, 11, 87, 136, 202
    action, 48, 59, 77, 80, 104, 231
    boil, 53, 57, 77, 98, 332
    damage, 141, 197, 201
    hardening, 151, 159, 189, 198
    resistance, 159, 196–201, 203–4,
        269, 341–2
fruticose and foliose lichen
    subformation, 33, 38, 47, 49
fruticose lichen (defined), 36
fruiting, 310, 314, 327, 333
Fuegian distribution, 28, 31
fugitives, 333
fumarole, 44–5, 69, 95, 326–7

G – see ground heat flux
gametangia, 316–23
gametophyte, 34, 278–9, 310–12,
    331
geese, 278–9
gemmae, 312, 314, 323, 327, 332
gene exchange, 323
genetic differentiation, 208–10
genetic recombination, 313, 336,
    339, 341
geographical isolation, 68
geological history, 1–7
geothermal heat, 41, 69, 95, 106,
    113, 326–7
glacial retreat, 4, 12, 80, 96, 336
glaciation, 2–3, 7, 22, 30, 47, 325
glycosides, 198
Gondwanaland, 4–5, 326
grass and cushion chamaephyte
    subformation, 33
grass heath, 14, 18, 20, 57, 260–1,
    263–4
grazing, 58, 254, 275–9
    by caribou and reindeer, 280–6
    by lemmings, 279–80
    pressure, 280, 285–6, 300
great growth period, 225, 240
greenhouse effect, 113
ground heat flux (*G*), 117–18,
    126–7, 130, 134
ground squirrel, 279
growing season, 12, 48, 62, 67, 192,
    213, 216, 239, 287, 318
growth, 11, 75–6, 78, 94, 103, 152,
    186, 192, 209–52, 294, 341
    annual increments, 213, 215,
        217, 229–45, 249, 329
    apical, 213, 215, 220
    climatic influence, 246–7
    curves, 220–1, 225–7, 232–3
    influence of climate, 246–7
    influence of nutrient availability,
        247–8
    inherent control, 245, 248–52
    intercalary, 220, 223, 230, 254
    marginal, 223
    methods of analysis, 229–33

growth (*contd.*)
  opportunistic, 248, 341
  relationship to NAR, 211–13
growth form, 13, 16, 19, 22, 32–8,
    69–75, 79, 123, 171, 249, 291
growth patterns
  in crustose and foliose lichens,
    223–9
  in fruticose lichens, 219–23
  in mosses, 213–15
  seasonal, 229, 242–6, 249
growth rate, 81, 143, 195–6, 201,
    243, 249, 305, 307, 333
  absolute, 220
  annual linear, 230, 238–40, 301
  radial, 224–6, 240–1
  relative (RGR), 220–1, 223,
    225–7, 260

H – see convective heat flux
hair points, 123
halophytes, 38, 57, 86
harvest techniques, 191, 253–4,
    256, 258
heat stress, 201–4
hemicellulose, 273
heterokaryosis, 313
herbfield, 15, 19, 57–8, 260–1, 267
herbivory, 275–86, 290
hiemal threshold, 139
high-Antarctic, 17–18
high-Arctic, 16–18
holocellulose, 273–5, 288, 290
human impact, 299–309
hydrogen sulphide, 67–8
hydroids, 72, 171, 217
hydrosere, 83
hygrophytic byrophyte
    communities, 62
Hyphomycetes, 52
hypsithermal period, 4

ice-wedge, 98–101
immigration, 26, 30–1, 68, 328,
    339, 341
inbreeding, 64, 336
industrial activity, 301–4, 307
infrared gas analysis (IRGA),
    141–2, 185, 189
innate markers, 229–30, 232–6
insect dispersal, 63–4, 333
insulation, 14, 75, 81, 132, 134–6,
    140, 196, 201
International Biological
    Programme, vii
International Geophysical Year, vii
introduced biota, 300
introgression, 336
IRGA – *see* infrared gas analysis
iron, 81, 280
isidia, 74, 313, 315, 323
isolichenin, 273
isostatic uplift, 7, 21, 241
isozymes, 207

K-selection, 334–5

laciniate lichen (defined), 36

lacustrine deposits, 7, 12
laminar sublayer, 120, 123–4, 323
large cushion (defined), 35
larch, 54, 91
latent heat, 93, 117, 122, 126–7
  flux (LE), 117, 121–2, 130–1
latitude, 7, 13, 16, 106–11, 131,
    190, 204–5, 268
Laurasia, 1, 326
LE – *see* latent heat flux
leachate, 93, 291, 299
leaching, 289–90, 294–5, 298
lead, 292, 303–4
leaf area index, 115, 334
Lejeuneaceae, 55
lemming, 64, 279–80
leprose lichen (defined), 36
leptoid, 217–18
lichen
  acid, 81, 189, 198
  factor, 240
  heath, 60–1, 76, 96, 263, 273–4,
    282, 289, 296
  resynthesis, 313, 325
  woodland, 54, 56, 61, 79, 91, 96,
    157, 263, 299
lichenin, 273
lichenometry, 233, 238–42, 256
life history strategies, 310, 331–5
light (*see also* PAR), 247, 249
  compensation, 64, 163–9, 186,
    203
  influence on NAR, 162–70
  penetration through snow,
    111–14, 190
  pentration through vegetation,
    114–16, 222
  quanta, 162
  saturation, 10, 163–9, 203–6
  screening, 203
  stress, 203–4
lignin, 269, 273–4, 287
limet shells, 41
linear phase, 223–6, 240
lipid, 218–19, 269, 279, 283, 312
lithophytes (*see also* epiliths), 174,
    249, 318–19, 332
longevity, 331–4
low-Antarctic, 17–18
low-Arctic, 16–18

magellanic moorland, 11, 17, 20,
    54, 63
Marchantiales, 314
marginal lobing, 228–9
marginal zonation, 39, 229–30
marine influence, 13, 38, 45, 57, 77,
    108, 248, 290
maritime Antarctic, 17–8
maturity index, 317–18, 320
meiosis, 310–13, 319, 334
membrane permeability, 196, 198,
    302
mesic communities, 14–15, 54–8,
    83, 263
Mesozoic, 2, 4
microclimate, 12, 93, 105–40, 256,
    311, 322
microconidium, 312

microtopography, 59, 79, 90, 97–8,
    139, 237
middle-Arctic, 16–18
mild-Arctic/Antarctic (defined), 19
mites, 278
monoclimax, 80
moraine, 7, 82–3, 232, 241, 316,
    332
moss balls, 104
muskox, 5, 60, 278
mycobiont, 34, 36, 52, 81, 162, 227,
    312, 329, 334

nematodes, 278
Neozealandic distribution, 28, 31
net assimilation (*see also*
    photosynthesis), 64–5, 75, 78,
    171, 193, 206, 212, 218, 254
net assimilation rate (NAR),
    141–2, 183–5, 193–5, 203,
    206–7, 228, 248, 304
  after freezing, 198–200
  after heat, 202
  diurnal variation, 198–200
  influence of snow cover, 189–91
  light responses, 162–70
  maximum, 143–52, 174–5, 179,
    186, 191, 204–5, 208, 211
  midday depression in, 160,
    184–6, 188, 213
  seasonal variation in, 143, 151,
    155–60
  temperature responses, 153–62
  water content responses, 172–83
  under field conditions, 185–91
nickel, 292
nitrate, 295–6, 303, 307
nitrogen, 13, 46, 50, 75, 78, 81, 93,
    248, 275, 283–4, 291–4, 299,
    302–3, 328
  fixation, 268, 275, 282, 285, 290,
    294–8, 302
nitrogenase activity, 296
nitrophiles, 64, 78–9
Nordenskjöld line, 20
norstitic acid, 52
nucleic acid, 109
nunatak, 1, 6–7, 22, 30, 49
nutrient
  availability, 67, 77, 92, 97, 191,
    247–8, 258
  cycling, 67, 90, 253, 290–9, 307
  enrichment, 15, 58, 63
  immobilisation, 295–6
  retention, 291, 294–5
  uptake, 217, 291–4, 296, 329

ocean barrier, 7, 21–2
octane derivative, 63
Ohm's law, 119
opportunistic responses, 203, 211,
    248, 341
origin of polar floras, 310, 335–42
ornithocoprophilous lichens, 30, 38
outbreeding, 331
oxalic acid, 81
oxygen, 51, 296
ozone, 109

palsa, 101, 125–6, 206–7
papillae, 170, 179
papules, 74
PAR – *see* radiation,
    photosynthetically active
particulates, 290, 293–4, 303, 305
pattern, in vegetation, 90, 94–105
patterned ground (*see also*
    polygons), 13, 57, 98
peat, 11–15, 42, 44, 53–4, 58, 63,
    87, 96, 135–6, 216, 278, 287,
    290, 292, 301
Peltigeraceae, 204
penguin, 46, 50, 58, 79
peristome, 63, 323, 331–2
perennial stayer, 331, 333
perithecia, 312
permafrost, 11–12, 42, 87, 97–8,
    101, 134–6, 258, 287, 289–90,
    295, 299, 301, 309
petrels, 58, 78
petroleum industry, 301–3
pH, 46, 61, 77, 97, 302, 304
phenolic compounds, 273, 275
phenotypic plasticity, 208–9, 250,
    340–1
physical disturbance, 301–2, 307
physiological buffering, 194, 196
phosphate, 67–8
phosphorus, 13, 46, 78, 248, 275,
    288, 290–4, 296, 299
photobiont, 36, 152, 154, 170, 180,
    223, 296, 313, 325, 329
photographic techniques, 232, 240,
    254
photoperiod, 67
photosynthesis (*see also* net
    assimilation), 10, 93, 106, 109,
    118, 122, 142, 171, 177, 179,
    181, 195, 203, 219, 225, 227,
    244, 248, 261, 297, 333
    gross, 153–4, 172, 188–9, 200,
        204, 214, 222, 248, 258
    influence of pollutants, 302–4
    influence of snow cover, 189–91
    light and dark reactions, 164
    variation with shoot age, 214–17
photosynthetic enzymes, 154
phytochrome photoequilibrium,
    114
phytomass, 253–68
    nutritional value of, 268–75
phytoplankton, 64
placodioid lichen (defined), 36
plakor, 16
Pleistocene, 3, 7, 12, 25–6, 30–1,
    68, 80, 325, 335, 340–1
Pliocene, 2
Podocarpaceae, 7
podsolisation, 13, 83
poikilohydry, 72, 74, 172, 203, 341
polar desert, 16–17, 22
pollution monitoring, 306–9
polyclimax, 80, 90
polygons, 13, 48, 54, 57, 61,
    97–101, 271
polyploidy, 335–41
polyol, 194–6
polytopic origin, 326

Polytrichaceae, 57, 77, 217–18, 337
post-linear phase, 227
potassium, 248, 275, 292–4
potential evapotranspiration, 20
Pottiaceae, 337
precipitation, 8, 11–13, 19, 23, 48,
    51, 54, 58, 62, 72, 81, 170, 174,
    240, 246, 290–2, 296, 307
prediction techniques, 256, 258
prelinear phase, 223–6
prenols, 218
primary consumption, 275–86
primary production, 65, 87, 92, 97,
    162, 191–2, 253–68, 279,
    286–9, 290–1, 295–6, 308
prions, 58
production : phytomass ratio, 260
propagule, 89, 94, 196, 324, 335
    asexual, 251, 312–15, 323, 327,
        331–2, 334
protein, 109, 194, 207, 227, 283
    content, 268–9, 275, 284–5
protocetraric acid, 283
protozoa, 278, 285
pseudocyphellae, 180
pteridophytes, 29, 279
pycnidia, 52, 312

Quaternary, 241

R_{air} – *see* resistance, air
R_{plant} – *see* resistance, plant
rabbits, 300
Racomitrium heath, 61
radiation
    diffuse, 108, 111, 115
    direct solar, 108, 123, 126
    infrared, 10, 108–9, 111, 114, 121
    longwave, 108–9, 113–14, 116,
        118–19, 138
    net, 10, 93, 116–19, 123, 126–7,
        130–1
    photosynthetically active (PAR),
        109, 111, 115, 121–2, 126, 131,
        152, 162–9, 202
    shortwave, 108–9, 111–12,
        115–16, 121, 134
    solar, 7, 10, 37, 106–16, 123, 132,
        162, 169, 184, 187
    ultraviolet (UV), 108–9, 325
    visible, 108, 111, 114, 121
radionuclide, 305–8
rain, 11, 19, 63, 126, 189, 195, 206,
    246, 294, 296
raised beach, 7, 38, 58, 61, 80, 86,
    96, 241
reciprocal transplantation, 208–9
recruitment, 329
reflectance, 114, 121–2, 126
refugia, 3, 7, 25, 26, 341
regeneration, 219, 312–13, 315
reindeer, 58, 219, 263, 280–86, 307
    lichen, 91
relative humidity (RH), 8, 10, 23,
    48, 51, 72, 75, 112, 119, 130–2,
    135, 192
    gradient, 127, 129, 291
renewal phase, 220, 230

reproductive biology, 31, 210, 312–
    42
    effort, 331–2
    phenology, 246, 316–20
    processes, 310–13
    strategy, 316, 334–5
resistance
    air (R_{air}), 119–25, 130–2
    boundary layer, 123, 179, 303
    canopy, 123, 125–7, 130, 171
    carboxylation, 179–80
    cell wall, 122
    cuticular, 120, 122–3
    plant (R_{plant}), 119–23, 131,
        171–2, 174
    stomatal, 120, 122
    to CO_2 diffusion, 156, 178–81
    to evaporation, 74, 119, 125, 127,
        184, 206
    to SO_2 uptake, 303
    to water loss, 72, 121, 197
respiration, 142–3, 151, 153–5, 157,
    159, 162, 171–3, 177–8, 184–5,
    190, 198–205, 211, 213,
    217–19, 222, 258, 286–7
    resaturation, 142, 189, 191,
        193–5, 206, 302
RGR – *see* growth rate, relative
RH – *see* relative humidity
rhizine, 36, 74, 81
rhizoid, 34, 36, 42, 50, 74, 82, 104,
    216, 312, 315
rhizome, 101, 214, 218, 272
rotifers, 64, 278
r-selection, 333–4
ruderals, 334
rumen content, 282–3
rugosity, 74
RWC – *see* water content, relative

scrub, 14, 20, 55, 255, 259, 274
seasonal response matrices, 178,
    183–5, 191, 246, 256
selection, 141, 203–10, 252, 334,
    336, 339–40
sensible heat – *see* convective heat
    flux
sex distribution, 87, 319–23
sexual dimorphism, 237
sexual reproduction, 312, 330–3
    causes of failure, 319–23
    frequency of, 313–15, 319–23,
        331
shoot
    density, 233, 237
    longevity, 213–14
short moss turf and cushion
    subformation, 33, 38, 41–2, 47,
        49–50
short turf (defined), 35
shuttle strategies, 332–4, 340
Siberian Province, 47
silver, 292
simulation models, 115, 191–2,
    194–5
small cushion (defined), 35
snow, 11, 14, 58, 70–1, 99, 132,
    219, 239, 280

snow (*contd.*)
  bank, 15, 50, 58, 62, 70, 75, 87,
    95, 136, 201, 267, 315, 324
  cover, 12, 48, 55, 61–2, 67, 75,
    96, 116, 157, 184–5, 190, 196,
    198, 201, 242–4, 300
  insulating effect of, 132, 140, 201
  melt, 71, 117, 140, 170, 246
sociation, 32, 38, 88
sodium, 46, 292
soil, 12–13, 17, 19–20, 23, 37–8, 41,
    46, 63, 74–8, 81–3, 98
  aeration, 66–8
  biota, 51
  erosion, 300
  heat flux – *see* ground heat flux
  moisture, 72, 96–7, 127, 131,
    179, 246, 299
  salinity, 23, 51, 59, 63, 78
  temperature, 134–6, 190, 299
solar constant, 106
solifluction, 13, 77, 87, 97–8
solstice, 106, 111, 116
solute leakage, 194
soredia, 313, 315, 323–4, 327, 330,
    335
southern beech, 54
southern temperate zone, 18, 327
spermatium, 312–13, 325
spore, 63, 279, 310, 314–15, 321–2,
    330, 333, 339
  dispersal, 323–8, 331
  germination, 311, 326, 332
  release, 319, 323–4
  size, 331–3
Splachnaceae, 63, 333
splash platform, 332, 324
sporophyte, 63, 278, 310–11,
    313–23, 327, 329, 331, 333
  abortion, 319–22
spotty tundra, 59
spring critical period, 140
spruce, 54, 56, 91, 115, 287, 299
squamulose lichen (defined), 36
stability, 65, 295, 299–300
starch content, 269
stationary phase, 224
Stefan–Boltzman law, 118
sterols, 218
Stictaceae, 204
stress, 141, 192–203
  tolerant ruderals, 334
  tolerators, 332–3
sub-Antarctic, 16–18, 21
sub-Arctic, 16–18, 21
subclimax, 286
subsistence lifestyle, 308
succession, 78–93, 96, 222, 296, 299
  auto-, 80, 87–9, 90
  cyclic, 80, 87, 89–90, 94, 99, 101,
    104
  directional, 87, 90, 94
  primary, 80, 290, 298
  secondary, 54, 80, 88, 91–3, 301,
    333
sugar content, 198, 269
sulphur dioxide and derivatives,
    302–4, 308
sunflecks, 115

supra-optimal
  irradiance, 163
  temperature, 156
  water content, 178–81
surface area : volume ratio, 74, 170,
    181
surface area : weight ratio, 127, 184
Surtsey, 83, 298, 327–8

T_{opt} – *see* temperature, optimum
    for net assimilation
tall moss turf subformation, 33,
    42–4, 47
tall turf (defined), 35
tardigrades, 278
temperate-disjunct distribution, 25
temperature, 8, 10, 12, 17, 19–23,
    47, 64, 75, 79, 81, 93, 118–19,
    177–8, 181, 183–5, 195–7, 200,
    203–4, 207, 219, 233, 240,
    246–7, 249, 289, 291, 296, 316,
    318
  compensation point, 156–7, 206,
    212
  diurnal fluctuation, 130–5, 139,
    153, 207
  gradient, 113, 118, 127–9, 134–5
  influence on NAR, 153–62
  influence on nitrogen fixation,
    296–7
  influence on respiration, 154–5
  influence on vegetation, 66–9
  interaction with irradiance,
    164–7, 203
  inversion, 118, 132, 300, 303
  microclimatic, 106, 108, 112,
    131–40, 158, 160–2, 204
  optimum for net assimilation
    (T_{opt}), 144–50, 153–62, 165,
    203–6, 208, 211
  plant, 118, 121–3, 125, 130, 138,
    184, 186–8, 192, 205, 207
  under snow, 132, 139–40, 190
Tertiary, 2, 7, 25, 80, 335, 339–41
thallose mat (defined), 35
thallus colour, 109, 121, 126, 184,
    190, 204
thaw lake cycle, 99–100
thermal
  conductivity, 134–5, 139
  diffusivity, 120
thermocarst, 13, 301, 309
thread (defined), 35
tolerance limit, 12
tomentum, 180, 214, 249, 329
topocline, 237
topography, 1–2, 12, 75, 79–80, 120
tourism, 301
tracked vehicles, 301
trampling, 286, 301
transfer cells, 217
translocation, 101, 217–19, 226–7,
    254, 268
transmittance, 114
tree line, 10, 280
triglycerides, 218–19, 269
tundra (defined), 1
turbulent air movement, 120, 123–4
turnover

  of nutrients, 295, 298
  of phytomass, 260, 263, 287, 289
tussock grassland, 15, 19, 58,
    260–1, 266–7

ultraviolet screening, 109
umbilicate lichen (defined), 36
upsik, 139
uronic acid, 292
usnic acid, 83, 109
UV – *see* radiation, ultraviolet

variation
  clinal, 223, 237–8
  genetic, 204, 208, 250–1, 339, 341
  in lichens, 206–8
  in mosses, 208–10
  interspecific, 202
  intraspecific, 185, 206, 250–1,
    339, 341
  plastic, 208–9, 250–1, 340–1
vegetation
  types, 13–15, 32–3, 37–65
  zones, 16–23, 66, 95–6
vegetative phenology – *see* growth
    patterns, seasonal
volcanic heat, 44, 314, 326–7
vole, 279

water
  availability, 11–12, 38, 62, 67–70,
    79, 162, 189, 192, 204, 219,
    233, 241, 246, 258, 266, 297
  holding capacity, 172, 174
  loss, 73–4, 170–2
  movement, 170–1, 175
  potential, 154, 175–7, 193
  relations and growth form, 69–75
  uptake, 72, 74, 170–2, 217, 291,
    329
  vapour absorption, 171, 189
water content, 72–3, 81, 97, 131,
    137, 154, 179–80, 184–6, 191,
    196, 206–7
  compensation point, 172, 174
  daily and seasonal variation,
    181–2, 192
  influence on NAR, 172–7
  influence on respiration, 177–8
  optimum for NAR, 142, 144–6,
    174–5, 204–5
  relative (RWC), 126, 130, 174–5,
    182, 194, 201
weathering, 52, 83, 290, 293
  chemical, 12, 81, 293, 295
  physical, 12, 77, 81–2
weevil, 278
weft (defined), 35
west wind drift, 326
weight per unit length, 233, 237,
    243, 247
wetland communities, 14, 52–4, 263
wet meadow, 14, 20, 52–3, 60,
    96–7, 264, 274
wetting and drying cycles, 194–6
willow, 22, 55, 59, 274
wind, 8, 10–12, 48, 50, 58, 61–2, 70,
    79, 87, 101–2, 232, 289

influence on dispersal, 323–8, 330
influence on evaporation rate, 123–5
influence on vegetation, 75–8
velocity, 120–1, 130, 303
winter kill, 213–14, 230, 329
Wisconsin glaciation, 3, 80, 341

woodland, 14, 19–20, 23, 54, 69, 88, 91–2, 180, 202–3, 206, 213, 217, 222, 237, 239, 255, 267, 274, 280, 289, 291, 297, 301, 333, 335
Würm glaciation, 3

xerix communities, 15, 58–9, 266

xerosere, 82

yeast, 49

zinc, 304
zonation, 19, 38, 51, 95–6, 233
zoospores, 313
zymograms, 207

DATE DUE

GAYLORD			PRINTED IN U.S.A.